Design Dimensioning and Tolerancing

by

Bruce A. Wilson

Publisher
The Goodheart-Willcox Company, Inc.
Tinley Park, Illinois

Library of Congress Catalog Card Number 94-18418
International Standard Book Number 1-56637-067-1

3 4 5 6 7 8 9 10   96   99 98 97

Library of Congress Cataloging in Publication Data

Wilson, Bruce A. (Bruce Allen)
   Design dimensioning and tolerancing/by Bruce A. Wilson.

   p. cm.
   Includes index.
   ISBN 1-56637-067-1
    1. Engineering drawings—Dimensioning.
    2. Tolerance (Engineering). 3. Engineering design. I. Title.
T357.W47 1995
604.2′43--dc20   94-18418
                   CIP

INTRODUCTION

A majority of commercial and military industries require their engineers and drafters to be knowledgeable in proper dimensioning and tolerancing methods. Machinists, inspectors, and industrial engineers are also required to understand dimensions and tolerances since they must work to the drawings created by the engineers and drafters.

The guidelines for consistent and clear application of dimensions and tolerances are defined by the standards of the American National Standards Institute (ANSI) as written by the American Society of Mechanical Engineers (ASME). The number of companies requiring compliance with national standards is continually growing.

Proper application of dimensions and tolerances is an important part of providing complete documentation of product requirements. ASME Y14.5M is the authoritative document for defining dimensioning and tolerancing symbols and application methods. **Design Dimensioning and Tolerancing** provides an expanded explanation of the material contained in ASME Y14.5M. Additional standards affecting calculation and application of tolerances are also explained and referenced within this textbook when appropriate.

The subject of this text ranges from the fundamentals of dimension application to extended principles of tolerance application. An explanation of dimensioning fundamentals is given since application of tolerances rely on the nominal size, orientation, and location dimensions being shown properly.

Tolerance application and interpretation explanations include all the categories of tolerances in ASME Y14.5M. Explanations are primarily based on the current version of the standard, ASME Y14.5M-1994. Past practices are briefly explained to provide information to people who must work on old drawings. Past practices are clearly distinguished from current practices.

Design Dimensioning and Tolerancing contains twelve chapters. Each chapter includes a list of major topics at the beginning of the chapter. A chapter summary is given at the end. Chapter review materials include questions and application problems. Additional problems are available in the **Study Guide for Design Dimensioning and Tolerancing.**

Generally, tolerancing terms are defined within the text when they are first introduced. Their definition is repeated in the Glossary to permit easy future reference. Other technical terms are italicized when they first appear. They may be defined within the text or in the glossary.

Dimensioning and tolerancing terms used within this textbook conform to those used in the ASME standards. Any additional terms are kept simple to make explanations easy to understand. Some general terms, such as centerline, are used to avoid complex terms that may be confusing. An explanation may refer to the centerline of a mating hole, rather than use a complex reference to the axis of the smallest mating perfect cylinder.

Illustrations are used extensively to clarify explanations. Each figure generally introduces only one new concept. Figures are not complicated with detail that is unrelated to the concept being explained. This makes it possible for the reader to quickly see the necessary information.

Color is used to separate explanation data from the main portion of the figures. Color is also used to highlight instructional information such as dimension spacing requirements and tolerance zone boundaries.

Interpretation figures are provided to show the tolerance zone that results from tolerance specifications. Tolerance zones and part variations in these figures are sufficiently exaggerated to make them visible.

Uppercase letters are used for notations that are part of the drawing. Lowercase letters are used for instructional notes and information.

Dimensioning and tolerancing of a drawing requires application of principles from the ASME Y14.5M standard. **Design Dimensioning and Tolerancing** shows how to apply the principles in the standard to many situations and offers suggestions regarding the extension of the principles to situations not shown in the standard. Care must be taken in the extension of principles, otherwise a violation of the standard or unclear requirement could result. Compliance with the standard is ultimately the responsibility of the person creating or working with a drawing.

ABOUT THE AUTHOR

Bruce A. Wilson, author and illustrator of **Design Dimensioning and Tolerancing,** has industrial and academic experiences that provide the expertise and practical knowledge necessary to write a dimensioning and tolerancing textbook that is technically correct and applicable to today's industrial needs.

Industrial experience includes positions as a drafter, designer, design manager, program manager, dimensional management technical leader, and industrial consultant. Prior to attending college, Bruce worked as a drafter for a manufacturer of pneumatic valves and automated pneumatic machines. Design engineering experience includes mechanical, electromechanical, optics packaging, and electronic packaging. Mechanical designs range from precision miniature gear drives to large sand castings. Electromechanical designs include mechanisms to position a missile guidance system camera that compensates for missile yaw and a redundancy switching system for the receiver optics in a laser communications system. Optics packaging includes an interchangeable system that splits a .003″ laser beam into four equal paths. Electronics packaging includes the first laser radar equipment launched into space.

Extensive design experience has provided a thorough exposure to dimensioning and tolerancing requirements as they apply to various applications. A wide range of applications have provided experience working to commercial and military documentation needs.

Mr. Wilson's background includes a Bachelor of Science and Master of Science degree from Southern Illinois University. It also includes teaching experience in a vocational center and in a major corporation. Bruce was a member of Phi Theta Kappa and Phi Kappa Phi fraternities, and a member of Iota Lambda Sigma professional fraternity.

Industrial training and consulting experience includes training program management, curriculum development, industrial training, GD&T implementation, and dimensional management implementation. Areas of concentration include dimensioning and tolerancing, computer-aided design, and a new designer training program. Dimensioning and tolerancing programs have been taught to designers, machinists, inspectors, industrial engineers, and managers.

Mr. Wilson has represented his company for many years at the ASME Y14.5 subcommittee meetings. He is currently a member of the ASME Y14.1, Y14.2, Y14.3, and Y14.5 subcommittees. He is chairman of Y14.3, vice-chairman of Y14.2, and section 5 leader for Y14.5. The subcommittees of the ASME Y14 committee write the national drafting standards. Bruce is also a member of the United States TAG/ISO/TC10/SC5-Dimensioning and Tolerancing Committee and the TAG/ISO/TC10/SC10-Engineering Symbology Committee.

ACKNOWLEDGMENTS

The author would like to express his appreciation to the following companies for their cooperation in the preparation of this book.

Figures in this textbook were created using AutoCAD software, a product of Autodesk, Inc.
Autodesk, Inc.
2320 Marinship Way
Sausalito, CA 94965

Figures in this textbook were plotted on a Summagraphics DMP-160 series plotter.
Summagraphics
8500 Cameron Road
Austin, TX 78754

Photographs were provided by the following company:
L. S. Starrett Company
Athol, Massachusetts 01331

CONTENTS

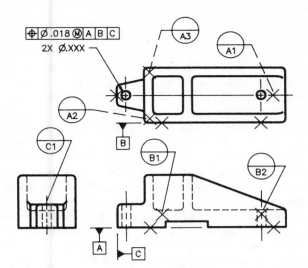

Chapter 1

Introduction to Dimensioning and Tolerancing

CHAPTER TOPICS

- ☐ The importance of accurately specifying dimensions and tolerances.
- ☐ The history and development of dimensioning and tolerancing methods.
- ☐ How teamwork can result in better definition of the dimensions and tolerances shown on a drawing or in a computer-aided design (CAD) file.

- ☐ Definition of the dimensioning and tolerancing skills needed for success in design- or production-related occupations.
- ☐ Probable industrial changes and possible impacts on dimensioning and tolerancing.

INTRODUCTION

This highly competitive, industrial world requires that full advantage be taken of all methods that help improve efficiency and reduce product cost. Application of dimensions and tolerances according to the current standards permits clearer definition of requirements than was ever possible before. Current standards define methods for increasing tolerance zones without reducing product quality or design function. Improving drawing clarity and increasing allowable tolerances are two means for improving competitive position.

A thorough knowledge of dimensioning and tolerancing methods helps to ensure clear application of part requirements. Tolerances can be maximized through careful dimension and tolerance calculations and application of the calculated values through proper utilization of standardized methods.

A clear definition of part size requirements is essential if parts are to be produced by more than one manufacturer. In the automotive industry, for example, parts are made by many companies and shipped to a factory for assembly into a functioning automobile. Requirements must be clearly defined if the parts are to fit together and properly function. Unclear dimensional requirements can result in parts that are incorrectly manufactured, and ultimately unable to be fastened to the assembly.

The number of small tolerances on a part should be reduced and all unnecessarily small tolerances should be eliminated. Calculating all tolerances is one step toward ensuring that tolerance values are maximized. Maximized tolerances permit the use of less expensive machine processes and less restrictive controls during the fabrication of parts.

THE DEVELOPMENT OF DIMENSIONAL CONTROL

There are two aspects to the development of dimensional control. One aspect is the establishment of standards that define how dimensions and tolerances are shown on a drawing. The other aspect is the establishment of a standard unit of measurement. The development of dimensioning and tolerancing standards has occurred only recently when compared to the time period over which units of measurement were developed.

Dimensioning and tolerancing standards have become relatively well defined during the past 100 years. Most of the progress has taken place since the mid-1940s. The greatest amount of change in these standards took place between the 1940s and 1973. Current standards have changed very little from the one issued in 1973. Although standard practices have existed for some time, many major corporations didn't make an effort to comply with the practices until the 1970s and 1980s.

Measurement standards have evolved over thousands of years, resulting in the two existing units of measurement–the inch and the millimeter. *The measurement unit that is used has no effect on the manner in which the dimensions are applied; it only affects the numerical values that are shown in the dimensions and tolerances on the drawing.* The present dimensioning and tolerancing standard is applicable to inch and SI (System International) metric dimensions.

Since either measurement system may be used, a note must be placed on each drawing to avoid any confusion as to which measurement system is in use. A drawing note such as the following can be used:

ALL DIMENSIONS ARE IN INCHES
or
ALL DIMENSIONS ARE IN MILLIMETERS

MEASUREMENT STANDARDS

The existing measurement standards have been developed over the past several thousand years. Units of measurement such as cubits, spans, palms, and digits were in use 5000 years ago. A *cubit* was the length of the Egyptian Pharaoh's forearm. A *span* was the distance from the tip of the thumb to the tip of the small finger with the hand spread. A *palm* was the distance across the hand, and a *digit* was the distance across a finger.

A permanent standard was made of black granite by the Egyptians. This was based on the cubit length (20.67″) and was referred to as the *Royal Cubit*. It was subdivided into small increments to allow relatively short distances to be measured accurately. The Royal Cubit was used as a reference against which measuring sticks were calibrated.

The Greeks established a unit of measurement called a *sixteen digit foot*. This distance standard (12.16″) was approximately two-thirds the length of a Greek Cubit (18.93″). The official name for the sixteen digit foot is the *Olympic Foot*. Its length is 12.16 inches.

The Romans established a unit of measurement called the *Roman Foot*. It was 11.65″ long. The *Anglo-Saxon Foot* was based on the Roman Foot and later became the standard for Great Britain.

The *yard* was established as a unit of measurement in the twelfth century when King Henry I of England decreed that the distance from his nose to the end of his thumb would be one yard. Three barley corns placed end-to-end were established as the standard for a one-inch increment in 1324 A.D. In 1855, a permanent standard for the yard was established in Great Britain and was called the *British Imperial Yard*.

In 1798, the standard meter was established. It is one ten-millionth of the distance from the north pole to the equator when measured along a line passing through Paris, France. In 1889, the *U.S. Prototype Meter number 27* became the standard against which all U.S. measurements were to be referenced. Although the inch system is generally used in the United States, the inch length is referenced to the standard meter. The length of Prototype Meter number 27 is established relative to the length of a specific wavelength of light.

Evolution of the units of measurement from the length of the Pharaoh's arm to the wavelength of light has brought about a phenomenal increase in the accuracy of distance measurements. The evolution of a fixed and accurate distance standard allowed for improvement in the equipment used to measure distance. See Figure 1-1. It would be impossible to produce machines that build and verify accurate distances without an accurate distance standard.

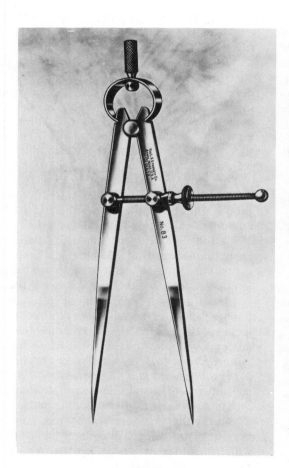

DIVIDER
FOR TRANSFERRING APPROXIMATE DIMENSIONS

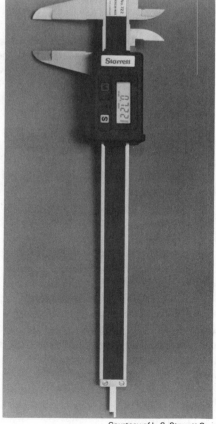

Courtesy of L. S. Starrett Co.

CALIPER
FOR ACCURATE DIMENSION MEASUREMENT

Figure 1-1. Accurate measurement tools are only feasible when an accurate measurement standard exists. As the measurement standards evolved, so did the measurement tools.

DIMENSIONING STANDARDIZATION

It was not possible to verify the dimensions of a produced part before measurement standards were well established and an accurate means for gaging distances was created. All parts were fitted together and assembled by hand, since it was not possible to verify by measurement that separately produced parts would fit together. Drawings did not show dimensions with allowable variations (tolerances), since only crude means for verifying size existed. There was no reason to show limits of allowable variation on any dimensions since there was no means for accurately verifying size.

The development of accurate measurement equipment made it possible to verify relatively small variations in part size. See Figure 1-2. This capability makes it reasonable to specify the limits of acceptable variation for a dimension.

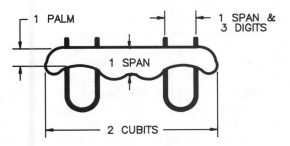

WOODEN YOKE
BUILT TO CRUDE DISTANCE
STANDARDS TO WHICH TOLERANCES
WOULD NOT BE LOGICAL

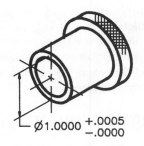

Ø1.0000 +.0005
−.0000

DRILL BUSHING
BUILT TO ACCURATE
DISTANCE STANDARDS TO WHICH
TOLERANCES ARE LOGICAL

Figure 1-2. Tolerances only make sense when there is an accurate unit of measurement.

If limits of variation are correctly calculated, and all parts made within the specified limits, mass production is possible. Efficient mass production is only possible when any part made within the limits shown on a drawing can be assembled with its mating parts. This is not to say that each part manufactured to specifications is exactly alike, because variations will be present. The important point is that all variations are within a specified allowable range.

The repeatable accuracy that resulted from establishing a fixed measurement standard helped to make interchange-able parts possible. Eli Whitney is credited with making the first interchangeable parts (1798). Repeatable accuracies that were achieved in the eighteenth century seem crude by today's capabilities, but they were adequate for making some interchangeable parts. Today, accuracies of .001″ inch are produced, and some machine processes can achieve accuracies in the millionths of an inch. Present industrial capabilities make most any design feasible.

It is the designer's and drafter's responsibility to master the dimension application guidelines. This allows for efficient utilization of today's industrial capabilities through properly specified dimensional control. However, there is a responsibility to utilize only the degree of dimensional control necessary to obtain the accuracy required for the part to function properly. Specifying accuracies smaller than needed increases both the cost and time required to produce a part.

The methods for completing design drawings have advanced along with the capabilities of production and inspection machines. Drawing instruments developed to the point where very accurate drawings could be produced, and computer-aided design (CAD) tools are now replacing drawing instruments. Computer-generated designs can contain dimensional data with accuracies to more than ten decimal places.

Although the computer data is extremely accurate, production machines still are not able to produce perfect parts from the CAD data. Since production machines introduce some errors, it is essential that the amount of acceptable error be defined. This is done through the application of tolerances. *Tolerances* are the acceptable dimensional variations that are permitted on a part.

Methods for showing dimensional requirements and permitted tolerances have advanced along with the measurement standards, drawing creation methods, and production machine capabilities. The dimensioning and tolerancing practices for showing size, form, orientation, runout, location, and profile requirements are better defined today than ever before. The dimensioning and tolerancing practices are standardized to provide both a clear and consistent set of guidelines.

There has been a great deal of effort made to standardize dimensioning and tolerancing methods. The voluntary efforts of the people serving on national committees for standardization have resulted in the current standards. Their efforts are an important part of making mass production possible for the complex and accurate products in today's industry.

DIMENSIONING STANDARDS

The main standard that defines dimensioning and tolerancing methods is ASME Y14.5M. Other standards have some impact on what is shown in the dimensions, such as thread specifications, but ASME Y14.5M is the standard that shows how the dimensions and tolerances are applied. The current standard has evolved over many years. Most of the dimensioning methods have remained the same for many decades. The area of greatest change has been in how tolerances are specified.

The methods for defining tolerances through what is commonly referred to as *positional tolerancing* was included in MIL-STD-8A in the 1940s. The methods have been revised several times. The following list shows the evolution of the dimensioning and tolerancing documents.

MIL-STD-8A

MIL-STD-8B

MIL-STD-8C

USASI Y14.5-1966

ANSI Y14.5-1973

ANSI Y14.5M-1982

ASME Y14.5M-1994

A new issue of the dimensioning and tolerancing standard is released periodically. New issues are released in an effort to keep the methods current with industrial needs. Changes are kept to a minimum in each new issue, and long-standing practices are not changed unless there is some compelling reason to do so.

There is no fixed period of time between issues of the standard. However, the last four issues were separated by approximately ten years. Very seldom are existing practices reversed by a new issue. The new issues primarily add new methods and clarify those already in effect. The small number of changes and the time between issues of the standard is sufficient to reduce concern about the need to learn a new system every few years. Although changes are generally minimized, it is important to learn the changes made in a new issue of the standard.

It is only necessary to know the methods defined by a previous issue of the standard when work is being completed on an existing drawing, or when creating a new drawing in fulfillment of an old contract. Contracts sometimes require compliance with one of the previously issued standards.

Since the current issue of ASME Y14.5M only includes minor changes from the previous issue, a person who understands current methods can also work to the previous issues with relative ease. This textbook includes descriptions based on the current dimensioning and tolerancing standard and highlights areas of significant change from previous issues of the standard.

It is not necessary to know the requirements of all previous standards since it is unlikely that many people are still working the initial standards such as MIL-STD-8. In the event that an old standard is a contractual requirement, a copy can be obtained from the contracting agency.

The most recent issue of the standard should be used when there is an option regarding which issue to use. If there is any doubt about whether a new issue of the standard has been released, either the American National Standards Institute (ANSI) or the American Society of Mechanical Engineers (ASME) can be contacted.

Standards are established to provide consistency in how things are done. However, a standard cannot show every possible situation that may arise; it can only show the principles and how they apply to various situations. It is up to the individual to apply the principles to specific situations as they are encountered.

Almost all figures in the current standard show metric

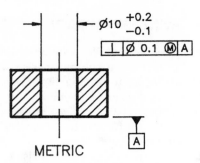

Figure 1-3. Illustrations in the current standard include metric dimensions.

dimensions. See Figure 1-3. The only exceptions are figures that show how dimension limits are applied to inch values. *The standard does indicate that either inch or millimeter dimensions are permitted on drawings.*

A large part of the industrial world still uses inch dimensions. The illustrations in this book are shown mainly with inch values. Changing the inch values to millimeters would have no effect on how the dimensions and tolerances are applied; it would only change the numerical values.

Principles covered by the current dimensioning and tolerancing standard may be applied to many situations that are not shown in the standard. If a complete understanding of the principles is gained, then the principles are adequate for meeting most design documentation needs.

Unique situations do occur where the principles of the standard do not meet the needs for a specific dimension or tolerance application. In these cases, it is possible to use drawing notes to explain the special requirement. However, notes should not be used for applications for which dimensioning and tolerancing methods are defined. The use of notes to explain tolerance requirements should be avoided since one of the purposes of the current standard is to provide a means for clear definition of dimensional requirements through symbology.

DESIGN THROUGH TEAMWORK

Product cost can be minimized by involving people from various disciplines in the design process. Many designers are very knowledgeable about manufacturing and inspection processes, however, they probably don't know all the specifics about the operations. Advice from industrial engineers, machinists, tool designers, tool fabricators, and inspectors is important in making sure the design can be produced at the lowest possible cost while still retaining the intended design function.

Advice from the previously mentioned disciplines can help in many ways. Basic design concepts can often be improved through ideas from people that must operate the machines to produce the parts. These people know the capabilities of the machines. Designers usually know machine capabilities and processes, but it is difficult to be as well informed as the machine operators. Technology changes too fast for one person to remain fully knowledgeable about everything in all fields that affect a design. People who work full time in a discipline can provide accurate and valuable advice.

Designers must determine how to use the advice to create the optimum design that meets the functional requirements of the product. Part of the design optimization process is to make decisions about how to show the dimensional requirements and the tolerances. Once again, input from the previously mentioned disciplines can be very helpful.

There are significant advantages to having all disciplines review drawings as they are completed. Comments can be acted upon before the drawing is finished. This prevents or reduces the need for changes after the drawings are released and production has been started. Another important advantage is the fact that everyone involved in the design process feels the design belongs partly to them. This helps develop a positive attitude among those who must produce the parts.

The most important advantage for involving various advisors in the design is that part cost will be reduced. The design will be producible, machine programming will be made easier through better applied dimensions, tool design will be facilitated because of better specification of functional relationships between features, and inspection will be better directed because of a clearer understanding of part requirements. These are only a few of the benefits that can be realized by designing through teamwork.

Sometimes, it is tempting to complete a design through an individual effort. The temptation is due mainly to the apparent increase in efficiency during the design process. However, this apparent improvement in efficiency may be misleading, especially when considering the time spent on changes that will occur after production has started. These changes will be necessary because of errors and oversights that an individual working alone is bound to make. It is important to stress that working with a design team does not always reduce efficiency. A design team can work together and be efficient. Efficiency can be achieved through close cooperation and a willingness to achieve a common goal. A close-working team will be efficient when compared to one individual trying to determine all design parameters without the assistance of experts.

PROFESSIONS AND TRADES AFFECTED BY DIMENSIONS

Designers are obviously affected by dimensions and tolerances since they must apply these requirements on the design drawings. A thorough knowledge of standard practices is required to make sure the information shown on the drawing "states" what is actually intended. Careless or uninformed application of tolerances can result in part requirements that are overly restrictive, forcing part cost to increase. Incorrect application of tolerances can also result in inadequate control, resulting in parts that will not assemble or will not function properly.

A complete understanding of how to apply tolerances results in design requirements that are applied in a clear manner, with meanings that are supported by a national standard. This is important no matter where parts are produced. However, the importance of clarity increases when parts are made by a company other than the one that created the design.

Industrial engineers and *machine planners* are often the people who decide how parts are to be produced. They must review drawings and make decisions based on the tolerances shown on the drawing. The dimensioning and tolerancing knowledge of the industrial engineer can have a significant impact on whether or not parts are made efficiently and correctly. Inadequate knowledge can result in an overly cautious approach; therefore, processes more accurate than needed might be used. This increases the product cost. Inadequate knowledge can also result in requirements that don't receive enough attention during fabrication.

Machinists must operate the equipment that produces the parts. A knowledgeable machinist is an important aspect of production. These are the people who take the idea from a piece of paper, or CAD file, and produce a real part. If they understand the requirements, they can produce the parts with the greatest amount of efficiency. If they don't understand the requirements, there is little chance of obtaining functional parts at the lowest possible cost. Misunderstood requirements can mean either excessive or inadequate care in the completion of fabrication steps.

Inspectors verify that parts are produced in compliance with drawing requirements, including dimensions and tolerances. If inspectors have a good understanding of dimensioning and tolerancing principles, they will accept good parts and reject the bad ones.

Tool designers must design the jigs, fixtures, and other machining tools that are required to produce parts. Proper understanding of the dimensions and tolerances is required to correctly design the tools. Tolerance specifications often dictate how a part must be located in a machining tool. If the tool designer doesn't understand the tolerances, incorrect location methods might be designed into the machining tool. This could result in ruining all parts produced with the tool. Correct understanding of the requirements improves the chances that the tool design will be correct.

Fabrication of a good-quality product depends on a knowledgeable team. It is important that all people involved in the design and fabrication process work together to develop a full understanding of the standard practices. A knowledgeable team can produce a product of which they can be proud.

Knowledge of standard principles and careful application of that knowledge can also reduce the risks of product liability. Product liability should be a concern to all who work on a design or its production. Individuals as well as their companies are sometimes held responsible for errors.

Government and military design personnel need to have a thorough knowledge of dimensioning and tolerancing principles. Nearly all government design and production contracts require that drawings be completed in compliance with the current dimensioning and tolerancing standard.

The requirement for compliance with the dimensioning and tolerancing standard on government and military contracts also indicates the need for people working in the defense industry to learn the principles. Drawings completed in compliance with the standard will be accepted with less need for revision than if there is incomplete compliance. Revisions made to drawings because of noncompliance with standards are usually made at the contractor's expense. If the expense of revising drawings must be paid by the contractor, that expense reduces the amount of profit made on the contract.

Proper application of the dimensioning and tolerancing principles will make part requirements more clear and result in tolerance zones that are determined by design function. Both of these factors reduce part production costs.

REQUIRED DIMENSIONING SKILLS

Dimensions and tolerances must be correctly applied to drawings. This requires that standard application methods and symbology be understood. The ability to interpret the symbols applied to a drawing, as well as their meanings, is necessary since the parts must be produced according to what is specified on the drawing. The ability to explain dimensioning and tolerancing requirements is needed when others who are less knowledgeable need assistance.

The ability to apply tolerances on a drawing is only part of the necessary tolerancing skills. It is also necessary to have the ability to calculate tolerances. Showing tolerances on a drawing is beneficial only if the tolerance values are correct. The only way to be sure tolerance values are correct is to determine the values through calculations.

APPLICATION

Past methods of showing tolerances was limited to tolerances on numerical values (such as size dimensions) and geometric controls that were explained in notes. It wasn't until the introduction of standard symbols that tolerancing methods progressed beyond the ambiguities of notations. The past practice of using notes to specify geometric requirements often resulted in uncertainty about the meaning of the note.

Symbols shown in the standard provide a means for defining requirements and remove the ambiguity that can result from using notes to define geometry. Figure 1-4 shows a perpendicularity requirement for a dowel pin. This tolerance specification has a well-defined meaning (as will be shown in the chapter on orientation tolerances). The same control expressed through a note takes more room and is subject to the language ability of the writer.

Proper application of dimensions and tolerances is usually the responsibility of the designer or drafter, therefore, they must have a good knowledge of dimensioning and

tolerancing application principles. Various others may review the drawing before it is completed to ensure the drawing is correct. The reviewers must also have an understanding of the principles applied on the drawing for the review to be valid and reliable.

Persons applying dimensions and tolerances must use only standard symbols in compliance with the methods defined in ASME Y14.5M. Nonstandard symbols should not be used; they result in confusion. Nonstandard symbols have no defined meaning. Therefore, various opinions can be developed about the meaning of any nonstandard symbols.

INTERPRETATION

Parts are usually manufactured by someone other than the person that completed the drawing. In large companies, many people will be involved in making a part. Each of them completes some portion of the part requirements. This requires that each person understand the meaning of the drawing. See Figure 1-5. People need to develop enough knowledge of dimensioning and tolerancing principles to be able to efficiently interpret and understand a drawing.

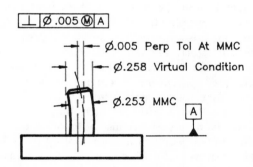

Figure 1-5. Permissible geometry variations can be determined if tolerances are properly interpreted.

Correct interpretation of dimensions and tolerances is required to produce a good-quality part at the minimum possible cost. Correct understanding of the drawing requirements makes it possible to match the machine processes to the specified accuracies. The correct processes ensure that excessive errors will not be made, nor will excessive control be used.

Competent designers produce drawings that accurately show the design requirements for their parts. Any person working with these drawings must interpret them. Since some designs are complex, it may take extensive knowledge of dimensioning and tolerancing principles to properly interpret the design drawings.

Every specification shown on a complex part may not be directly supported by a specific figure or paragraph in the standard. A person must be able to interpret dimensions and tolerances when they are applied to parts that have greater complexity than the figures shown in the standard. This means that the principles in the standard must be understood well enough to extend them to complex situations.

Some industry drawings contain dimensioning and tolerancing errors, and, therefore, are not in complete

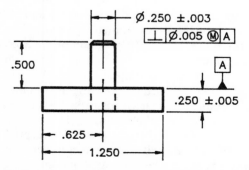

Figure 1-4. Symbology is used to define tolerances.

compliance with the standard. One reason for incomplete compliance is that not all designers have a thorough knowledge of dimensioning and tolerancing principles. Many have started to learn the subject, but have not yet mastered it.

When an incorrect drawing is identified, a decision must be made–work can be completed to the drawing, or drawing corrections can be requested. Sometimes, it is necessary to work with an incorrect drawing. This is a risky practice, however, requiring that a person attempt to guess what requirements the drafter meant to show on the drawing. If a portion of the drawing is inadequate, then parts made to the assumed requirements may not function correctly. Working with an incorrect drawing puts at risk the full cost of parts that will not be acceptable due to someone else's assumption of what the drawing states.

The amount of risk taken when working on the basis of an incorrect drawing is dependent on the type of errors on the drawing. A judgment must be made as to the severity of errors and the potential cost of incorrectly interpreting the drawing. It is always better to ask for the drawing to be corrected than risk manufacturing a large quantity of bad parts.

When there is doubt about the meaning of a drawing, it is a good idea to discuss the drawing with the person that made it. If the verbal requirements and the drawing requirements do not correspond, the drawing should be corrected to reflect the actual design needs.

The ability to interpret dimensions and tolerances is only an ability to understand what a drawing means when it is completed in compliance with the standard. *The ability to interpret a drawing is not an ability to guess what a drawing means when it is not in compliance with an established standard.*

DISCUSSION OF PRINCIPLES

Understanding the dimensioning and tolerancing methods well enough to apply and interpret them is very important. It is also important to be able to discuss the principles with other knowledgeable people. Care should be taken to discuss the principles and methods that are defined by the standard, and not to create new methods that conflict with those already standardized.

There is a wide range of dimensioning and tolerancing capabilities throughout industry. Many companies have utilized the principles covered in ASME Y14.5M for years. Others have just started to learn them, and a few have yet to begin. Even within companies that have started to use the methods, there is a wide range of capabilities.

Not everyone will have the same opportunities to work with the principles, and, therefore, will not learn them at the same rate. Each person will encounter different applications, and may understand how to apply the principles to some applications, and not fully understand others. Regardless of the reasons, various levels of capability do exist. It is likely, however, that anyone using the ASME Y14.5M principles will need to be able to explain what they have done. This is especially true when it comes to working with complex parts. It is also necessary to be able to understand the explanations of others as they describe new situations.

CALCULATIONS

Properly calculated tolerances take advantage of the maximum amount of part variation that can exist, yet still ensure proper function of the part. Many tolerance calculations are simple to complete; others require a significant amount of thought. An explanation of various tolerance calculation methods is given in the following chapters of this book.

Properly calculating tolerances is an important part of ensuring product quality and minimum cost. If tolerances are properly calculated, parts made within specified values will fit together correctly. If tolerances are calculated, the values are known to be the maximum permissible amount of variation. Large tolerances typically reduce part cost.

When design schedules are rushed, there may be a tendency to assign tolerance values without completing calculations. This is a risky situation. Parts made to "assumed" tolerances will not function correctly if the tolerances are larger than those resulting from calculations. If the assumed tolerances are smaller than necessary, the parts will cost more than expected. Product liability and the competitive industrial market should deter anyone from guessing at tolerance values. Records of calculations should be maintained to provide evidence of a good design.

Calculated tolerance values give the designer confidence the design will work. Anyone that has ever assigned tolerance values by "feel" or "guessing" knows the type of anxiety that is involved. A designer's career is directly impacted by how well that person's designs work. If tolerances are guessed, the designer must wait until parts are fabricated to determine if they will function properly. If the guesses are incorrect, then the chances of the design working correctly are significantly reduced. The wait and the uncertainty can be avoided by properly completing the tolerance calculations.

Designs do not always function as intended by the designer. When the first set of units are built, the designer must figure out what to do if the design fails to assemble or operate correctly. If tolerances are calculated, there is one less possible cause for the problem.

FUTURE TRENDS

A significant change now taking place in industry is the increased use of computers. Even small companies can now afford to use *computer-aided design (CAD)*. CAD is often referred to as *computer-aided design and drafting (CADD)*.

The minimum requirements for any CAD system is a computer and the CAD software. See Figure 1-6. A variety of computer models and software are available. The capabilities and prices of CAD systems vary. CAD systems range from inexpensive personal computers with some type of general CAD software packages to costly dedicated design systems utilizing specialized CAD software. The software used to develop the illustrations in this textbook is used by a large number of businesses.

A CAD system usually includes more than just the computer and software. It can also include pen plotters or other peripherals used to make *hardcopies* (paper copies) of the

Courtesy of Autodesk, Inc.

Figure 1-6. Computer-aided design (CAD) is an important part of our future.

CAD design files. Software for operating *numerically controlled (NC)* machines can also be utilized on the same computers as the CAD software. These software packages are usually considered separate from the CAD system.

Software for creating the computer files for running NC machines is commonly known as *computer-aided machining (CAM)* software. CAM files can be generated from CAD files. The designer creates CAD design files that define part requirements, and an industrial engineer or machine planner uses the CAD file to create an NC machining file.

Some companies are now attempting to eliminate the need for hardcopies. The intent is to have the designer create the computer file that defines the part requirements. Someone else then copies the design file and creates the NC machine file that runs the production machines. The NC machine file is created without the use of a paper drawing.

A design file can also be used to create a NC machine file to operate an inspection machine. One type of inspection machine is called a *coordinate measuring machine (CMM)*. One type of CMM is driven by a computer file that can be created from a design file. No paper drawing is needed to create the CMM file. Creation of the design file, CAM ma-

chining file, and CMM inspection file can be completed without the generation of any paper. This type operation is part of what is being referred to as the *paperless factory*.

How will the paperless factory affect the application of dimensions? The answer isn't yet clear. People may always want to see a drawing that they can look at to determine the required dimension values. The drawing may be on paper or on a computer display. At the present time, dimensions and tolerances are shown, regardless of whether a drawing is printed on paper or whether it is shown on a computer display. There may be a time when dimensions are not shown in the CAD file.

People only need to read dimensions if they are going to manually input machining data for fabrication or if inspection is to be completed manually. When all companies have the ability to create CAD files, drive NC machines with CAM generated files, and complete inspection of parts with CMM, then there may no longer be a need to dimension drawings. However, for the present time, it is still necessary to create drawings and to show the dimensions. The necessary technical advances and expense of those advances will result in dimensioned drawings remaining a part of our world for a long time.

How will the paperless factory affect the application of tolerances? Regardless of whether the NC and CMM machines get dimensions from a computer file or from manual input off a paper drawing, there must be some method for "telling" the machines how much variation is permitted. This means that tolerances must be assigned in some manner.

Even if dimensions are omitted from CAD design files someday, it is quite likely that some means for showing tolerances will always be required. One possibility is that CAD software may be written so that tolerance specifications are assigned as parameters associated with the features in the design file.

Future changes will probably eliminate or greatly reduce the number of paper drawings created. It will most likely eliminate the need to show dimensions on the CAD file, but will not eliminate the need to indicate tolerance requirements and specifications.

The changes that the future holds will require a continual learning process for people to remain competent in their professions. Learning the information in this book will provide the knowledge necessary to work to the requirements defined by the current dimensioning and tolerancing standard.

CHAPTER SUMMARY

Development of an accurate distance standard was an important part of making possible the accurate production of parts.

Compliance with dimensioning standards makes possible a common understanding of drawing requirements.

A team consisting of people from various disciplines can produce a better design than an individual working alone.

Many professions and trades are affected by dimensions and tolerances. The more knowledgeable people are about di-

mensioning and tolerancing, the more accurately and efficiently they can complete their daily jobs.

Required dimensioning and tolerancing skills include application, interpretation, and calculation. It is also an advantage if you are able to explain the meaning of various specifications on a drawing.

CAD is certain to be part of the industrial future, as is the need to define dimensional and tolerance requirements.

REVIEW QUESTIONS

Answer the following questions on a separate sheet of paper. Do not write in this book.

MULTIPLE CHOICE

1. The current distance standard for the meter is based on the _____.
   A. distance from the north pole to the equator
   B. wavelength of a certain color of light
   C. distance between the earth's poles
   D. distance around the equator
2. The _____ measurement unit must be used when working to the current dimensioning and tolerancing standard.
   A. inch
   B. millimeter
   C. inch or millimeter
   D. None of the above.
3. A(n) _____ must be able to understand dimensions and tolerances since he/she is responsible for making sure the produced parts meet the requirements of the drawing.
   A. inspector
   B. machinist
   C. design manager
   D. None of the above.
4. Producibility of a design can be improved the most by _____ working to optimize the design.
   A. a single designer
   B. a design team
   C. two designers
   D. None of the above.
5. Tolerance values should be _____.
   A. assigned on the basis of what seems good
   B. minimized to ensure proper function
   C. calculated to ensure proper function
   D. None of the above.

TRUE/FALSE

6. Measurements can be made more accurately when using the inch dimensioning system than when using the system international. (A)True or (B)False?
7. When the note "ALL DIMENSIONS ARE IN INCHES EXCEPT WHERE SHOWN OTHERWISE" is used, any metric dimensions must have the abbreviation "mm" applied as a suffix. (A)True or (B)False?
8. If a symbol does not exist for a special need, it is a good idea to create one and use it on the drawing. (A)True or (B)False?
9. Standard tolerance application methods are required since inconsistent application methods would result in unclear requirements being shown. (A)True or (B)False?
10. Working to a drawing with incorrectly applied tolerances is likely to result in disagreements about whether or not the parts are properly made. (A)True or (B)False?

SHORT ANSWER

11. Why is it necessary to have an accurate distance standard in industry?
12. Cite one example of how a machinist can help a designer show dimensions and tolerances in the best manner.
13. Show a note that would be included on a drawing that primarily has millimeter dimensions on it.
14. Explain why an incorrectly applied tolerance can be a problem.
15. Why is it necessary for a tool designer to be able to interpret dimensions and tolerances?

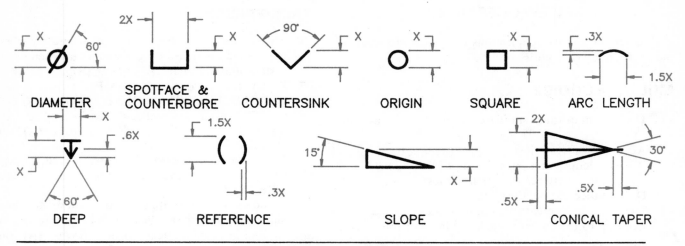

DIAMETER   SPOTFACE & COUNTERBORE   COUNTERSINK   ORIGIN   SQUARE   ARC LENGTH

DEEP   REFERENCE   SLOPE   CONICAL TAPER

DIMENSIONING SYMBOLS

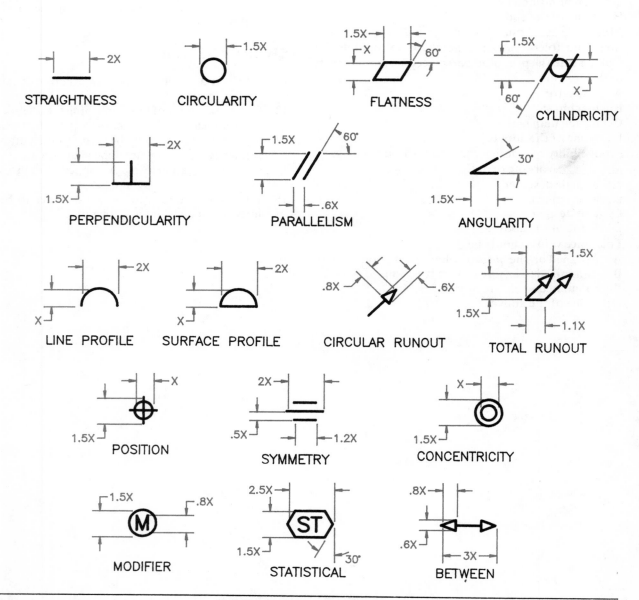

STRAIGHTNESS   CIRCULARITY   FLATNESS   CYLINDRICITY

PERPENDICULARITY   PARALLELISM   ANGULARITY

LINE PROFILE   SURFACE PROFILE   CIRCULAR RUNOUT   TOTAL RUNOUT

POSITION   SYMMETRY   CONCENTRICITY

MODIFIER   STATISTICAL   BETWEEN

TOLERANCING SYMBOLS

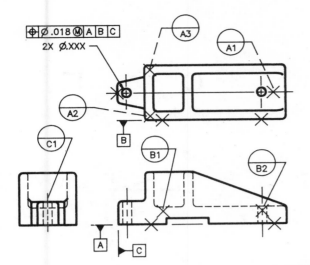

Chapter 2

Dimensioning and Tolerancing Symbology

CHAPTER TOPICS

☐ General dimensioning symbols and their common applications.
☐ Symbols used for application of tolerances.

☐ Proper composition of feature control frames and the effect of placement within a drawing.
☐ An introduction to datum identification and basic dimensions.

INTRODUCTION

Symbology is used to express dimensioning and tolerancing requirements on engineering drawings. The shapes of dimensioning and tolerancing symbols provide a logical connection between the symbol and the characteristic it represents. The easily identifiable shapes of the symbols permit a standard symbol set to be used internationally without the problems caused by differences in language.

It is important to create engineering drawings that have clearly defined dimensional requirements. The symbology defined by ASME Y14.5M makes a clear and consistent expression of requirements possible. The same level of clarity and consistency is difficult to achieve when using notations. The difficulty with notations is caused by various language mastery levels and personal writing styles.

GENERAL SYMBOLS AND ABBREVIATIONS

Some dimension values shown on a drawing require clarification through the use of symbols. As an example, a dimension applied to a circular part is clarified by a diameter symbol. The diameter symbol makes it clear that the specified value is a diameter, and not a radius or spherical diameter.

Standard dimensioning symbols, for characteristics such as a diameter, are used to provide consistency regardless of who creates a drawing. See Figure 2-1. The symbols currently defined by ASME are shown. Abbreviations are no longer illustrated in the standard. The omission of abbreviations from the standard is a clear indication that symbols are preferred on the drawing. However, abbreviations are still required for use in notes. Symbols are not to be used in notes.

Symbol size is established relative to the character height of dimension numerals. See Figure 2-2. The character height

is to be substituted for the "X" in each of the symbol size formulas. The formulas shown in the figure are recommended symbol sizes and are not absolute requirements.

GENERAL DIMENSIONING SYMBOLS		
CURRENT PRACTICE	ABBREVIATION IN NOTES	PARAMETER
∅	DIA	DIAMETER
S∅	SPHER DIA	SPHERICAL DIAMETER
R	R	RADIUS
CR	CR	CONTROLLED RADIUS
SR	SR	SPHERICAL RADIUS
⊔	CBORE or SFACE	COUNTERBORE SPOTFACE
∨	CSK	COUNTERSINK
⊽	DP	DEEP
○	—	DIMENSION ORIGIN
□	SQ	SQUARE
( )	REF	REFERENCE
X	PL	PLACES, TIMES
⌒	—	ARC LENGTH
▷	—	SLOPE
▷	—	CONICAL TAPER

Figure 2-1. Symbology is used in place of many abbreviations previously used for dimensioning.

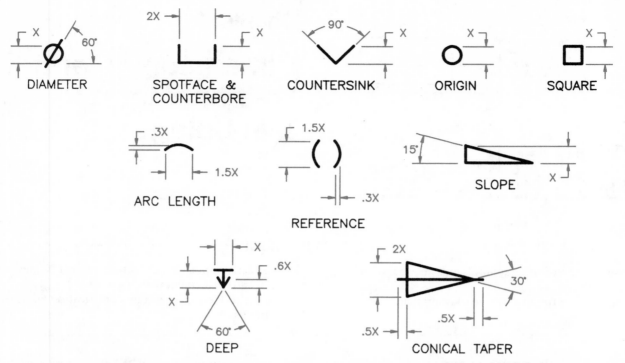

Figure 2-2. Dimensioning templates make it easy to control symbol proportions. The proportions shown in this figure can be used if a dimensioning template is not available.

The following example shows how a dimension for a symbol can be calculated.

Drawing Character Size	.125″
Illustrated Symbol Dimension	1.5X
Symbol Size	1.5 x .125″ = .188″

Symbols can be drawn with a template when manually creating drawings. Dimensioning symbol templates are available for standard character heights of .125″, .156″, 3mm, and 4mm.

Many CAD systems include a library of dimensioning symbols, and automatically size the symbols when they are applied to a drawing. A library of dimensioning symbols should be created by the operator of any CAD system that does not already contain them. A library of symbols permits the appropriate symbol to be pulled into a drawing without recreating the symbol each time it is needed.

The symbols in Figure 2-2 should not be used in general notes or text. Abbreviations are used in place of the symbols for these applications.

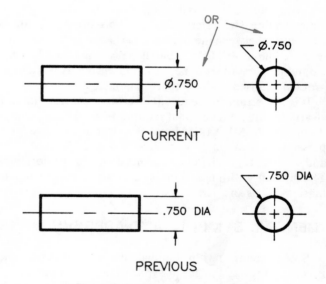

Figure 2-3. The diameter symbol.

SYMBOL APPLICATION

Symbol application requirements *do* exist. Correct dimensioning practices require placement of each symbol in a specific location relative to the associated numeric value. In most cases, the symbol precedes the value.

The *diameter symbol* is a circle with a line drawn through it. It always precedes the numeric value. See Figure 2-3. A full space is not left between the symbol and the number. An excessive amount of space should not be used or the symbol and number may appear detached. Zeros should not

have a line drawn through them to differentiate between them and the letter O. A line through a zero can cause it to be mistaken for a diameter symbol.

The abbreviation **R** is used as the *radius symbol*. This is not a change from previous practice. However, the R now precedes the numeric value. See Figure 2-4. Previous practice applied the R as a suffix to the dimension value. A full space is not permitted between the R and the number value. An excessive amount of space should not be used because the abbreviation and number would appear detached.

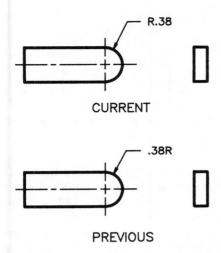

CURRENT

PREVIOUS

Figure 2-4. The radius abbreviation.

A *controlled radius* is specified in the same manner as a radius, but the letters **CR** are used as a prefix to the dimension value.

The *spherical diameter symbol* combines the letter **S** and the diameter symbol. It is placed in front of the dimension value. See Figure 2-5. There is no space between the **S** and diameter symbol.

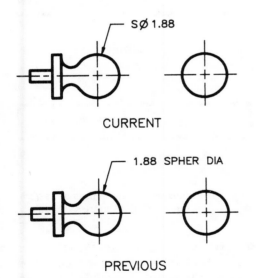

CURRENT

PREVIOUS

Figure 2-5. Spherical diameter is indicated with an S placed in front of the diameter symbol.

A *spherical radius* is indicated by the letters **SR** preceding the dimension value. No space is placed between the two letters. See Figure 2-6. A small space between the letters and dimension value is acceptable if the letters and numbers do not appear detached. A full letter space is not permitted.

Counterbores and *spotfaces* are indicated by the same symbol. See Figure 2-7. The symbol is shaped like the bottom of a counterbore. The counterbore symbol precedes the diameter symbol in the specification.

Depth can be specified in a dimension notation by placing the depth symbol in front of the depth value. The depth symbol is a downward-pointing arrow that extends from a horizontal line. See Figure 2-7. To avoid confusion, the

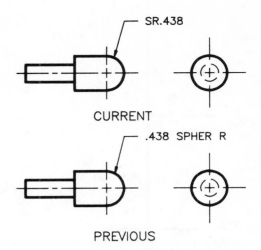

CURRENT

PREVIOUS

Figure 2-6. The spherical radius abbreviation.

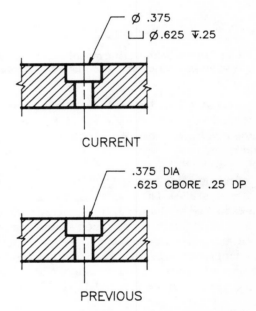

CURRENT

PREVIOUS

Figure 2-7. Application of the counterbore and depth symbols.

depth symbol should not be placed too close to the dimension value that precedes it.

The letter **X** can be used to indicate the number of *times*, or *places*, something is required. See Figure 2-8. The countersunk hole shown in this figure is required in six places; the notation for the hole is preceded by **6X** to indicate the required number of holes. There must not be space between the **6** and the **X**; an **X** has a different meaning if there is a space between it and the preceding number.

The letter **X** can mean "BY". It is used in this manner when noting chamfer sizes and countersinks. See Figure 2-8. The countersink shown in this figure is noted to have a size .438 diameter by 100°. *An X has the meaning of "BY" when there is a space on each side of the letter.* In the shown countersink notation, the letter **X** is used in one location to mean six times and in another location to mean .438″ BY 100°. Care must be taken to control the spaces around the letter **X** to ensure the proper meaning.

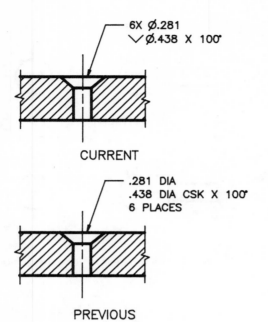

CURRENT

PREVIOUS

Figure 2-8. Two applications and meanings for the symbol X. Also an application of the countersink symbol.

A *countersink* is indicated by a V-shaped symbol. See Figure 2-8. This symbol has the appearance of a countersink, and therefore, a logical connection between the symbol and the required part geometry is easy to make. The diameter symbol follows the countersink symbol.

A dimension value enclosed in parenthesis is a *reference value*. See Figure 2-9. The proper usage of reference dimensions is explained in Chapter 3. If the reference dimension is a diameter or radius, the associated symbols may also be included inside the parenthesis.

A symbol is now available to indicate that a given dimension defines an *arc length*. The symbol is an arc drawn above the dimension value. See Figure 2-10. There was no provi-

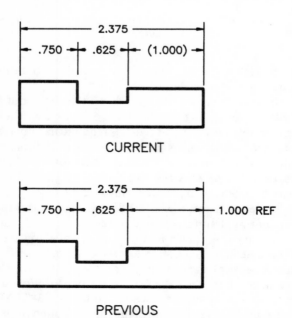

CURRENT

PREVIOUS

Figure 2-9. Indication of a reference dimension.

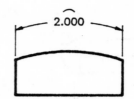

Figure 2-10. The arc length symbol applied to indicate the length of an arc.

sion for indicating arc length prior to the 1982 standard. Since previous practices did not define a method for indicating arc length, it was necessary to apply a notation to the dimension to explain the special requirement.

A small *square* is used as the symbol for dimensions that apply across the flats on a square shape. See Figure 2-11. The square symbol precedes the dimension value. A space is not used between the square and the dimension value. Previous practice used the abbreviation **SQ** as a suffix to the dimension.

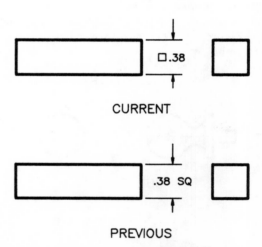

CURRENT

PREVIOUS

Figure 2-11. The distance across the flats of a square shape is indicated by using a square symbol.

An *origin symbol* is available for the rare occasion when it is necessary to indicate the origin of a dimension. A small circle is placed at the terminating point of the dimension line that is to serve as the origin. See Figure 2-12. No arrowhead is drawn at the end of the dimension line where the origin symbol is located.

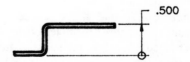

Figure 2-12. The origin symbol is used to indicate the surface from which measurements are to be made.

The *slope* on a flat surface can be specified using a symbol and the slope value. See Figure 2-13. The slope dimension is the amount of change in surface height per unit of distance along the base line from which height measurements are

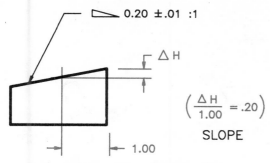

Figure 2-13. Slope symbol application.

made. If the change in height is specified in inches, then the unit length is one inch. The slope symbol is placed before the slope value. The complete dimension includes the amount of slope and tolerance on the slope per inch along the base line. The slope specification is written as a ratio. The specified slope value in Figure 2-13 indicates a rise of .20″ plus or minus .01″ per inch of distance (.20 ± .01 :1).

Several methods exist for dimensioning tapered parts. When the amount of taper per unit of length is specified, a *taper symbol* is used. See Figure 2-14. The symbol appears as a cone on an axis. The taper symbol precedes the taper value and its tolerance. The specified taper value in Figure 2-14 indicates a change in diameter of .18″ plus or minus .02″ per inch along the axis (.18 ± .02 :1).

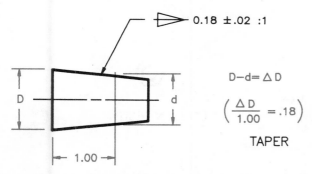

Figure 2-14. Taper on a diameter dimensioned using the taper symbol.

GEOMETRIC TOLERANCING SYMBOLS

Parts made in production must have tolerances applied to define the allowable variations from the nominal dimension values. Tolerances are required since it is not possible to produce every part to an exact size. The type and amount of tolerance depends on many variables, many of which are discussed in the following chapters.

The utilization of symbology for specification of tolerances has been evolving since MIL-STD-8 was issued in the 1940s. Initially, control of form, orientation, and position was achieved through notations placed on the drawing. As the variations in notations became an obvious problem, the utilization of symbols began to replace notations. Symbology is utilized for a majority of controls at this time. Only occasionally is a control needed that requires notations to supplement the existing symbology.

SYMBOL SHAPE AND SIZE

Tolerancing symbols, like general dimensioning symbols, have shapes that logically connect to the associated control. Each symbol has one clearly defined meaning. Correct utilization of standardized symbols provides a clear tolerance specification. Utilization of nonstandard symbols or misapplication of standard symbols can cause tolerances to be ambiguous.

Standardized symbols exist for specifying *form, orientation, runout, profile,* and *location* tolerances. There are additional symbols for clarifying and modifying the listed tolerance types.

Each symbol has a recommended size that is related to the general character height used on the drawing. See Figure 2-15. Illustrated symbol proportions include dimension values related to the variable "X." The value of "X" represents the character height on the drawing where the symbol is to be used.

Form tolerance symbols are used for specifying requirements that apply to individual features. The application and interpretation of form tolerance specifications is explained in Chapter 5.

The form tolerance symbols are *straightness, circularity, flatness,* and *cylindricity.* A straight line is used to indicate a *straightness* requirement. Since the symbol has the same shape as the desired control, it is easily identified. A circle is used to indicate a *circularity* requirement. As with the straightness symbol, this symbol resembles the indicated control. A parallelogram is used to indicate a *flatness* requirement. This symbol can be thought of as an oblique view of a flat surface. The *cylindricity* symbol is a circle with two tangent lines.

The symbols for specification of orientation tolerances are *perpendicularity, parallelism,* and *angularity.* See Figure 2-15. A *perpendicularity* tolerance is indicated by perpendicular lines. *Parallelism* is indicated by two inclined parallel lines, and the *angularity* symbol is two lines drawn to form a 30° angle.

There are two types of profile tolerances. See Figure 2-15. The *profile of a line* symbol is a semicircle. The *profile of a surface* is also indicated by a semicircle, but it has a horizontal line drawn across the bottom to distinguish it from the profile of a line symbol.

There are two types of runout tolerances. See Figure 2-15. A single arrow indicates *circular runout.* The arrowhead can be filled solid, or it can be left unfilled. *Total runout* is indicated by two arrows connected by a horizontal line. The total runout symbol that is shown in the figure was first made a requirement in the 1982 issue of the dimensioning standard.

Position tolerances are indicated by a symbol composed of two lines crossing at the center of a circle. See Figure 2-15. It is symbolic of the center lines used on a drawing to indicate the desired location for a hole. Another type of location tolerance is *concentricity.* It is indicated by two concentric circles.

A third type of location tolerance is *symmetry.* Symmetry may be indicated by either of two types of symbols, depending on the type of control desired. A position symbol may

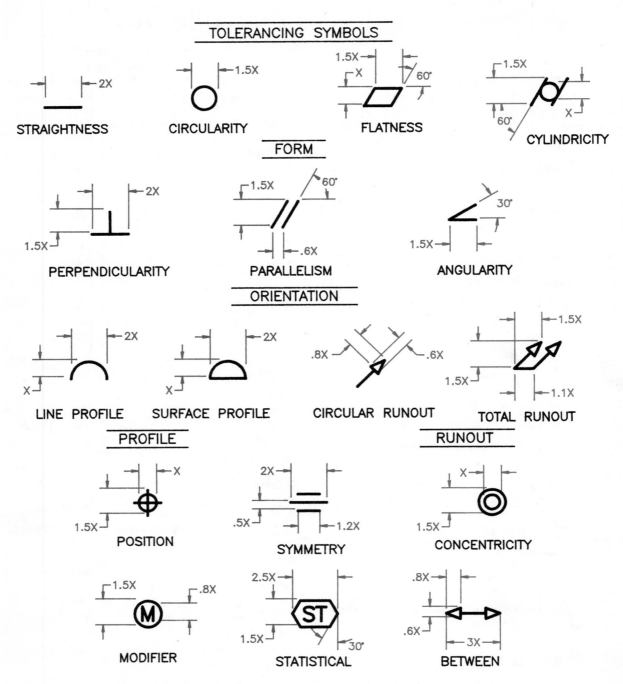

Figure 2-15. Tolerance symbols are shown grouped by tolerance categories.

be used when specifying the symmetrical location of features of size at MMC. A symmetry symbol that consists of three horizontal lines may be used for other symmetry tolerances. The symmetry symbol containing three lines was not included in the 1982 standard.

Modifiers are often applied to tolerance specifications. There are several modifiers, and the proportions shown in Figure 2-15 apply to all of them.

A tolerance calculated by statistical methods is identified with a special symbol. The symbol for a *statistical* value is an "elongated hexagon" containing the letters **ST**.

A double ended arrow is used in place of the word *between*. It Is used for tolerance specifications that extend be-

tween two points.

Other Symbols

A material condition modifier applies to every geometric tolerance that is attached to a feature of size. The modifier can be drawn in the tolerance specification, or one can be assumed on the basis of rules to be defined in later chapters.

There are two material condition modifiers illustrated in the standard. See Figure 2-16. If neither modifier is shown in a tolerance specification, the tolerance is assumed to apply regardless of feature size. It is still permissible to show the RFS symbol, but it is not illustrated in the standard. The preferred practice is to omit the RFS symbol.

Material condition modifier symbols are used only on the field of the drawing. They are not used in drawing notes or other written text. If it is necessary to reference a material condition modifier in written text, the term is spelled out or the appropriate abbreviation is used.

Some tolerance applications require the specified tolerance zone to project outside the object. When this is necessary, a *projected tolerance zone* symbol is used. See Figure 2-16. The projected tolerance zone symbol is the letter P within a circle.

Tolerances intended to apply in the *free state* are indicated with a letter F inside a circle. Tolerances that apply to a *tangent plane* instead of to the surface are identified by a letter T inside a circle.

Profile tolerances can be applied to extend all around the profile of the view to which the tolerance is applied. Multiple tolerance specifications can be avoided on a view by using an *all around* symbol. See Figure 2-16. The all around symbol is a circle placed at the corner in the tolerance application leader. It is similar to the all around symbol used in welding specifications.

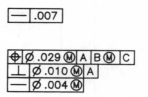

Figure 2-17. A feature control frame can be composed of one or more lines.

COMPOSITION OF THE FEATURE CONTROL FRAME

Every feature control frame follows the same order of specification, reading from left to right. See Figure 2-18. The tolerance symbol (characteristic) is always shown first at the left end of the feature control frame. It is followed by the tolerance value. Datum references, if required, follow the tolerance value.

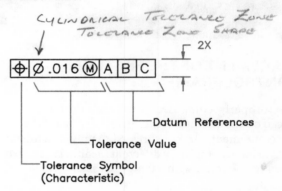

Figure 2-18. Feature control frames are always read from left to right.

Words are not generally written inside a feature control frame. Some special applications do include the abbreviation MAX, but symbols, numbers, and datum reference letters are generally adequate for specification of standardized tolerance controls.

The tolerance value in a feature control frame may be no more than a number indicating the size of the tolerance zone. It is also possible for the tolerance value to include a diameter symbol, a material condition modifier, a statistical tolerance symbol, and a tangent plane symbol. See Figure 2-19. When used, the diameter symbol precedes the tolerance value. It is required if the tolerance zone is round or cylindrical. If a material condition modifier is to be shown, it follows the tolerance value.

Datum references are required in all feature control frames, except those specifying form or profile tolerances. The number of referenced datums depends on the tolerance being specified and the control that is desired.

Datum references are always written left to right with the primary datum shown first. See Figure 2-19. It is followed by the secondary and tertiary datum references. Material condition modifiers can be applied to the datum references.

Figure 2-16. Modifier symbols are used only within feature control frames.

Ⓜ	MAXIMUM MATERIAL CONDITION (MMC)
Ⓛ	LEAST MATERIAL CONDITION (LMC)
Ⓟ	PROJECTED TOLERANCE ZONE
Ⓕ	FREE STATE
Ⓣ	TANGENT PLANE
⌀⟋	ALL AROUND

FEATURE CONTROL FRAMES

Geometric tolerance requirements are specified in *feature control frames*. They can contain no more than a tolerance symbol and a tolerance value, or they can contain multiple lines of requirements that include tolerance symbols, tolerance values, modifiers, and datum references. See Figure 2-17. The amount of information contained in a feature control frame depends on the desired level of control to be established.

Determining how to show the required information in a feature control frame is a primary subject of this book. The methods for completing each type specification are described in several chapters. The following segments of this chapter describe the required composition for a feature control frame.

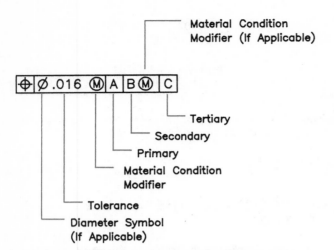

Figure 2-19. Whether a diameter symbol and material condition modifier are used, or omitted, depends on the desired tolerance specification and the type of feature being controlled.

Modifiers are only required on references to datum features of size, and then the modifiers are applied in compliance with rules explained in following chapters.

PLACEMENT OF THE FEATURE CONTROL FRAME

A properly composed feature control frame must be placed on the drawing in the correct manner if the appropriate requirement is to be expressed. Applying a tolerance to a surface has a completely different meaning than associating the same tolerance with a feature of size.

Application to Surfaces

Placing the feature control frame on an extension line from a surface indicates the specification of a requirement for that surface. Feature control frames can be placed on either side of extention lines. The side of the extension line on which it is located makes no difference. See Figure 2-20.

A leader can also be used to connect a tolerance specification to a surface. The use of a leader has exactly the same effect as attaching the specification to an extension line. The

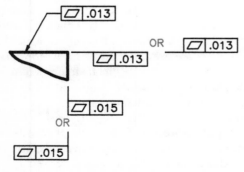

APPLICATION TO SURFACES

Figure 2-20. The end to which a leader or extension line attaches on a feature control frame does not affect the interpretation of the requirements.

leader can be extended from either end of the feature control frame.

Application to Features of Size

Tolerances can be associated with a feature of size by placing the feature control frame adjacent to a dimension or attaching it to the dimension line. See Figure 2-21. When a feature control frame is placed adjacent to a dimension value, care must be taken that confusion doesn't exist regarding other dimensions in the vicinity. Always make sure that the feature control frame is clearly associated with only one dimension.

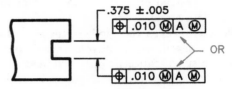

APPLICATION TO FEATURES OF SIZE

Figure 2-21. Placement of a feature control frame on the dimension line or near the dimension value indicates a control on the feature of size.

Application to Threads

A geometric tolerance applied to a thread is either placed adjacent to the thread specification or is attached to the threaded feature by a leader. See Figure 2-22. *Any tolerance applied to a thread and shown in a feature control frame applies to the pitch diameter unless noted otherwise.* A majority of the tolerances applied to threads should apply to the pitch diameter.

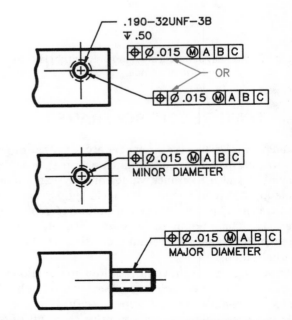

Figure 2-22. A notation must be placed under the feature control frame to control a feature other than the pitch diameter of a threaded part.

Sometimes it is desirable to specify a geometric tolerance that controls the major or minor diameter of a thread. This can be done by noting the applicable diameter beneath the feature control frame. See Figure 2-22. In the given examples, the minor diameter is controlled on an internal thread, and the major diameter is controlled on an external thread.

Application to Gears and Splines

Geometric tolerances applied to gears and splines are not assumed to control a specific diameter. Control of any feature requires a notation. When control of the pitch diameter is desired, a notation of PITCH DIAMETER is placed under the feature control frame to prevent any possible confusion about the feature that must be controlled.

DATUM IDENTIFICATION

Datum features are the surfaces and features of size used to establish the locations for datum planes and datum axes. Standard symbols are used to identify datum features.

DATUM FEATURE SYMBOL

The *datum feature symbol* is a square drawn around a letter. See Figure 2-23. The point of application is shown by a datum feature triangle. The triangle may be filled or unfilled. The datum feature symbol and triangle may be used to identify a surface or feature of size as a datum feature. The given example only shows the symbols as they apply to surfaces. Datum letters are assigned alphabetically, omitting letters, I, O, and Q. No two datums on a single part should have the same datum letter.

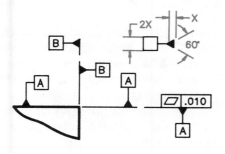

APPLICATION TO SURFACES

Figure 2-23. Datum feature symbols can be associated with the feature in any of several ways.

Three methods are used to show that a datum feature symbol applies to a specific surface. The triangle can be placed on an extension line from the surface. The side of the extension line to which the symbol is applied is not important. A second means for applying the symbol is to attach the triangle directly to the surface with a leader. The triangle can also be used to attach the datum feature symbol to a feature control frame.

DATUM TARGET SYMBOL

Surface irregularities or part size can prohibit using an entire feature to locate a datum. Datum targets can be identified on a part to require specific locations on the datum features to be used for establishing datum locations. See Figure 2-24. Datum target symbols are used to identify each datum target. These symbols are not drawn at the target location, but are connected to target locations with a leader. No arrowhead is drawn on the datum target symbol leader.

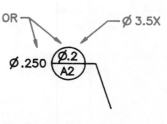

CURRENT PRACTICE

Figure 2-24. Datum targets must each be identified with a datum letter and number.

The *datum target symbol,* implemented in 1982, is a circle with a horizontal line drawn across it. The datum identifying letter and a number are written in the lower half of the circle. The letter identifies the datum. The number is the target number for the particular datum. Targets for each datum are numbered beginning with 1.

If datum target areas are used, the target diameter can be written in the top half of the circle. Unless single digit decimals are being used, the diameter values will not fit in the symbol. If two- or three-place decimals are being used, the diameter can be placed adjacent to the symbol.

DATUM TARGETS

Datum target symbols can be attached to any of three types of *datum targets*. The target type shown on the part depends on the design function and the characteristics of the part. The proper application of targets is explained in the chapter on datums.

Datum target points are indicated with two perpendicular lines oriented to appear as the letter X. See Figure 2-25. The size of the symbol prevents it from being mistaken for an **X**. *Datum target lines* are indicated with a phantom line. *Datum target areas* are outlined with a phantom line, and filled with cross hatching.

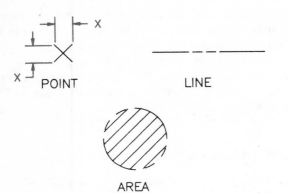

POINT

LINE

AREA

Figure 2-25. There are three types of datum targets

BASIC DIMENSIONS

Every feature of size and every feature location on a drawing is assigned a dimension value. Perfect fabrication processes do not exist, therefore, an amount of permissible variation must be defined for every feature of size and every location.

Permissible variation for any one dimension can be assigned with a general note, title block tolerance, tolerance shown on the dimension, or through the use of a feature control frame. If a feature control frame is used to specify a tolerance, then it is not desirable to have confusion about whether or not other tolerances apply. To avoid confusion, dimensions must be identified as basic when the tolerance is shown in a feature control frame.

A *basic dimension* is a theoretically exact value for which there is normally a tolerance shown in a feature control frame. A basic dimension is not an indication of zero tolerance for a dimensioned feature. Anytime a basic dimension is seen, it is known that a feature control frame, or other control, identifies the amount of tolerance for the dimensioned feature.

BASIC DIMENSION SYMBOL

A rectangle placed around a dimension on the field of the drawing identifies that dimension as basic. See Figure 2-26. The title block tolerances do not apply to that dimension, and a feature control frame, or other control, expresses the amount of tolerance for the dimensioned feature.

NOTED BASIC DIMENSIONS

Drawing notes can be used to indicate that dimensions are basic. See Figure 2-27. If all dimensions on the drawing

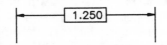

Figure 2-26. A rectangle around a number identifies it as a basic dimension.

ALL DIMENSIONS ARE BASIC

OR

ALL UNTOLERANCED DIMENSIONS
ARE BASIC

Figure 2-27. Notes can be used to indicate that dimensions are basic.

are basic, a general note to that effect can be used. Should it be necessary to use plus or minus tolerances on some of the dimensions, a general note can be used to state that untoleranced dimensions are basic. Title block tolerances do not apply to any dimension if either of the two notes shown in Figure 2-27 are used.

PAST PRACTICES

Existing drawings and those being completed under continuing contracts may require knowledge of previously used methods.

PAST SYMBOLS

The symbology used for tolerance specification has gone through some changes as the dimensioning and tolerancing standard evolved. Knowledge of past practices is needed for those situations in which an old drawing must be used or when a continuing contract still references an old version of the dimensioning and tolerancing standard.

Prior to 1982, the total runout symbol with two arrowheads was not used in the United States. See Figure 2-28. The previously used total runout symbol looked like the symbol for circular runout, but the word TOTAL was written beneath the feature control frame.

A separate symbol existed for symmetry tolerances prior to the 1982 standard. The 1982 version of the standard eliminated the symmetry symbol, and the position symbol was defined to be correct for specification of symmetrical location requirements. The 1994 standard reinstated the symmetry symbol.

A regardless-of-feature-size modifier was included in the standard prior to 1994. Effective in 1994, tolerances are assumed to apply RFS except when indicated otherwise. Since tolerances are now assumed to apply at RFS, the symbol is no longer needed.

The datum feature symbol prior to 1994 was a rectangle. The rectangle included a datum identifying letter with a dash at each side of the letter.

Previous practice required the utilization of a circle divided into quadrants for the datum target symbol. See Figure 2-29. The datum letter was placed in the upper left quadrant, and the number was placed in the lower right quadrant.

PAST FEATURE CONTROL FRAME FORMAT

Feature control frames can look different, depending on which of the previous standards was in effect when the drawing was completed. The difference in appearance may be caused by previously permitted options on the datum reference location within the feature control frame. Another difference is that rules regarding the usage of material condition modifiers have changed.

Current practices require that datum references be shown last in the feature control frame. In accordance with the 1973 standard, datum references were permitted in either of two locations—they could be shown between the tolerance symbol and the tolerance value, or they could be shown in the same manner as defined for present practices. See Figure 2-30. Prior to 1973, datum references were required to be located between the tolerance symbol and tolerance value.

Prior to 1982, the diameter symbol was not required and the diameter abbreviation (DIA) was used. DIA followed the tolerance value.

A significant difference in the feature control frames involves material condition modifiers. Prior to 1982, position tolerances were assumed to apply at the maximum material condition (MMC) if no modifiers were shown in the feature control frame. Assumptions **were not permitted** for positional tolerances in 1982. The current standard assumes the RFS modifier on all tolerances.

PAST BASIC DIMENSIONS

Old drawings might include the word BASIC or the abbreviation BSC adjacent to some dimensions. These notations were permissible under previous issues of the dimensioning and tolerancing standard. As part of the effort to replace words with symbology, the notations are no longer used.

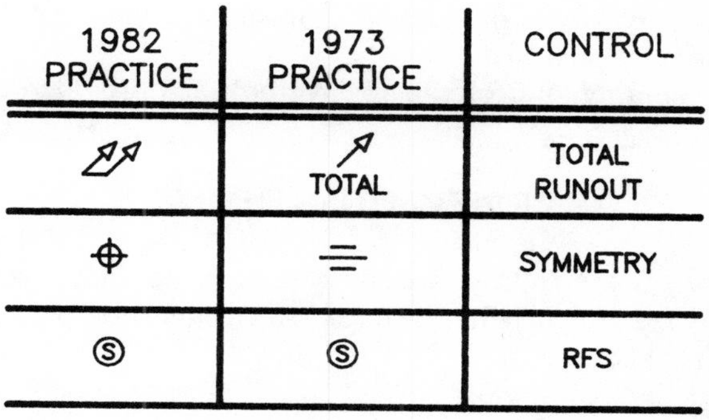

1982 PRACTICE	1973 PRACTICE	CONTROL
↗↗	↗ TOTAL	TOTAL RUNOUT
⊕	⸗	SYMMETRY
Ⓢ	Ⓢ	RFS

Figure 2-28. Some tolerance symbols have changed as the dimensioning and tolerancing standard evolved.

1973 PRACTICE

Figure 2-29. Prior to 1982, the datum target symbol had a vertical line and a horizontal line that divided the symbol into quadrants.

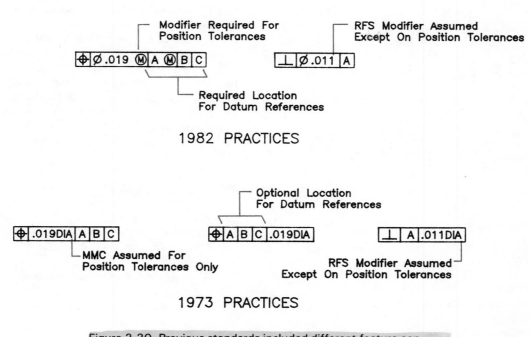

Figure 2-30. Previous standards included different feature control frame format requirements.

Symbology is replacing the previously used notations and abbreviations that were applied on engineering drawings.

Symbology for general dimensioning and tolerance application is well defined.

Dimensioning and tolerancing symbols can provide better drawing clarity than can be achieved through the utilization of notes.

There are rules for applying dimensioning and tolerancing symbols. These rules must be followed properly if the full advantage of symbology is to be achieved.

REVIEW QUESTIONS

Answer the following questions on a separate sheet of paper. Do not write in this book. Accurately complete any required sketches.

MULTIPLE CHOICE

1. _____ reduce the number of words used on a drawing.
   A. Notes
   B. Symbols
   C. Dimensions
   D. Tolerances

2. Symbols are _____ abbreviations.
   A. preferred in place of
   B. being gradually replaced by
   C. not defined as well as
   D. no longer used in place of

3. The proportions for dimensioning symbols as shown in the Appendix of the standard are _____.
   A. based on a fixed character height of .125″
   B. based on the character height used for dimensions
   C. related to the character height of the drawing title
   D. modified depending on the drawing scale

4. Dimensioning symbols are usually placed _____ the number to which the symbol is applied.
   A. under
   B. over
   C. after
   D. in front of

5. There are _____ form tolerancing symbols.
   A. no
   B. two
   C. four
   D. five

6. Perpendicularity is a type of _____ tolerance.
   A. form
   B. orientation
   C. position
   D. angularity

7. The first cell in a feature control frame _____.
   A. must always show the tolerance symbol
   B. sometimes includes a diameter symbol
   C. shows either a tolerance symbol or datum reference
   D. None of the above.

8. A feature control frame can include a diameter symbol in the cell with the _____.
   A. tolerance value
   B. datum references
   C. tolerance symbol
   D. None of the above.

9. Position tolerances applied to threads control the location of the _____, unless noted otherwise.
   A. mating part
   B. pitch diameter
   C. major diameter
   D. minor diameter

10. A datum feature is identified with a _____.
    A. letter inside a square
    B. number inside a rectangle
    C. note
    D. Either A or B.

11. One dimension on a drawing can be made basic by _____.
    A. drawing a circle around it
    B. using a note
    C. drawing a rectangle around it
    D. All of the above.

TRUE/FALSE

12. The shapes of tolerancing symbols do not have any relationship to the indicated control. (A)True or (B)False?

13. The letter X with a space on each side of it means BY. (A)True or (B)False?

14. Flatness is a form tolerance. (A)True or (B)False?

15. Material condition modifier symbols may be used in drawing notes. (A)True or (B)False?

16. Some feature control frames are not required to include datum references. (A)True or (B)False?

17. A dimension value can be identified as basic. (A)True or (B)False?

FILL IN THE BLANK

18. The symbol **X** has _____ meanings.

19. How many orientation tolerances exist?

20. Form and orientation tolerances are specified in a _____.

21. A rectangle drawn around a number indicates the number is _____.

22. The abbreviation for maximum material condition is _____.

SHORT ANSWER

23. Describe what the letter X means when no space is placed between it and a preceding number.

24. List the form tolerances.
25. Sketch each of the following tolerance symbols.
    A. Straightness
    B. Perpendicularity
    C. Total runout
    D. Line profile
    E. Position
26. Sketch each of the following dimensioning symbols.
    A. Diameter
    B. Counterbore
    C. Spotface
    D. Countersink
    E. Slope
27. Prior to 1982, the datum references in a feature control frame could appear in what location(s)?
28. Show the two material condition modifier symbols and label each of them.

APPLICATION PROBLEMS

Some of the following problems require that a sketch be made. All sketches should be neat and accurate. Each problem description requires the addition of some dimensions for completion of the problem. Apply all required dimensions in compliance with dimensioning and tolerancing requirements.

29. Show a 1.125″ diameter for the given shaft.

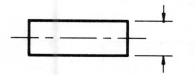

30. Show a .625″ radius on the inside corner.

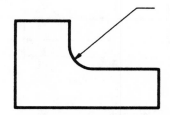

31. Use symbology to specify the counterbore and hole size in a note attached with a leader. The hole has a .282″ diameter. The counterbore is .438″ diameter and .375″ deep.

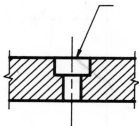

32. Use symbology to specify the countersink and hole size in a note. The hole has a .218″ diameter. The countersink has an 82° angle and a .438″ diameter.

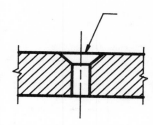

33. Make the 1.25″ dimension a reference value.

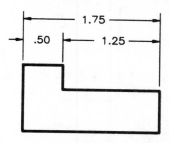

34. The given flatness tolerance specification is to be drawn so that it applies to only the top surface of the given part.

⬜ .013

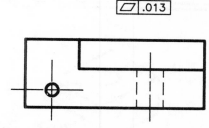

35. In accordance with the 1973 standard, the given tolerance specification is assumed to have what material condition modifier?

⊕ ⌀ .016 A B C

36. In accordance with the current standard, the given tolerance specification is assumed to have what material condition modifier?

⊕ ⌀ .016 A B C

37. Make all hole location dimensions basic.

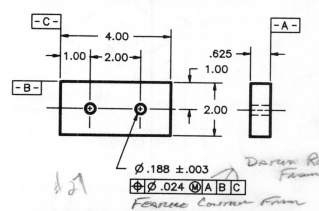

Ø .188 ±.003

⊕ ⌀ .024 Ⓜ A B C

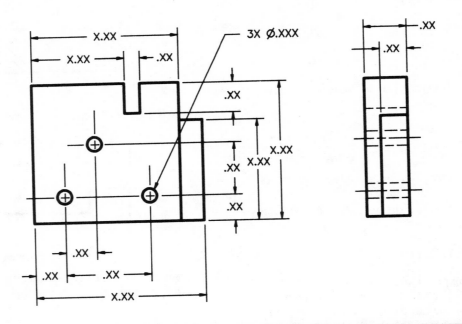

3X Ø.XXX

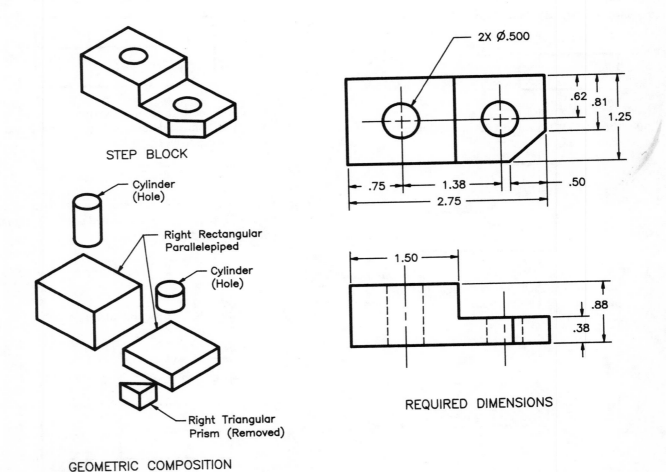

STEP BLOCK

Cylinder (Hole)

Right Rectangular Parallelepiped

Cylinder (Hole)

Right Triangular Prism (Removed)

GEOMETRIC COMPOSITION

2X Ø.500

.62
.81
1.25

.75
1.38
.50
2.75

1.50

.88
.38

REQUIRED DIMENSIONS

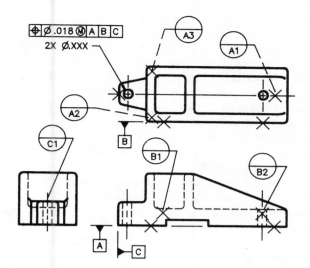

Chapter 3

General Dimensioning Requirements

CHAPTER TOPICS

□ Dimensioning methods including standard line types, lettering sizes, arrowhead form, and the proper utilization of these parameters for dimensioning.

□ General dimension systems including how to apply chain, baseline, rectangular coordinate, and polar coordinate dimensions.

□ Guidelines for dimension placement to provide clear part requirements specification.

□ General and specific drawing notes.

□ Definition of general categories of fit between mating features of size.

INTRODUCTION

Dimensions applied to multiview drawings provide the sizes and locations of features. Proper application of dimensions make size and location requirements easier to understand and increases the probability that the parts will be properly produced. A standard published by the American Society of Mechanical Engineers (ASME)—ASME Y 14.5M-defines the guidelines for application of dimensions.

Proper application of the size and location dimensions, supplemented by tolerance specifications as defined in the following chapters, provides complete dimensional control for a part. Dimensional specifications must be shown on any multiview drawing that is to be used for the production of parts. This is true regardless of whether the multiview drawing is produced by manual means or by a computer-aided design (CAD) system.

It is possible that a CAD-generated design will be fabricated directly from the design file data without the generation of a multiview drawing. If a CAD design file is used without the generation of a multiview drawing, dimensions defining locations, sizes, and tolerances may still need to be added to the file. Although all entities in the design file have dimensional data associated with them, this data does not indicate allowable variations. The allowable variations are defined through the application of dimensions and tolerances.

CAD file entities may have attributes assigned to them to indicate allowable tolerances at some point in the future. However, the methods for identifying allowable variations on design file entities have not been standardized at this time.

DIMENSIONING METHODS

Standard dimensioning methods are used to provide maximum clarity and consistency in the presentation of product requirements. Dimensioning systems are universally understood throughout the United States since a single national standard defines dimensioning methods.

A single standard is necessary so that information can be easily interchanged throughout the United States. If companies were to establish their own dimensioning methods, reading drawings from various companies would require understanding each company's guidelines. Since this would cause confusion and waste time, a national standard that defines most dimensioning applications is necessary.

It is impossible to write a single document that addresses every possible dimensioning situation. Therefore, the national standard addresses common situations and requirements through a set of principles that can be extended to most applications. Since there may be an occasional special situation that can't be addressed by extension of the standard principles, individual companies might want to establish their own drafting and dimensioning requirements for special applications. Care should be taken not to create guidelines that conflict with the national standard.

LOCATION AND SIZE DIMENSIONS

All dimensions applied to a part show either size or location. See Figure 3-1. A *location dimension* (specified with an L in the figure) describes *where* a feature is, and a *size dimension* (specified with an S in the figure) indicates how large a feature is. Location and size must be controlled in all three axes.

Most feature locations are specified with one or two dimensions. This is possible since the remaining location dimension is defined by other features on the part. The location of a drilled hole is an example; two coordinates locate its center, while the surface into which it is drilled locates the start of the hole.

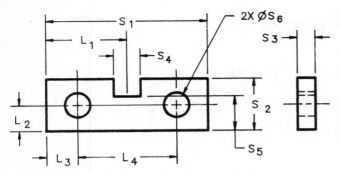

Figure 3-1. The letter S indicates size dimensions, and the letter L indicates location dimensions.

Size can be completely dimensioned through several means. A rectangular prism is sized by dimensioning the height, width, and depth. A hole can be sized by giving its diameter and depth. If a hole goes through a part, its size is specified by giving only the diameter. The part thickness defines the length of a through hole.

LINE USE

The three line types used for dimensioning are extension lines, dimension lines, and leader lines. All of them are drawn with thin, dark lines. The thin, dark lines provide enough contrast with the wide object lines to avoid confusion between the object outline and the dimensions. Good control of line quality is essential to provide maximum readability of a dimensioned drawing.

Extension Lines

Extension lines are used to extend features on an object to allow the application of dimensions. See Figure 3-2. ASME Y14.5 requires that a visible gap exist between the object and extension line. It is generally a good practice to begin an extension line approximately .031″ to .062″ from the feature. These lines extend approximately .125″ beyond the last dimension line. CAD systems often have these predetermined distances as system defaults, but many of the CAD systems also permit adjustment of the preset values.

Extension lines may cross object lines or other extension lines. No breaks are made in an extension line when it crosses an object line or another extension line. Breaks in extension lines cause a discontinuity that makes the drawing harder to read. It is only acceptable to break an extension line if it crosses or is sufficiently close to an arrowhead as to cause confusion.

Extension lines are perpendicular to the dimension lines to which they extend, and are *normal* to the feature from which they extend. An extension line from the center of a circular feature starts adjacent to the centerline cross and extends to the dimension line. CAD limitations may require that the extension line start at the center point.

Sometimes, there is a need for a special treatment of extension lines. Special treatments include an offset (dogleg) inserted in an extension line and oblique extension lines. These special treatments are defined in the section of this chapter regarding ordinate dimensioning.

Dimension Lines

Dimension lines show the direction and magnitude of a dimension. The direction of a dimension line is parallel to the distance being specified. The dimension magnitude is shown by the value inserted in the dimension line. Arrowheads are placed at each end of a dimension line to show the point of application.

There are four arrangements in which the dimension line and values can be shown. See Figure 3-3. The preferred arrangement is with the arrowheads and value inside (between) the extension lines. This arrangement is used when sufficient space exists between the extension lines to show the arrowheads, dimension line, and value.

When the distance between extension lines is limited, one of the following arrangements is used:

1. The arrowheads can be placed inside and the value outside.
2. The value can be placed inside and the arrowheads outside.
3. Both the arrowheads and value can be placed outside.

When a combination of dimension line and value arrangements is used, it is acceptable for one arrowhead to be used for two dimensions as shown in Figure 3-3.

Dimension lines must be adequately spaced to make them easy to read. The minimum recommended space between the outline of the part and the first dimension line is .40″ (10mm). The minimum recommended distance between succeeding dimension lines is .24″ (6mm). Preferable spacing is .50″ from the object and .38″ between succeeding dimension lines. These distances are general guidelines; variation is permitted, provided the dimensions do not run together when reduced-size copies of the drawing are made. The spacing used between dimension lines should appear constant to provide an easy-to-read drawing.

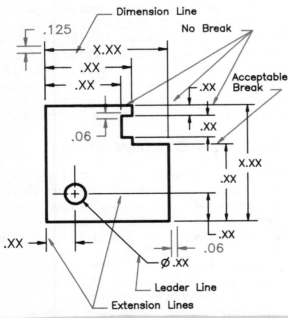

Figure 3-2. Extension lines are used to extend features for the application of dimensions.

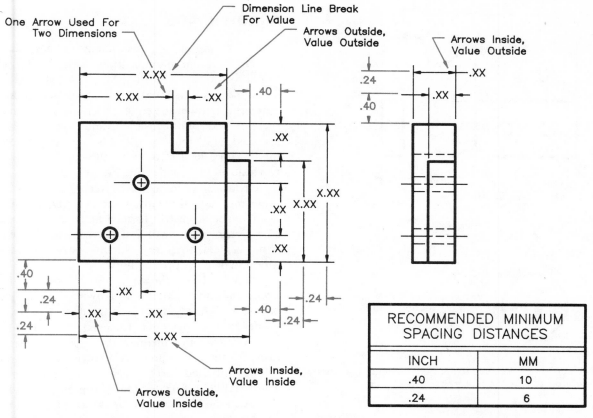

Figure 3-3. Proper dimension line spacing allows each dimension to be seen clearly. Minimum recommended spacing is shown.

RECOMMENDED MINIMUM SPACING DISTANCES	
INCH	MM
.40	10
.24	6

The distance to the first dimension line is measured from the outline of the object. If there is an offset in the part outline, it is preferable to place the first dimension outside the outline of any adjacent features. See Figure 3-4. A remotely located feature does not affect the location of the first dimension.

Leader Lines

Leader lines, or *leaders,* are used to connect information, such as a note or symbol, to a specific feature on the part. See Figure 3-5. The leader line extends from the first or last character in the note. A short horizontal line normally extends from the first or last character, and a leader extends from the horizontal line to the noted feature. Some companies omit the horizontal line. The leader is terminated with an arrowhead or dot. An arrowhead is used if the leader terminates on the profile of the feature, and a dot is used if the leader terminates on a surface.

The horizontal bar on the leader is drawn with a visible gap between it and the character. The size of an arrowhead is proportional to the line width. Dot terminators are .062″ in diameter.

Leader lines are not drawn horizontally or vertically; they are always inclined. Horizontal or vertical leaders could be confused with dimension and extension lines. Leaders can cross object and extension lines and are not broken at the

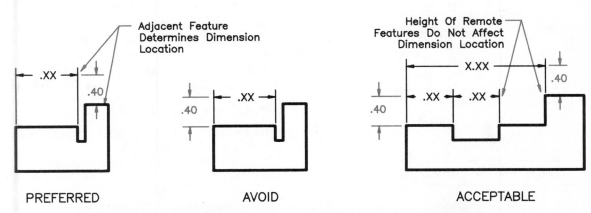

Figure 3-4. Adjacent features must be considered when locating dimensions.

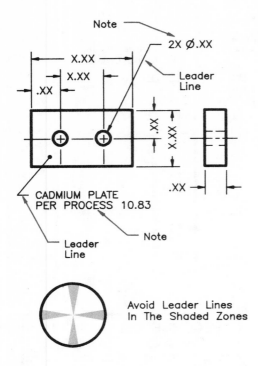

Figure 3-5. Leader lines are used to connect notes and drafting symbols to specific features on a drawing.

ARROWHEADS

Arrowheads are used on dimension lines and leader lines to show the point of application. Arrowhead size is proportional to the line width (ASME Y14.2M-1992). However, a common practice that results in an acceptable size is to make the arrowheads equal in length to the character size (height) used for dimensioning and notes. If .12″ characters are used, .12″ long arrowheads are used. Arrowheads are made with a length-to-width ratio of 3:1. A .12″ long arrowhead is .04″ wide. See Figure 3-6.

Arrowheads can be drawn freehand, by template, or by other mechanical means, such as CAD. Freehand arrowheads must be clearly formed and relatively consistent in size. Conformance with exact arrowhead size requirements is not critical, therefore estimated sizes can save drawing time and increase productivity.

All arrowheads shown in ASME Y14.5 are filled. Many companies do not fill their arrowheads. The time required to fill arrowheads is not well spent, and the extra graphite on pencil drawings tends to smear. The filled arrowheads in CAD drawings increase data storage space and slow response time on some CAD software. A two- or three-stroke unfilled arrowhead can save time. ASME Y14.2M-1992 illustrates four acceptable arrowhead styles. Although the filled arrowhead is recommended by Y14.2M, all four styles are acceptable.

intersections. Intersections between leader lines and dimension lines are to be avoided. A leader line should be broken if it crosses an arrowhead.

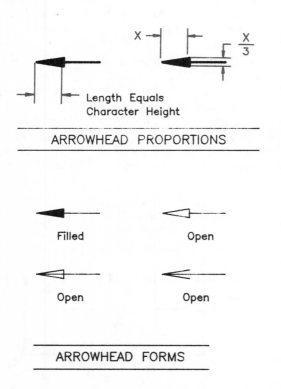

Figure 3-6. Arrowheads may be open or filled.

CHARACTER SIZE AND STYLE

The height of all numbers and letters used for the dimensions and notes on a drawing must be the same. Character size for all drawings sizes is standardized at 3mm (approximately .125″). The 3mm character is adequate for making microfilm copies of drawings if the characters are clearly formed. Characters this small will not reproduce well if they are poorly formed. Gothic and microfilm lettering styles and sizes for engineering drawings are defined in ANSI Y14.2M.

Past practice established character size on the basis of drawing sheet size. Drawing sheets that were 17″ x 22″ or smaller required a minimum character height of .125″ (1/8″). Drawing sheets larger than 17″ x 22″ required a minimum character height of .156″ (5/32″). The larger character size of large sheet sizes permitted clear reproduction of the characters on reduced copies.

Only one letter style should be used on a drawing. Even when an existing drawing is revised, the letter style used for the revision should match the letter style already on the drawing.

Hand-drawn numbers and letters should always be drawn using guidelines to control character height and line spacing. Guidelines should locate the center of the dimension values on dimension lines. See Figure 3-7. A break in the dimension line is made at the dimension value. The break is large enough to leave a visible gap between the dimension line and number.

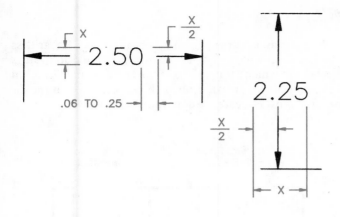

Figure 3-7. Dimension values are centered on the dimension lines.

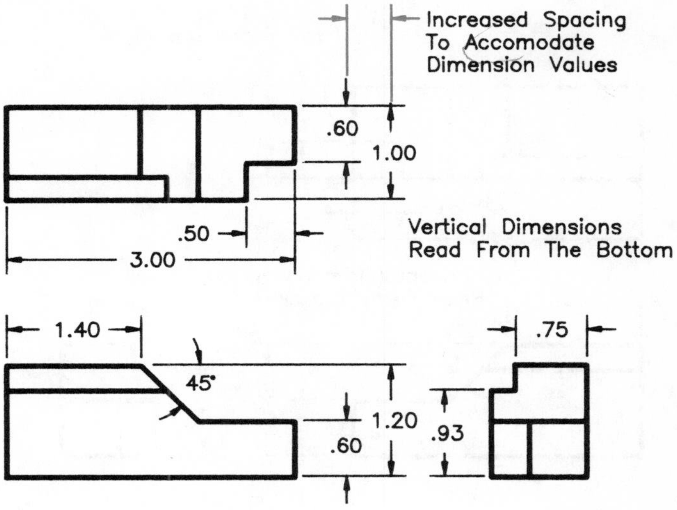

UNIDIRECTIONAL DIMENSIONS

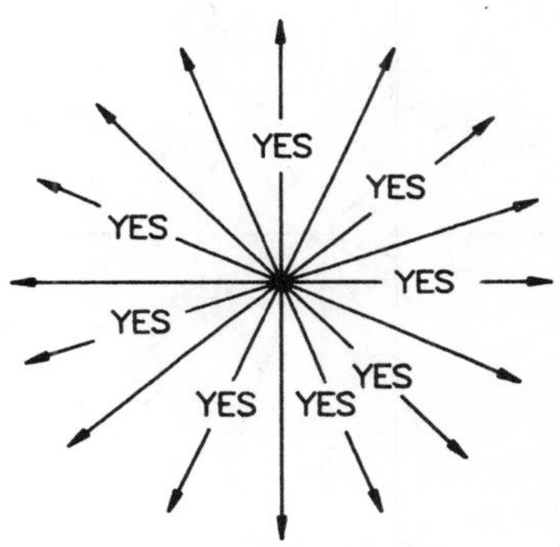

UNIDIRECTIONAL DIMENSIONS
CAN BE USED IN ANY ORIENTATION

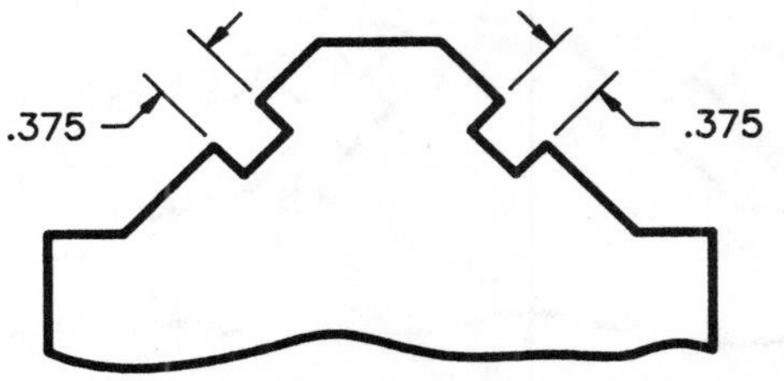

ALL UNIDIRECTIONAL DIMENSIONS
ARE READ FROM THE BOTTOM

Figure 3-8. All dimensions are read from the bottom of the page when using unidirectional dimensions.

Decimal points included in dimensions must be dark and clearly formed. Only clearly formed decimal points are certain to be seen on reproduced drawings. Any decimal point that cannot be easily seen on a drawing can cause an error when producing the part.

UNIDIRECTIONAL DIMENSIONS

The dimension value is always written horizontally when using *unidirectional dimensions*. All dimensions are readable from the bottom of the page. Unidirectional dimensions are used exclusively in ANSI Y14.5M.

Unidirectional dimensions do require more space between vertical dimension lines than is normally required for horizontal dimensions. The increased space for vertical dimensions is required to prevent dimension values from overlapping or crowding adjacent dimension lines.

All values are applied in the same orientation regardless of the dimension line orientation. See Figure 3-8. Increased dimension line spacing for vertical and inclined dimensions requires an increased amount of area for a drawing.

ALIGNED DIMENSIONS

Dimension values are parallel to the dimension lines when the *aligned dimensioning* method is used. See Figure 3-9. Horizontal dimensions are written from left to right, and read from the bottom of the page. Vertical dimensions are written from bottom to top, and read from the right side of the page. Aligned dimensions are no longer illustrated in the dimensioning and tolerancing standard, but they continue to be used by some companies. The use of aligned dimensions is not recommended.

Aligned dimensions must be read from the bottom or right side of the drawing sheet. The orientation of some dimension lines makes it difficult to achieve the required orientation of the dimension text. See Figure 3-9. When an undesirable orientation of dimension lines is required, the dimension values can be rotated to the horizontal orientation. Any dimension entered in this way does not conform to

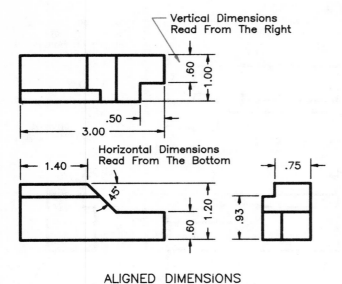

Vertical Dimensions Read From The Right

Horizontal Dimensions Read From The Bottom

ALIGNED DIMENSIONS

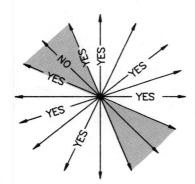

ALIGNED DIMENSIONS CANNOT BE CORRECTLY SHOWN IN THE SHADED AREA

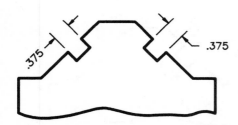

DIMENSIONS IN THE SHADED ZONE ARE ORIENTED TO READ FROM THE BOTTOM

Figure 3-9. Aligned dimensions cause difficulty in application for some orientations.

the aligned dimensioning method, but is preferable to a value that is read from the left side or top of a drawing.

The unidirectional and aligned dimensioning methods are not to be mixed within a drawing. One method or the other must be selected. Normally, only one method is used on drawings within a company.

SCALE

All mechanical drawings are drawn to scale. Regardless of the scale at which views are drawn, the dimension values are those to which the part is to be made. See Figure 3-10. An 8″ feature drawn at half scale is shown 4″ long, but the dimension must indicate the 8″ requirement.

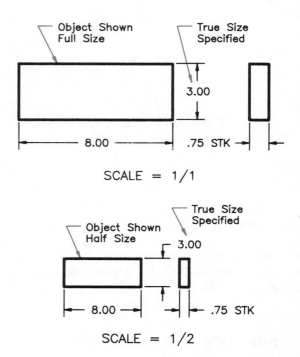

Figure 3-10. The part size to be produced is shown in the dimension regardless of the drawing scale.

The scale of a drawing is always noted in the title block. Scale may be noted in fractional form, such as 2/1, 1/1, 1/2, and 1/4. The scale notation is not provided to make it possible to scale the print to determine dimensions; rather, it is noted to give the reader a quick reference for mentally forming a true-size image of the part. The noted scale also informs everyone of the scale to use if the drawing original must be revised.

A reproduced copy of a drawing should never be scaled (measured) to determine the size of a feature. Scaling a reproduced copy is a risky practice. First, if a dimension is missing from the drawing, the drawing is incomplete. An incomplete drawing is proof that a mistake has been made, and it is possible that more mistakes exist. The drawn location or size of the undimensioned feature may also be a mistake. Second, the missing dimension may be a critical value that was calculated and then accidentally omitted from the drawing. If the dimension is critical, the gross measurement from a copy of the drawing will not reflect the accuracy needed. A third reason for not scaling a drawing is the actual process of making a reproduced copy. Most reproduction processes either cause a slight reduction or enlargement in at least one direction along the copy. Inaccurate reproduction processes guarantee that any measurements made from the drawing are incorrect.

MEASUREMENT SYSTEM

The two measurement systems used today are the inch and SI (metric) systems. The inch system is used extensively in the United States, but the metric system is becoming more widely accepted. ANSI Y14.5M is illustrated using metric values.

Most countries outside the United States use the metric system. The application of dimensions is affected very little by which measurement system is used. Only the representation of the dimension value is different.

The measurement system used on a drawing should be noted in or near the title block. This is especially true of drawings completed using the metric system. The note should state:

UNLESS OTHERWISE SPECIFIED, ALL
DIMENSIONS ARE IN MILLIMETERS
or
UNLESS OTHERWISE SPECIFIED, ALL
DIMENSIONS ARE IN INCHES

The measurement system noted in the title block is used for all dimensions. Any exceptions are noted by showing the unit of measurement beside the dimension value. See Figure 3-11. When the measurement system noted in the title block is inches, any metric dimension shown on the drawing is followed by the abbreviation for millimeters (mm). When the noted measurement system is millimeters, any inch dimension shown on the drawing is followed by the abbreviation for inches (IN). The unit of measurement is not shown with dimensions that are based on the measurement system noted in the title block.

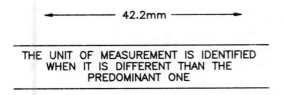

THE UNIT OF MEASUREMENT IS IDENTIFIED
WHEN IT IS DIFFERENT THAN THE
PREDOMINANT ONE

Figure 3-11. The unit of measurement is noted on dimensions that are different than the predominant measurement system used on the drawing.

Regardless of which unit of measurement is being used, it is sometimes desirable to show both the inch and metric values. When both values are shown, one of them must be a reference value. If the inch value is the firm requirement, the metric dimension is placed in parentheses to indicate that it is reference information. See Figure 3-12. The reference value is placed directly below the firm dimension requirement, and both values are placed in the same break in the dimension line. A note such as the following is added to the drawing that states:

VALUES SHOWN IN PARENTHESES ARE
METRIC AND FOR REFERENCE ONLY.

This note makes it clear that all production and inspection measurements are to the firm dimension values.

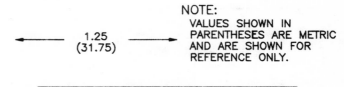

WHEN BOTH MEASUREMENT SYSTEM VALUES
ARE SHOWN FOR ONE DIMENSION, ONE
VALUE IS IDENTIFIED AS REFERENCE.

Figure 3-12. A reference value can be shown in a unit of measurement different than the firm requirement.

The same methods can be used on metric drawings to show a reference inch value. The metric value is shown as a firm requirement, and the inch dimension is shown as a reference value in parenthesis. The reference inch value is placed below the metric value. A note such as the following is added to the drawing that states:

VALUES SHOWN IN PARENTHESES ARE IN
INCHES AND ARE FOR REFERENCE ONLY.

Conversion of dimension values from one system to the other must be approached with a great deal of caution. The two values must be equal. This not only includes the nominal value, but also any tolerance applied to the nominal value. Conversion from one measurement system to the other requires the use of an accurate conversion factor. Rounding off converted values must be avoided until the final calculation is completed. Rounded-off values must not be used in the calculation of other values. The accumulation of round-off errors can cause unacceptable errors in part dimensions. Accurate conversion from one measurement system to the other should be based on 25.40 millimeters per inch. Inches are converted to millimeters by multiplying the inch value by 25.40 millimeters per inch. For example:

$$.875'' \times 25.40\text{mm/inch} = 22.23\text{mm}$$

Inch Measurement System

Dimension values shown in inches should be shown as *decimal inches*. The application of fractional dimensions is not defined by the current dimensioning standard. The following explanation of how to use fractions is based on commonly used practices. This explanation is provided to recognize the fact that many small companies continue to use fractions. However, this explanation should not be interpreted as a recommendation to use them.

Fractional values are limited in the degree of size control that can be achieved. Designs that do not have critical size requirements can be adequately dimensioned with fractions, but the limitations must be recognized. Any size or location that does not fall on a 1/64″ increment cannot be dimensioned using common fractions. The acceptable amount of variation (tolerance) on any dimension value must be 1/64″ or more when using fractions. This is not a high degree of accuracy.

Applying fractions to a drawing is relatively easy. See Figure 3-13. The fraction bar for horizontal dimensions is drawn parallel to, and aligned with, the dimension line. A

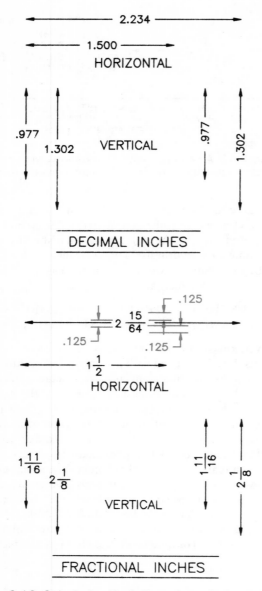

Figure 3-13. Only decimal inch dimension values are shown in the current dimensioning system, but the application of fractional inch values is still used by some companies.

diagonal line is not used for fractions in dimensions. The direction of the fraction bar on vertical dimensions is determined by whether aligned or unidirectional dimensions are used. The fraction bar is aligned with the dimension line when using aligned dimensions. It is always horizontal when using unidirectional dimensions.

The character height for the whole number, numerator, and denominator must meet the minimum character height requirements for the drawing. This means that a fraction is more than twice the height of a whole number since both the numerator and denominator must be clear of the fraction bar.

Applications in today's industry often require that a high degree of accuracy be maintained in the manufacture of products. The required degree of accuracy makes fractions impractical. As an example, installation of some bearings requires that shaft sizes be produced to within .0005″ (five

ten-thousandths) of diameter. Specifying dimensional control of this magnitude does not permit the use of fractions. The use of decimal dimensions permits specification of allowable part variation to any necessary degree of accuracy. Of course, the specified tolerance should never be smaller than what is needed to achieve the intended design function.

The decimal system not only makes control of the size easier, it also provides a system on which mathematical operations are easier to perform. Adding, subtracting, multiplying, and dividing operations are all easier to perform on decimal numbers than on fractions.

Total freedom in size increment is achieved by using decimal measurements. Many designers attempt to maintain fractional or .10″ size increments, even when using the decimal system. The practice is mainly due to standard sizes for vendor-supplied items (screws, seals, bearings, etc.) being based on nominal fraction sizes.

A zero is not placed in front of a decimal inch measurement less that one inch. The *number of decimal places* normally shown is two or three; but one, four, or more may be used. The number of decimal places does not affect the accuracy of the specified value. The decimal values .23, .230, and .23000 are exactly the same.

There is a common practice of specifying *title block tolerances* that apply according to the number of decimal places shown in the dimension value. Tolerances are shown as in the following example:

.XX	± .02
.XXX	± .010
ANGLES	± .5°

The required accuracy for producing a .23″ dimension is ± .02″ because of the tolerances shown in the example. The required accuracy for producing a .230″ dimension is ± .010″. The required accuracy for producing a .2300″ dimension is unknown since the title block does not show a tolerance for a four-place decimal. A tolerance for four-place decimals could be added to the title block, or a specific tolerance could be applied directly to the dimension.

Proper use of decimal values requires an understanding of how to *round-off* numbers. When using two- or three-place decimal equivalents for common fractions, some of the decimal values must be rounded off. An example is a feature that is 3/32″ in size. The decimal equivalent of 3/32″ is .09375″. The two-place decimal equivalent is .09″, and the three-place decimal equivalent is .094″.

When a number is to be rounded off, specific rules are to be followed. Rules for rounding off numbers are defined by ANSI/IEEE Std 268. The rules used to round-off decimal values are:

1. If the value following the last figure to be retained is less than 5, do not change the last figure. (.3125 rounds off to .31, since the figure following the 1 is less than 5.)
2. If the value following the last figure to be retained is greater than 5, the last figure is increased by 1. (2.1261 rounds off to 2.13, since the figure following the 2 is greater than 5.
3. When the value following the last figure to be retained

is exactly 5, the last significant figure is made an even value.

.31250 rounds off to .312

.43750 rounds off to .438

Rounding off values is necessary when using two- and three-place decimals for dimensioning. When a number is rounded off, it is no longer an exact equivalent of the fraction value. The allowable tolerance is applied to the rounded-off value, not to the fractional value.

A 3/32″ value rounded off to .09″ is the same as .090000″. Specifying the dimension as .09 or .090 only affects which of the title block tolerances is applicable.

Part of the inch measurement system is the foot. In machine drafting, dimensions under 72″ are shown in inches. No symbols are used to indicate the unit of measurement when a dimension only shows inches. Some companies show dimensions greater than 72″ in feet and inches. Symbols are used when both feet and inches are shown. A dash is used to separate the number of feet from the number of inches.

Example:

9′-3.50″ or 111.50

Metric Measurement System

The standard unit of measurement for drawings using the metric measurement system is the millimeter. One thousand millimeters equal one meter, and 25.4 millimeters equal one inch.

Metric values are entered in dimensions the same way as decimal inch values. See Figure 3-14. Either the aligned or unidirectional method of dimensioning can be used. All metric dimensions are shown using decimals. One-place decimals are common. A zero is placed in front of the decimal point for values smaller than one. This is different than the inch system, where a leading zero is not shown. The abbreviation for millimeters (mm) is not shown except when a metric dimension is applied on a drawing that is predominantly based on the inch system.

DIMENSION SYSTEMS

Several systems exist for the application of dimensions. Each system has specific applications for which it is best suited. The systems are often used in combination to achieve maximum control over the size of a part.

Chain, rectangular coordinate, ordinate, and tabulated dimensioning systems are used to show size and locations. These systems control the manner in which dimensions are applied and affect the location tolerances for features. If several features are dimensioned from one line, each feature's location varies in relationship to that line. If features are located by a chain of dimensions, the location tolerance for each feature in the chain is added together.

CHAIN DIMENSIONING

Chain (point-to-point) dimensioning is a system in which all features are located from adjacent features. See Figure 3-15. It is the most direct method for applying dimensions, but it has a definite disadvantage related to the accumulation of tolerances. The location of every feature depends upon the location of the feature from which it is dimensioned. If one feature is out of position by the maximum amount allowable, the location of the next feature is affected.

The object shown in Figure 3-15 is dimensioned using chain dimensions. This type of part would not normally be dimensioned in this manner, but the example emphasizes the effect of *tolerance accumulation*. As shown in the figure, the title block tolerances are applicable to each dimension in the chain.

RECTANGULAR COORDINATE DIMENSIONING

All dimensions originate from three mutually perpendicular planes in the *rectangular coordinate dimensioning* system. The three planes establish a coordinate system. Three methods of showing rectangular coordinate dimensions are baseline dimensioning, ordinate dimensioning, and tabular dimensioning.

Baseline Dimensioning

Baseline dimensioning reduces the tolerance accumulation that is caused by chain dimensioning. The accumulation is reduced because all dimensions extend from baselines. The example in Figure 3-16 shows the application of baseline dimensions.

Two features on the part are used to establish the baselines. The left surface establishes the baseline for X coordinates (width dimensions) and the bottom surface establishes the baseline for Y coordinates (height). All dimensions extend from these surfaces.

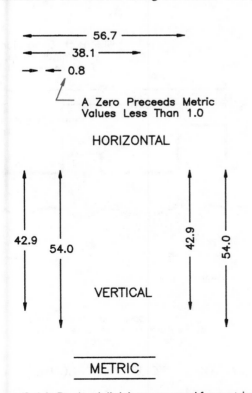

Figure 3-14. Decimal divisions are used for metric measurements.

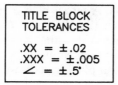

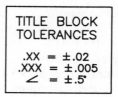

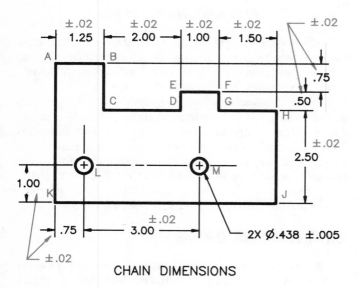

CHAIN DIMENSIONS

With All Features At The Nominal Size, The Part Is 5.75 Inches Long.

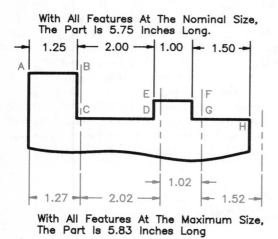

With All Features At The Maximum Size, The Part Is 5.83 Inches Long

EFFECT OF TOLERANCE ACCUMULATION

Figure 3-15. Chain dimensioning results in a large tolerance accumulation when it is applied to multiple features.

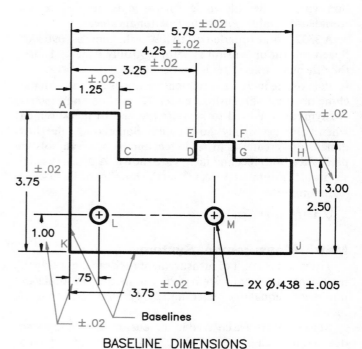

BASELINE DIMENSIONS

The title block tolerances apply to all dimensions in the given figure since no specific tolerances are used with the dimensions. All shown dimension values are two-place decimals, so their values can vary ± .02″ from their respective baselines. Since all features vary in location to the baselines by ± .02″, their locations relative to each other can vary by ± .04″.

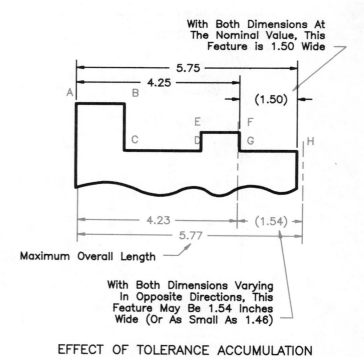

EFFECT OF TOLERANCE ACCUMULATION

Figure 3-16. Baseline dimensioning can reduce the amount of tolerance accumulation on a drawing.

Ordinate Dimensioning

In the *ordinate dimensioning system,* dimensions are shown on a drawing without using any dimension lines. The dimension values are written adjacent to the extension lines. See Figure 3-17. The dimension values shown in the figure are rectangular coordinates referenced to the origin. The origin for the coordinate dimensions is identified by zeros placed on extension lines from the origin location. Features located to the left of the origin have a -X value, and those to the right of the origin have a +X value. Features located below the origin have a -Y value, and those above the origin have a +Y value. Although not required by the national standards, it is common practice to draw arrows from the 0,0 extension lines to show the +X and +Y directions.

When using aligned dimension values in the ordinate dimensioning system, all dimension values are written above the extension lines. Sufficient space must be provided between the extension lines to allow the dimension value to be entered and clearly associated with only one extension line. A *dogleg* (offset) may be required in some extension lines to provide sufficient space for the dimension value.

The dimension values may also be placed at the end of the extension lines. This reduces the required space between extension lines.

Combining ordinate dimensions and unidirectional di-

mension values requires more space between extension lines to prevent the dimension values from running together. More doglegs must be inserted into the extension lines to provide adequate clearance, and many of the doglegs will be offset by a greater distance. The increased amount of offset in the doglegs makes unidirectional dimensions hard to read when using the ordinate dimensioning system.

Hole sizes can be noted on the part or referenced to a hole size table. When a table is used, the hole on the part is labeled with a letter that corresponds to a size shown in the table. For example, hole A in Figure 3-17 is .156″ in diameter and the diameter tolerance is ± .003″.

Tabulated Dimensions

Locations and sizes for a large number of holes can be conveniently shown in a table. See Figure 3-18. A table showing hole locations and sizes provides a well-organized and precise system for dimensioning a large number of holes. All hole locations in the table must be given in relationship to one point.

Tables used to show hole locations and diameters are not standardized. Any arrangement that clearly shows all data is

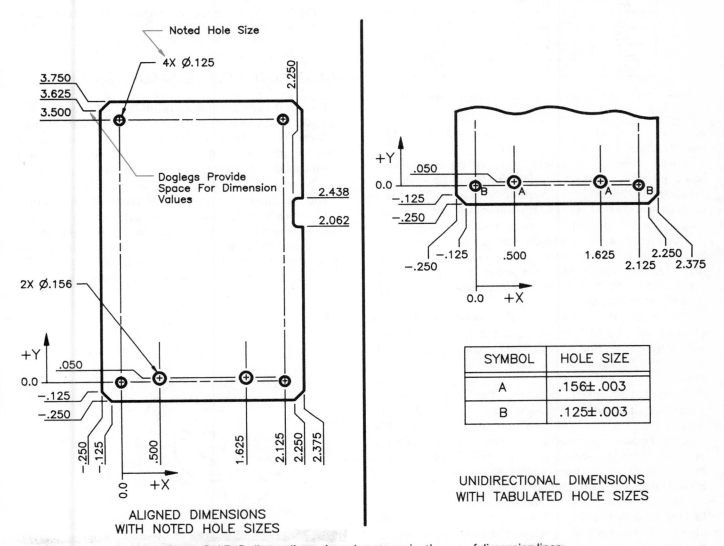

SYMBOL	HOLE SIZE
A	.156±.003
B	.125±.003

UNIDIRECTIONAL DIMENSIONS
WITH TABULATED HOLE SIZES

ALIGNED DIMENSIONS
WITH NOTED HOLE SIZES

Figure 3-17. Ordinate dimensions do not require the use of dimension lines.

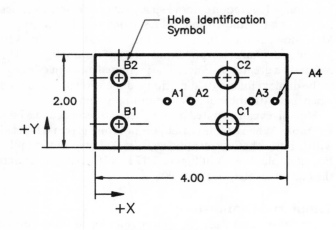

Hole Identification Symbol

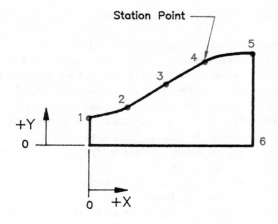

Station Point

DRILL TABLE

| SYMBOL | LOCATION | | SIZE | TOL |
	+X	+Y		
A1	1.50	1.00	.125	+.005 −.000
A2	2.00	1.00		
A3	3.25	1.00		
A4	3.75	1.00		
B1	.50	.50	.312	+.005 −.000
B2	1.00	1.50		
C1	2.75	.50	.438	+.006 −.000
C2	2.75	1.50		

Figure 3-18. Hole locations and sizes can be shown in a table.

| | STATION | | | | | |
	1	2	3	4	5	6
+X →	0	.40	.80	1.20	1.70	1.70
+Y ↑	.30	.42	.67	.91	1.00	0

Figure 3-19. Station points can be used when outline dimensions are shown in a table. Dimensions in a table are known as tabulated dimensions.

acceptable. The table used as an example in Figure 3-18 is one acceptable format.

A *hole identification symbol* is shown in one column to identify a hole on the part. The locating X and Y coordinates are shown in the column adjacent to the symbol. The nominal hole size is listed in the next column, and the size tolerance for the hole diameter is shown in the last column.

The hole identification symbol in the example includes a letter and number. The letter represents all holes of a particular size. The number identifies a specific hole of the size represented by the letter. It is not required that this system of hole identification be used.

A geometric symbol can be used to represent a hole size in place of the letter and number shown in the figure. A circle with a horizontal line across it might represent each .093″ hole, while hexagons might stand for the .125″ diameter holes.

Tabulated dimensions can also be used to dimension the outline of an object. If a table is used, features along the outline are identified with symbols. See Figure 3-19. The identified points are called *station points*. Letters or numbers are normally used as symbols for identifying station points. On curved outlines, station points are located close enough together to obtain the required accuracy for the curve. The station points are shown in a table along with the located coordinates for each point. A curved surface can also be dimensioned by specifying the mathematical formula that defines the curve.

POLAR COORDINATE DIMENSIONING

The *polar coordinate dimensioning system* requires that the location definition include an angular dimension. Angular location of related features is sometimes more correct from a functional standpoint than the rectangular coordinate location. Angular relationships are often important for rotating parts.

The part shown in Figure 3-20 includes several features that are dimensioned through specification of polar coordinates. The two holes are dimensioned by an angle showing the distance from the edge of the part to one hole. Another angle is specified for the distance between the two holes. The location of the two holes is further defined by a radius

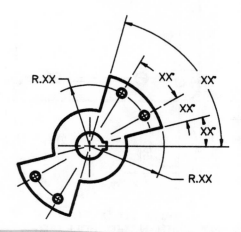

Figure 3-20. Polar coordinate dimensions combine angular and distance dimensions to define sizes and locations.

dimension. Omission of the radius dimension or either of the two angle dimensions would result in incomplete definition of the hole locations.

DIRECT (COMBINED) DIMENSIONING SYSTEMS

The combined use of baseline dimensioning and chain dimensioning systems allows the tolerance accumulation on a drawing to be minimized. The sheet metal bracket in Figure 3-21 shows the combination of the two systems. The bracket is designed to position a machine block at a specific height and orientation. The block height is controlled by the height of the attachment screw, and its orientation is controlled by an alignment pin. The bracket has two mounting holes in the bottom flange.

The mounting surface of the bracket is used as a baseline from which the attachment screw hole is located. This determines the height of the block relative to the mounting surface. The same baseline is used to show the overall height of the bracket. The left side of the bracket is used as a baseline from which the attachment screw hole is located in the width dimension. The same side of the bracket is also used as a baseline for locating one of the mounting holes and to dimension the width of the bracket.

The back of the bracket is used as a baseline from which the mounting holes are located and to show the overall depth of the bracket. The back surface of the bracket is selected as a baseline to ensure a controlled location of the machine block relative to the mounting holes. The back surface location relative to the mounting holes can only vary by ±.02″ from the .75″ specified.

The alignment pin hole is located from the attachment screw hole. This ensures a minimum tolerance between the two holes. Minimizing tolerance helps ensure that the alignment pin can go through both the bracket and machine block while the block is attached to the bracket.

One hole within a hole pattern is often located and then each hole in the pattern is located from the first one. This is in effect using the first hole as an origin for the other holes. The attachment screw hole and alignment pin hole in the given bracket can be thought of as a hole pattern containing two holes.

Parts should be dimensioned using functional features as baselines. See Figure 3-22. On the given part, the 1.000″ location dimension for the rail is from a surface that must be cleared by the part that slides on the rail. The use of this surface as a baseline reduces the tolerance accumulation between it and the rail. Clearance between the baseline surface and the sliding part is only affected by one dimension on each of the shown parts.

The rail width is dimensioned from the located side of the rail, rather than by a second dimension from the rail locating baseline. This reduces the tolerance on the rail width to the variation applicable to the one dimension. Reducing the tolerance on the rail helps to ensure that it will fit into the mating slot.

Locating features from functional baselines reduces the tolerance accumulation on their locations. Using chain dimensions to define feature size reduces tolerance accumulation on the individual feature.

DIMENSION PLACEMENT

General guidelines must be followed when applying dimensions. These guidelines ensure that size and location information are presented in a clear and consistent manner. Following dimension placement guidelines further clarify a drawing and make it easier to read.

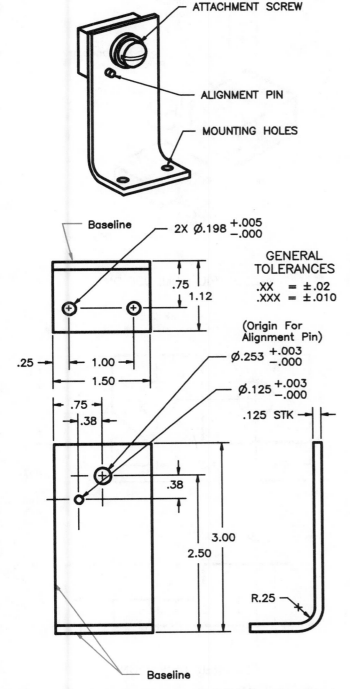

Figure 3-21. A combination of baseline dimensioning and chain dimensioning maximizes the dimensional control that can be achieved on a part.

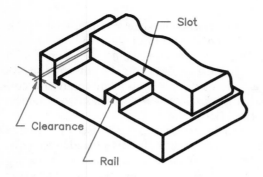

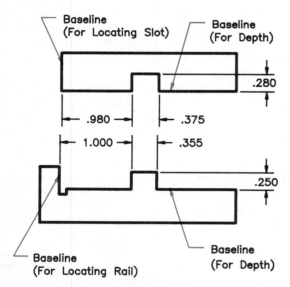

Figure 3-22. Functional baseline locations provide the locations of related features.

Regardless of the dimensioning system being used, such as chain or rectangular coordinate systems, the following general guidelines are applicable.

1. Dimension where the feature contour is shown.
2. Dimension between the views.
3. Dimension off the views (outside the object).
4. Dimension with consideration given to how the parts are assembled.
5. Consider the fabrication processes and capabilities.
6. Consider the inspection processes and capabilities.
7. Create a logical arrangement of dimensions.
8. Stagger dimension values.
9. Avoid dimensioning to hidden lines.

All dimensioning guidelines are based on common sense and industrial processes. As knowledge in machine shop and inspection practices is gained, the dimensioning guidelines become more natural and require a minimal amount of effort to remember.

Correct dimension application requires that the location and size for every feature be shown. This means that the height, width, and depth must be given for each feature and that the location of the feature on the part must also be given. To ensure that a part is completely dimensioned, the part can be mentally broken down into the geometric shapes that form the part.

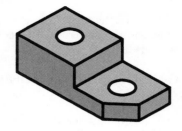

STEP BLOCK

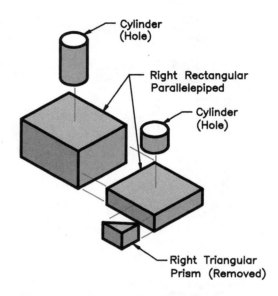

GEOMETRIC COMPOSITION

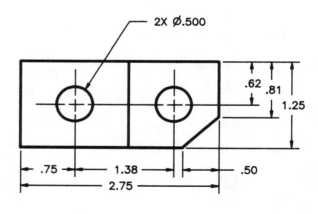

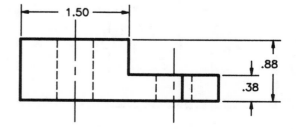

REQUIRED DIMENSIONS

Figure 3-23. Mentally breaking a part into the geometric shapes that form the part is one method for determining the dimensions needed on a drawing.

A step block is shown in Figure 3-23. The step block consists of five geometric shapes. Two right rectangular parallelepipeds form the basic shape of the part. Two cylinders and a right triangular prism are removed from the part to complete the step block. The height, width, and depth dimensions for the two parallelepipeds are given to outline the object. The locations and diameters for the two holes are specified. The size of the right triangular prism is specified by locating two corners. The third corner of the triangle is the removed corner of the step block.

DIMENSIONING FEATURE CONTOUR

Dimensions are applied where the contour of a feature is best shown. See Figure 3-24. This permits the dimension and the feature contour to be viewed simultaneously. The front view of the given object shows a distinct offset in the height of the part. The offset appears as a line across the top view and does not show any distinct change. The offset is dimensioned in the front view as 1.00″ from the left side of the object. The .38″ height of the offset is also dimensioned in the front view.

An angled surface on the part removes one of the square corners. The location of the angled surface is dimensioned in the top view since the change in contour is clearly seen there, and the corner appears only as a line between two surfaces in the front view.

Dimensions applied to the wrong views force the reader to look away from the dimension to see a view where the contour is shown. This is time consuming and can result in errors. Placement of each dimension on the contour of the feature allows the size or location to be determined without as much chance for error.

DIMENSIONING BETWEEN VIEWS

Dimensions of size and location should be placed between views when practical. See Figure 3-25. This general

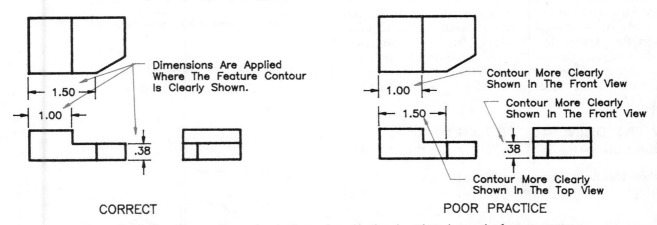

Figure 3-24. Features are most clearly dimensioned in the view that shows the feature contour.

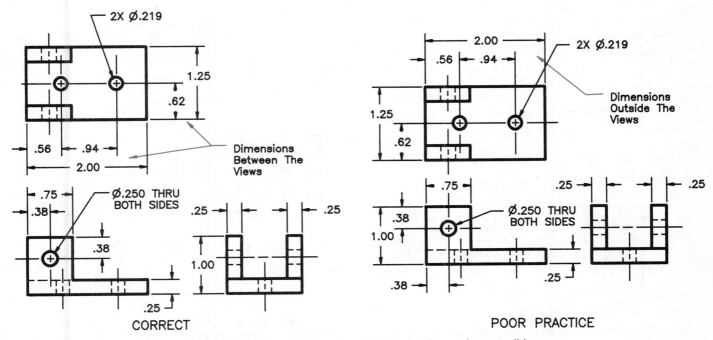

Figure 3-25. Dimensions are placed between views when possible.

guideline is followed for all dimensions that can be applied between the views without causing a loss of drawing clarity. A dimension should not be placed between views if it requires running extension lines across several features within a part.

The given example shows a part for which all dimensions are placed between the views. The extension lines used to show the dimensions between the views do not cause any loss in drawing clarity. Dimensions applied between two views allow the dimensions to be related to both views. Extension lines from any one dimension generally extend to only one view, but the proximity of the dimension allows it to be related to the second view. If dimensions are not placed between the views, the view on which the dimension is applied separates the dimension from the second view.

DIMENSIONING OFF THE VIEWS

Dimension lines are not to terminate on object lines if it is possible to clearly dimension the feature using extension lines. Extension lines extend the feature outside the object to allow placement of the dimension lines on the outside of the object. See Figure 3-26. The example shows the correct way in which extension lines are used to locate dimensions off the views. Incorrect use of extension lines results in dimensions that are inside the object and dimension lines that terminate on object lines.

DIMENSIONING ACCORDING TO RELATED PARTS

Parts on a machine are dimensioned according to the way in which the parts fit together. Consideration must also be

given to the function of the parts. See Figure 3-27. Four parts from a machine are shown. They are a base plate, slide, latching solenoid, and solenoid bracket. The slide moves on the base plate and is guided by a rail. The latching solenoid is mounted on the solenoid bracket, and the solenoid bracket mounts on the base plate.

The solenoid bracket must locate the solenoid in a position that will allow the solenoid plunger to enter the latching holes in the slide. This will occur only if the latching holes

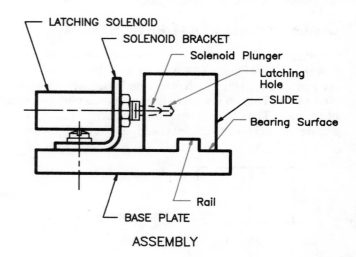

ASSEMBLY

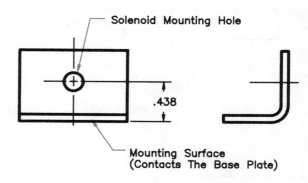

SOLENOID BRACKET

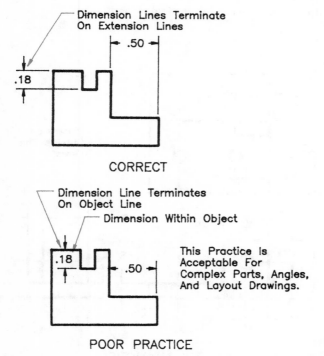

CORRECT

POOR PRACTICE

Figure 3-26. Dimensions are normally placed off the object. Avoid placing dimensions on the object.

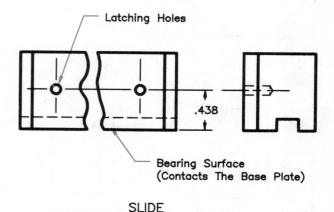

SLIDE

Figure 3-27. The location of the solenoid mounting hole is dimensioned from the same baseline as the latching holes. Related features on separate parts should be dimensioned from a common baseline.

are located the same distance from the base plate as the solenoid plunger. To accomplish this, the solenoid mounting hole is dimensioned from the bottom surface of the solenoid bracket. The latching holes are dimensioned from the bearing surface (bottom) of the slide. The surfaces used as baselines on the two parts are both in contact with the base plate when the parts are assembled. Locations for the holes on two different parts therefore originate from a common baseline in the assembly. Use of a common baseline for two parts ensures a minimum amount of tolerance accumulation between the two parts.

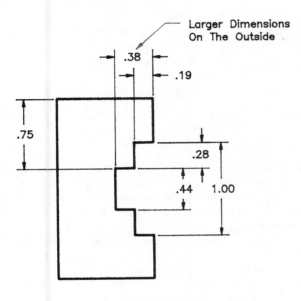

CORRECT

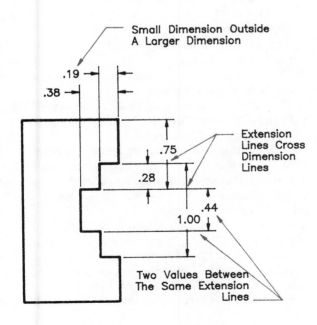

POOR PRACTICE

Figure 3-28. Dimensions are arranged to avoid crossing dimension and extension lines. This normally requires that the large dimensions be placed outside smaller ones.

GENERAL ARRANGEMENT OF DIMENSIONS

Dimensions are arranged to minimize crossing of extension and dimension lines. See Figure 3-28. Placing large dimensions farther from the object than small dimensions is one way to minimize crossing lines. In the given example, the illustration labelled as "correct" shows the effect of placing large dimensions outside the smaller ones. All size and location dimensions for the stepped groove are clearly shown.

When the same stepped groove is dimensioned with small dimensions outside larger ones, the size and location information is not clear. Dimension and extension lines cross because of the poor dimension arrangement. It is difficult to determine the point of application for a dimension if the dimension line crosses several extension lines. Placement of two or more values between the same extension lines is necessary if small dimensions are placed outside larger ones, causing confusion as to which dimension value is applicable. Confusing dimensions are avoided by using a general arrangement where small dimensions are placed closer to the object than large dimensions.

STAGGERED DIMENSION VALUES

Offsetting dimension value locations helps to more clearly show the values when several rows of dimensions are used. See Figure 3-29. Each value is placed near the center of

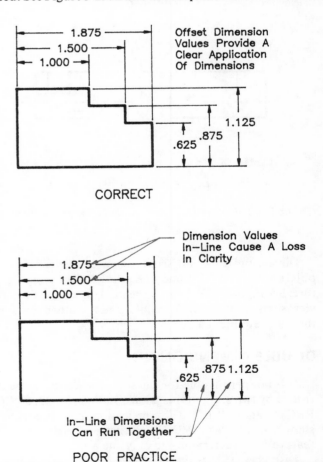

CORRECT

POOR PRACTICE

Figure 3-29. Staggered positions for dimension values make it easier to read the dimensions.

its dimension line, but enough offset is used to avoid putting the values directly in-line with one another. Failure to offset vertical dimensions can result in dimension values that run together.

DIMENSIONING HIDDEN FEATURES

Generally, it is not acceptable to dimension a feature where it is shown using hidden lines. It is preferable to dimension the feature where it is drawn as an object line. See Figure 3-30. If the feature appears hidden in all views, a section view is required. The section can be a broken out section that shows the feature with object lines, or it can be one of the other types of section views.

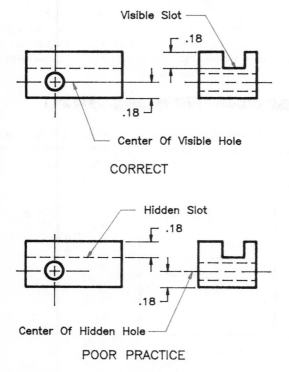

CORRECT

POOR PRACTICE

Figure 3-30. Features are dimensioned where they are shown as visible.

Dimensions in a half section view can extend from an object line to a hidden line when showing the size of a feature. See Figure 3-31. This is acceptable since one side of the view shows the feature as visible, and the other side shows the feature as hidden.

DOUBLE DIMENSIONING

It is not permissible for the size of a feature to be controlled by two sets of dimensions. In Figure 3-32, a "CORRECT" and "WRONG" application of dimensions is shown. In the "CORRECT" illustration, the size of each feature can be determined in only one way.

Distances A,B; B,E; and A,F have dimensions applied directly to them in the illustration labelled as "CORRECT." Title block tolerances specify the acceptable variation for

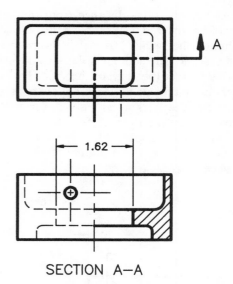

SECTION A—A

Figure 3-31. Dimensions can extend to the hidden portion of a feature shown by a half section.

each dimension. Distance E,F is not directly specified in the "CORRECT" example; it does not require a dimension. When all dimensioned features are produced to the sizes specified, distance E,F will be made to a definite size. The permissible range of sizes at which distance E,F may be produced is determined by the tolerances on the other features (which are dimensioned), and can be calculated through the shown procedure.

The accumulation of tolerance on distance E,F can be controlled by dimensioning the part differently if the dimensions in the given figure are unacceptable. However, no more than three dimensions may be used, regardless of the desired result. Adding a fourth dimension creates a double-dimensioned feature.

The illustration labeled "WRONG" shows the features double dimensioned. *Double dimensioning* results in a condition where the size or tolerance for a feature can be determined in more than one way.

Four dimensions are used to define distances A,B; B,E; E,F; and A,F in the "WRONG" example. This is one dimension too many. A title block tolerance of ± .02″ is applied to each dimension. This tolerance applied to each individual feature conflicts with the tolerance accumulation that occurs when all other features are considered.

Examination of distance A,F shows how a conflict in tolerances occurs when double dimensions are used. A dimension of 2.25″ with a tolerance of ± .02″ is directly applied to distance A,F. This provides one requirement for distance A,F. A second requirement for distance A,F is determined by adding the dimensions applied to distances A,B; B,E; and E,F. The sum of these dimensions is 2.25″ with a total tolerance accumulation of ± .06″ for distance A,F.

The directly applied dimension for distance A,F requires a tolerance of ± .02″ and the derived requirement from the other features is ± .060″. Since these values are not equal, there is a conflict in the two requirements. It is not acceptable to place two conflicting requirements on a feature.

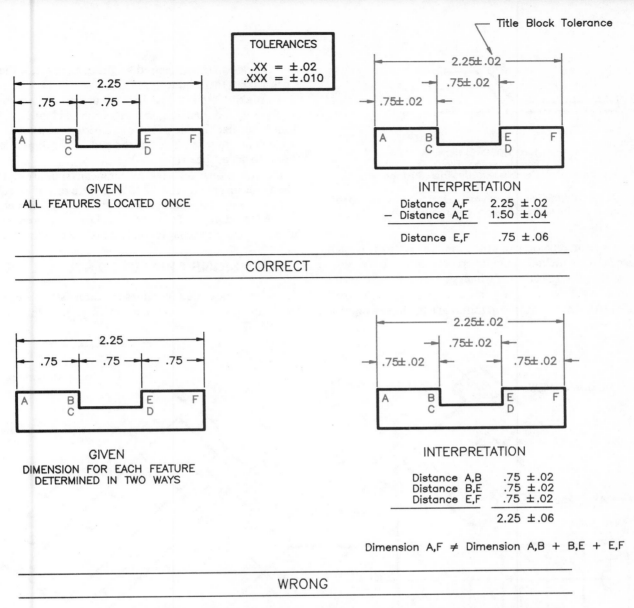

Figure 3-32. Double dimensioning causes an unacceptable conflict between the tolerance accumulations on a part.

REFERENCE DIMENSIONS

It is possible to apply a *reference dimension* to a feature that is already controlled by other dimensions. See Figure 3-33. Reference dimensions are used only for what the name implies; they are used as a reference that shows a nominal size or location, but they do not indicate the accuracy of the dimension. Since no tolerance is implied by a reference dimension, its use does not indicate double dimensioning. Tolerance for the affected feature is determined from the other dimensions on the drawing.

Figure 3-33 shows the same object that was shown in Figure 3-32. Distance E,F in Figure 3-33 is controlled by the dimensions on distances A,F; A,B; and B,E. The nominal size of distance E,F is shown as a reference dimension. The dimension is only provided to show the approximate size of distance E,F. It is not meant to be a number that must be used while making the part. If a person does use a reference dimension when making a part, the other dimensions must be produced within the acceptable variations allowed by specified tolerances.

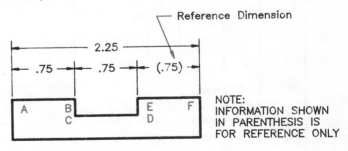

Figure 3-33. A reference dimension shows a nominal size or location. No tolerance is applicable to a reference dimension.

INDICATING A DIMENSION ORIGIN

Identification of the *dimension origin* is sometimes required when variations on a part could cause a difference in how features are measured for acceptance or rejection of produced parts. See Figure 3-34. The given example shows a height dimension between offset surfaces. One of these

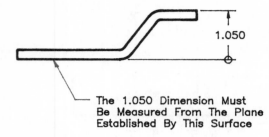

The 1.050 Dimension Must
Be Measured From The Plane
Established By This Surface

Figure 3-34. A dimension origin may need to be indicated to
ensure the desired design requirement is met.

surfaces is relatively long and the other one is short. If each
of these features includes angular errors relative to the vertical leg of the part, selection of which surface acts as an origin
can impact whether or not parts are acceptable. A small
angular error on the short horizontal leg can have a signifi-

cant impact on the measured height of the part, if the short
leg is used as the origin.

To make certain the correct feature is used as the origin
for the height dimension, the origin can be identified. This is
done by replacing one of the dimension arrowheads with a
circle. The circle is placed at the end of the dimension line
that is to act as the origin.

Identification of an origin by this means should only be used
when measurement relative to a specific origin is required,
and then only if the part is not controlled by tolerances referenced to a datum reference frame. Datum reference frames
are described in the chapter on datums and datum references.

DIMENSIONS APPLIED TO SPECIAL VIEWS

The methods and guidelines for dimensions are applicable to all views in a multiview drawing, including auxiliary

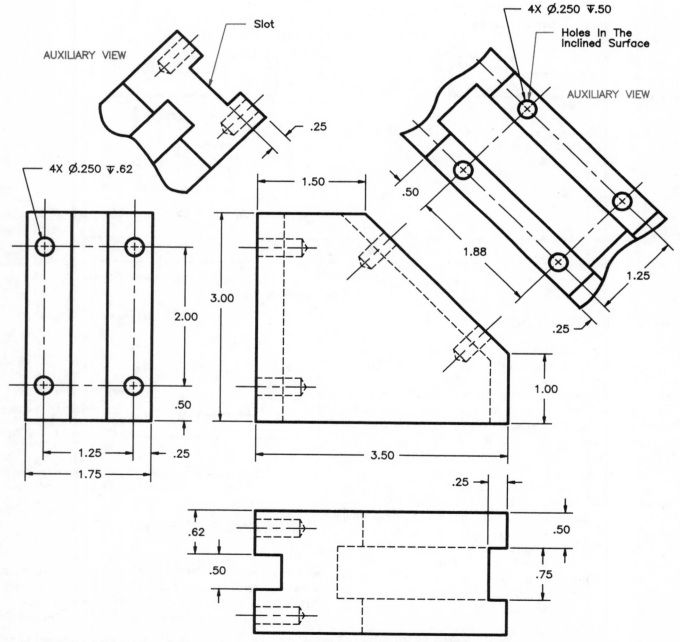

Figure 3-35. Dimensions are shown on an auxiliary view only when they cannot be shown clearly on the principal views.

and section views. Pictorial drawings are not part of a multi-view drawing, and require special dimension application guidelines.

DIMENSIONING AN AUXILIARY VIEW

Auxiliary views are used to show the true shapes of features on inclined and oblique surfaces. It is required that dimensions be applied where the feature is seen in true size. Features shown true size only by an auxiliary view must be dimensioned in the auxiliary view. See Figure 3-35. Features shown true size by principal views should be dimensioned on a principal view.

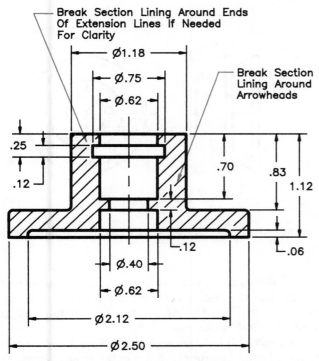

Figure 3-36. Internal features are dimensioned in section views.

The given example has an inclined surface with four drilled holes. A slot is cut into the inclined surface. True shape and size for the inclined surface and slot is shown in auxiliary views.

Dimension lines in the auxiliary views are parallel to the distances being specified. Extension lines are perpendicular to the dimension lines.

DIMENSIONING A SECTION VIEW

Section views are dimensioned in accordance with all dimensioning requirements previously explained for principal views and auxiliary views. Section lines (crosshatch lines) sometimes create a need for special consideration when dimensioning. Extension lines and section lines are drawn with the same line type. To avoid confusion between the two lines, a break in the section lining may be made near the end of an extension line. See Figure 3-36. Section lines may also be broken around arrowheads that fall within a sectioned area. However, breaking section lines is not always possible on CAD drawings.

DIMENSIONING PICTORIAL DRAWINGS

Pictorial drawings are not normally used to show all the dimensions necessary to make a part. Exceptions do exist, such as in plans for home project kits and drawing assignments in textbooks. Dimensions on pictorial drawings are normally limited to those necessary to indicate overall size.

Dimensions on pictorial drawings are clearest when all the dimensions running in one direction are shown parallel to a single plane. See Figure 3-37. All width dimensions in an oblique drawing can be shown on either the horizontal or frontal planes. Only one of the two planes should be used. Depth dimensions can be shown on either the horizontal or profile planes. Height dimensions can be applied on the frontal or profile plane. All width dimensions in the given oblique drawing are shown parallel to the frontal plane, height dimensions parallel to the frontal plane, and depth dimensions parallel to the horizontal plane.

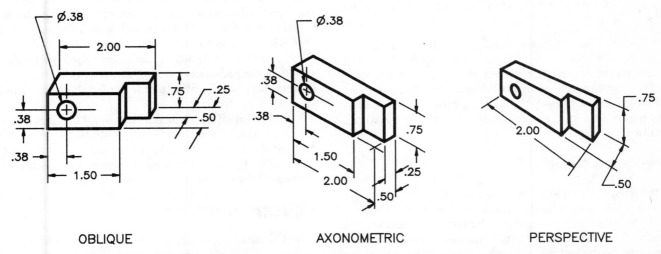

OBLIQUE   AXONOMETRIC   PERSPECTIVE

Figure 3-37. Pictorial drawings are usually not dimensioned. When dimensions are used, they are applied to appear in the same plane as the dimensioned feature.

Axonometric drawings are dimensioned following the same general rules as are applied to oblique drawings. All dimensions in one direction should lie parallel to one plane.

Perspective drawings are seldom dimensioned with more than overall dimensions. The dimension lines in a perspective view recede toward the vanishing points used for the object lines.

DIMENSIONING WITH NOTES

Information that cannot be clearly shown through the orthographic views, dimensions, or symbology must be put into written form. Any written information that applies to a drawing is considered a drawing note. Notes can be applied on the field of the drawing or put on separate note sheets.

The type information given in a note may be general and apply to an entire drawing. It is also possible for a note to be very specific and apply only to a single feature on a part.

Notes are an important part of any drawing. Compliance with the specifications given in drawing notes is as much a requirement as is meeting the dimensions on the part. Failure of the manufacturer to meet any noted requirement is grounds for part rejection. For this reason, the manufacturer must follow the notes, and the designer must not overspecify the requirements.

The method in which information is given in a note depends partly on the use of the drawing. If the drawing is used within the designing company, standard processes can be referenced. Drawings used outside the designing company generally do not reference any of the designing company's standard processes. Instead, they describe procedures completely, or reference federal standards and industrial standards. Typical notes for installing fasteners are—
For use by the design company:

INSTALL FASTENERS PER SP663

For use by other companies:

INSTALL FASTENERS PER MIL-STD-XXX
or
INSTALL FASTENERS PER DETAILED VIEW SHOWING CORRECT FASTENER STACKUP. TORQUE FASTENERS TO THE VALUES SHOWN IN TABLE ONE.

The first note references a standard process manual that shows correct fastener installation and torque values. The second note is an example of a reference to a military standard that contains the necessary information. The third note gives all required information. Its use requires detail views showing fastener stackup and a table of torque values within the drawing.

SPECIFIC NOTES

Information that applies to a specific feature is given in a specific note. See Figure 3-38. A specific note can be written on the drawing and connected to the applicable feature by a leader. If a separate notes list is being used, a note number is connected to the feature by a leader. The note number refers to a note in the list. Note numbers are placed inside a symbol

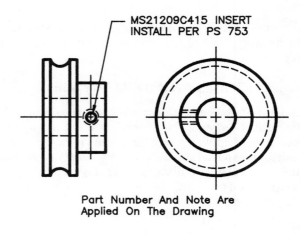

Part Number And Note Are Applied On The Drawing

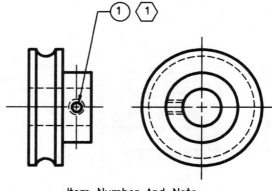

Item Number And Note Number Refers To A List

Figure 3-38. A specific note shows information applicable to a specific feature.

that identifies it as a note number. The symbol must be different than the one used to identify item numbers on an assembly drawing. Use of one symbol for both purposes would result in confusion.

Two methods for showing a note callout that defines how to install a thread insert is shown in the given example. The first method shows the noted part number and installation process attached by a leader that points to the insert. The second method shows an item number and note number in place of the noted part number and process. The item number identifies the insert in a parts list, and the note number identifies the specific note that is applicable to the insert. The referenced note states the same requirement as was shown in the previous example. The note is located in a list of notes that may be either on the field of the drawing or in a separate notes list.

Application of the same note to several features is simplified by using note numbers. The note is only written one time in the notes list, and the number is applied to each applicable feature. Drawing space and time are saved through this procedure.

GENERAL NOTES

Information that applies to an entire drawing is contained in a general note. See Figure 3-39. If a drawing is a single detail drawing, then information that affects the

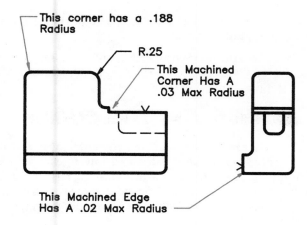

NOTES:

1. ALL FILLETS & ROUNDS R.188 UNLESS SPECIFIED OTHERWISE

2. ALL MACHINE CUTTER CORNER RADII .03 UNLESS OTHERWISE SPECIFIED

3. REMOVE ALL SHARP EDGES AND BURRS PER SP 750 R.02 MAX

4. INFORMATION IN PARENTHESIS IS REFERENCE INFORMATION

Figure 3-39. General notes affect the entire drawing.

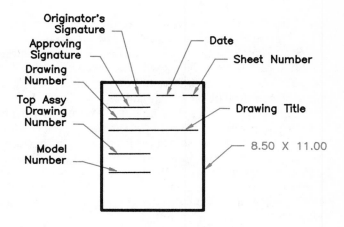

COVER SHEET

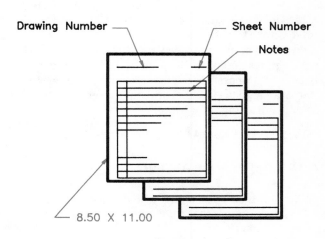

NOTE SHEETS

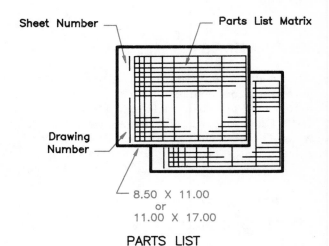

PARTS LIST

Figure 3-40. If a separate notes and parts list is used, it is as much a part of the drawing as if the information was shown on the field of the drawing.

entire part is a general note. If a drawing is a multidetail drawing, information that affects all parts is a general note.

A fillet radius that is the same for most corners on a part can be given in a general note. The note specifies the radius one time and prevents the need for repeated radius dimensions on the part. Only radii that differ from the noted radius must be dimensioned. A standard machining process for removing sharp edges and burrs is a common general note.

A general note is very precise in its requirements. It is not general in the sense of being vague. Vague notes are not of any use on a drawing. General notes are only general in the sense that they apply to the entire drawing.

SEPARATE NOTES AND PARTS LISTS

Preprinted forms are commonly used to document the drawing notes and parts list data for a drawing. Placing notes and parts data on separate notes and parts lists allows the drawing sheets to be used entirely for the views, title block, and revisions block. Information in a separate notes and parts list is considered part of the drawing.

Some CAD systems automatically generate a notes list and parts list based on data entered as the design model is produced. The manner in which the notes and parts data is output from the CAD system depends on the particular software being utilized. It may be plotted on the field of the drawing, plotted on a separate drawing sheet as a first page in a multisheet drawing, or printed on a line printer.

The use of preprinted forms saves time when manually producing separate notes and parts lists. There may be three or more standard forms used within one separate list.

A *cover sheet* is commonly used to show general information. See Figure 3-40. It is normally an 8.5″ x 11″ sheet. It shows the drawing number, signatures, and the drawing title. Other information can also be shown, such as the top assembly drawing number, model numbers, and serial numbers. The sheet number and total number of pages in the list are shown, such as: 1 of 6. The drawing number and title are required to associate the list with the correct drawing. The originator's signature and an authorizing or approving signature is required by most companies.

Note sheets are used for the notes. A column for note numbers is provided. A space for the sheet number is also provided. A symbol of some type is placed at the beginning of each specific note to indicate that the corresponding note number is called out on the drawing or in the parts list. The symbol can be a reduced version of the symbol used to enclose note numbers on the drawing. This symbol is not placed adjacent to general notes.

Parts list sheets are used for entry of information about parts and materials to be used in producing the parts or assembly. Spaces for the drawing number and sheet number are provided. Columns and headings are typically preprinted on the forms.

One note should be placed above the title block when a separate notes and parts list is used. The note indicates the existence of a separate list. This ensures that anyone looking at the drawing sheets will know that a separate list forms part of the drawing. For example:

SEE SEPARATE PARTS LIST

NOTES ON THE DRAWING

When notes are shown on the field of a drawing, all notes are placed in one area. Common locations are either the right or left margins of the first sheet. The first sheet is used entirely for notes if there is a large amount of written information. The notes list is headed by the word **NOTES**. The notes are numbered consecutively to make each one easy to locate.

ABBREVIATIONS

Abbreviations are used extensively to reduce the space required for dimensioning and notes. Standard abbreviations for technical terms are defined in ANSI Y1.1 and MIL-STD-12. Only standard abbreviations are to be used on a drawing. The use of nonstandard abbreviations is confusing and may result in production errors. An example of the space saved by abbreviations can be seen in the following example.

Abbreviated example:

2X .250-20UNC–3B THD X .38 DP

Nonabbreviated example:

.250-20 UNIFIED NATIONAL COARSE–3B
THREAD X .38 DEEP 2 PLACES

Acronyms are also used to reduce the space required for notes. An acronym is a word made from the beginnings of the words in a phrase. The acronym does not always look like a word. An acronym can be created for any name that is used repeatedly in a notes list or report. The first time it is used, the name is spelled out and the acronym is shown in parentheses. For example:

The Laser Altitude Sensor (LAS) must be assembled in a class 100,000 clean room. After assembly, seal the LAS in its shipping container (P/N 11304).

CATEGORIES OF FIT BETWEEN PARTS

The relationship between mated parts falls into two major categories–clearance fits and interference fits. See Figure 3-41. A slip renewable drill bushing has a clearance fit that permits easy installation and removal. A drill bushing liner has an interference fit that provides a semipermanent installation.

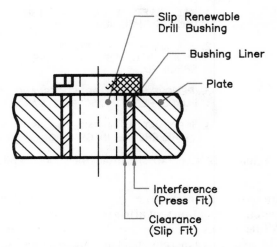

Figure 3-41. Clearance fits permit parts to move, and interference fits hold parts in place.

CLEARANCE FIT

A *clearance fit* exists when an internal part is smaller than the mating external part. Clearance fits are used for moving parts. The amount of clearance provided between the moving parts is based on the amount of movement, rate of movement, finish on the parts, lubrication requirements, and nominal feature size.

Clearance fits are also used for location of parts relative to one another. See Figure 3-42. The figure shows a slip renewable drill bushing that slips into a drill bushing liner. The renewable bushing must clear the liner for easy installation and removal. The maximum amount of clearance is based on the amount of location error that can be tolerated between the bushing liner and drill bushing. The minimum clearance is based on the amount of clearance necessary to ensure removal of the drill bushing.

INTERFERENCE FIT

An *interference fit* exists when an internal part is larger than the mating external part. See Figure 3-43. The amount of interference depends on the function of the interference fit. A small interference is all that is required for most loca-

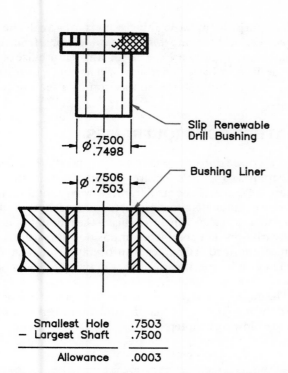

Smallest Hole	.7503
– Largest Shaft	.7500
Allowance	**.0003**

Figure 3-42. Proper calculation of a clearance fit will result in limits of size that provide clearance between the features for all possible size combinations.

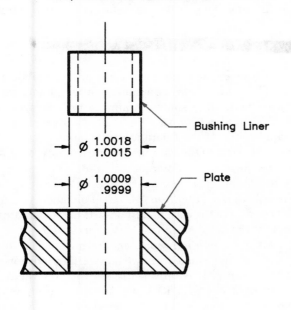

MAXIMUM INTERFERENCE

Smallest Hole	.9999
– Largest Shaft	1.0018
Allowance	**–.0019**

MINIMUM INTERFERENCE

Largest Hole	1.0009
– Smallest Shaft	1.0015
	–.0006

Figure 3-43. A negative allowance value between two features indicates that an interference fit exists.

tion interference fits. A larger amount of interference is required if the fit must hold parts in place.

Too much interference between parts can cause damage. Hollow parts pressed into a hole can collapse, and thin external parts can rupture. Care must be taken to use only the amount of interference that is required for the design function. The stresses developed in an interference fit can be calculated using formulas contained in most engineering handbooks.

The bushing liner shown in the given example is pressed into the drill template (plate). The interference fit holds the liner in the plate and at the same time locates the liner. The amount of interference required for this design is relatively small since no substantial forces are applied to the liner.

TRANSITION FIT

A *transition fit* permits size variations that can result in either an interference or clearance between two mating parts. See Figure 3-44. An interference exists when the smallest permissible hole is mated with the largest permissible shaft. A clearance exists when the largest permissible hole is mated with the smallest permissible shaft.

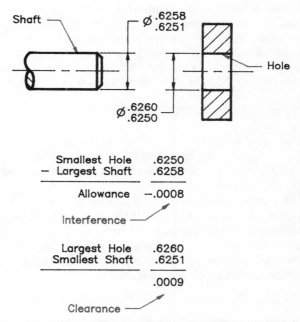

Smallest Hole	.6250
– Largest Shaft	.6258
Allowance	**–.0008**
Interference	
Largest Hole	.6260
Smallest Shaft	.6251
	.0009
Clearance	

Figure 3-44. A transition fit results in a clearance fit at one extreme of the applied tolerance limits, and an interference fit at the other extreme.

LINE FIT

A *line fit* has a zero allowance between mating parts. See Figure 3-45. A *zero allowance* means that the difference between the maximum shaft and the minimum hole is equal to zero. A zero allowance is assumed to create an interference fit, since two parts with the same diameter do not have a clearance. In the given figure, an interference fit only exists when both parts are at the .250″ diameter. All other conditions for these two parts result in a clearance fit.

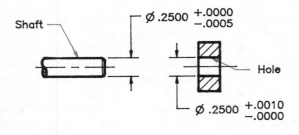

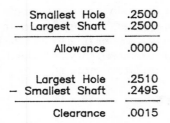

Smallest Hole	.2500
– Largest Shaft	.2500
Allowance	.0000
Largest Hole	.2510
– Smallest Shaft	.2495
Clearance	.0015

Figure 3-45. A line fit results in line-to-line contact when both parts are at MMC.

GEOMETRIC TOLERANCES

Control of form, orientation, location, runout, and profile are all achieved through the specification of geometric tolerances. Some of the symbols in Chapter 2 are used for geometric tolerances, and the topics in following chapters define how geometric tolerances are applied to drawings. The application of dimensions as defined in this chapter provides the size and location specifications from which the above mentioned tolerance controls can be specified.

BASIC DIMENSIONS

A *basic dimension* is a theoretical dimensional value applied to the size or location of a feature. Title block tolerances or general tolerances specified in notes do not apply to basic dimensions.

A basic dimension is identified in one of two ways. Probably the most common method for identifying a basic dimension is to draw a rectangle around the dimension value. See Figure 3-46. The second means for identifying basic dimensions is to place a general note in the notes list. A typical note used to indicate basic dimensions is:

ALL UNTOLERANCED DIMENSIONS ARE BASIC

When this general note is used, each dimension is considered basic unless it has a specific tolerance associated with the dimension value. A dimension of 2.500″ is basic if the above note is on the drawing. A dimension of 2.500″ ± .008″ is not basic whether or not the above note is on the drawing.

A past practice for indicating basic dimensions was to place the word BASIC or the abbreviation BSC adjacent to the dimension value. This method is no longer defined in the dimensioning standard.

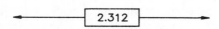

Figure 3-46. A basic dimension can be identified by placing the dimension value inside a rectangle.

The purpose for basic dimensions is to provide the theoretical values for feature sizes and locations from which geometric tolerances are specified. Any feature size or location defined by basic dimensions will normally have a tolerance applied through the application of a feature control frame.

FEATURE CONTROL FRAMES

A *feature control frame* is used to specify the tolerance that applies to specific features or patterns of features. See Figure 3-47. Feature control frames always contain at least a tolerance symbol (characteristic) to indicate the type of control and a tolerance value to indicate the amount of acceptable variation. The complete composition of feature control frames is defined in Chapter 2. Information on methods for applying these specifications to a drawing are also briefly described in that chapter.

The definition of the achievable control that can be exercised through the use of feature control frames is explained in the following chapters.

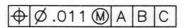

Figure 3-47. A feature control frame contains tolerance requirements that apply to a feature or a pattern of features.

MATERIAL CONDITION MODIFIERS

Tolerances specified on features of size include a modifier. The modifier may be assumed, in compliance with rules in the standard, or shown by application of a symbol. The modifiers indicate at what material condition the tolerance is applicable. If the tolerance must be met at all possible produced sizes of the feature, it is said to apply *Regardless of Feature Size (RFS)*. Tolerances that must be met only when the most material exists in the part are said to apply at *Maximum Material Condition (MMC)*. Tolerances that must be met only when the least material exists in the part are said to apply at the *Least Material Condition (LMC)*.

Material condition modifiers are included in tolerance specifications to indicate at what material condition the tolerances apply. It is only necessary to show the MMC or LMC modifiers since RFS is assumed when no modifier is shown. See Figure 3-48. The modifiers are only used in feature control frames. When referencing material conditions within a note, the applicable term is to be spelled out or an abbreviation may be used.

Figure 3-48. Material condition modifiers are applied to tolerances to indicate at what material condition the tolerance is applicable.

Size description is defined through dimensions.

Location dimensions define where a feature is located.

Size dimensions indicate how large a feature is.

Extension lines are used to extend features for the application of dimensions.

Break extension lines only where they cross an arrowhead.

Dimensions are placed inside a break in the dimension line.

The minimum recommended space from an object to a dimension line is .40″ (10mm). The minimum recommended space between dimension lines is .24″ (6mm).

All dimensions on a drawing are based on one measurement system. Exceptions are identified by showing the units of measurement beside each dimension that is an exception.

A combination of dimensioning systems can be used to control tolerance accumulation.

Size and location for every feature on a part must be given.

Dimensions are applied where the feature contour is visible.

Dimensions are applied between views when practical.

Dimensions are applied off the views.

Dimensions are applied according to how parts fit together.

Large dimensions are generally placed outside small ones.

Stagger dimension values to avoid confusion.

Avoid dimensioning to hidden features.

No feature should have two methods for defining its size. Do not double dimension.

Avoid using pictorial drawings to define size for production.

Notes are written to give exact information.

Notes are part of a drawing, even when written on a separate list. Notes define part requirements that must be met.

REVIEW QUESTIONS

Answer the following questions on a separate sheet of paper. Do not write in this book.

MULTIPLE CHOICE

1. Dimensions are used to specify _____.
   A. size and volume
   B. size and location
   C. size and shape
   D. shape and location

2. _____ lines are used to extend features for the application of dimensions.
   A. Dimension
   B. Leader
   C. Extension
   D. Object

3. An extension line should be broken when it crosses _____.
   A. another extension line
   B. an arrowhead
   C. an object line
   D. All of the above.

4. It is preferable to show dimension values and dimension lines _____.
   A. outside the object outline
   B. inside the object outline
   C. Both A and B.
   D. Neither A nor B.

5. Leader lines should not be drawn _____.
   A. horizontal
   B. vertical
   C. at a 45° angle
   D. Both A and B.

6. The dimension values are oriented to be read from _____ when using unidirectional dimensioning methods.
   A. the right or left side
   B. the bottom and right side
   C. the bottom
   D. any convenient direction

7. The dimension values are oriented to be read from _____ when using aligned dimensioning.
   A. the right or left side
   B. the bottom and right side
   C. the bottom
   D. any convenient direction

8. When dimensioning in inch units, it is preferable to use _____ values.
   A. full inch
   B. fractional inch
   C. decimal inch
   D. Both B and C.

9. A zero is placed in front of dimension values less than 1 when using the _____ measurement system.
   A. metric
   B. inch
   C. pound
   D. Both A and C.

10. The _____ dimensioning system uses a single feature from which other dimensions originate.
    A. chain
    B. combined
    C. baseline
    D. Both A and C.

11. Polar coordinates include a(n) _____ dimension.
    A. angle
    B. latitude
    C. longitude
    D. Both B and C.

12. Dimensions should be applied to the view in which the _____ of the feature is clearly shown.
    A. texture
    B. centerline
    C. profile
    D. height

13. Placing dimensions _____ permits easier association of the dimension to multiple views.
    A. inside the object outline
    B. outside the object outline
    C. close to the object outline
    D. between the views

14. Dimension values in vertical dimensions are usually _____ to make them easier to read.
    A. aligned
    B. centered
    C. offset
    D. underlined
15. Features are generally dimensioned where they are _____.
    A. hidden
    B. visible
    C. Either A or B.
    D. Neither A nor B.
16. A _____ dimension is provided for general information only, and is not a requirement that must be met.
    A. radius
    B. reference
    C. limit
    D. polar

TRUE/FALSE

17. An extension line is not broken where it crosses another extension line. (A)True or (B)False?
18. A dimension line can be drawn with both arrowheads on the outside of the extension lines, and the dimension value shown between the extension lines. (A)True or (B)False?
19. Dimension lines may be drawn at any angle without causing any difficulty in orienting the dimension value when using unidirectional dimensions. (A)True or (B)False?
20. The scale of a drawing determines the dimension value that is applied to a feature. (A)True or (B)False?
21. A metric dimension must show the abbreviation for millimeter even if all dimensions on the drawing are in millimeters. (A)True or (B)False?
22. An origin must be indicated by showing 0,0 values when using the ordinate dimensioning system. (A)True or (B)False?
23. Only aligned dimensions can be used in the ordinate dimensioning systems. (A)True or (B)False?
24. Tabulated dimensions can't be used to define dimensions for any features other than holes. (A)True or (B)False?
25. It is not always possible to place every dimension between views. (A)True or (B)False?
26. It is preferable to avoid extending a small dimension across a larger one. (A)True or (B)False?
27. It is permissible to dimension to a hidden line when the hidden line is part of a feature in a half section view. (A)True or (B)False?
28. It is permissible to dimension a part so that its size or location can be determined in more than one way. (A)True or (B)False?
29. Notes provide supplementary information and are not considered to be part of the drawing requirements. (A)True or (B)False?
30. A basic dimension indicates that title block tolerances do not apply. It also indicates that the tolerance on the feature is probably provided in a feature control frame. (A)True or (B)False?

FILL IN THE BLANK

31. A _____ is used to connect a notation to a specific feature on the drawing.
32. A dimension applied to a 6″ feature that is drawn at half scale must show a dimension of _____ inches.
33. Arrowheads are typically drawn with a length-to-width ratio of _____.
34. The _____ dimensioning system places dimensions in line, and is also known as the point-to-point dimensioning system.
35. If all features on a part are located by baseline dimensions, the maximum tolerance accumulation is equal to the tolerance on _____ of the dimension values.
36. _Double_ dimensioning results in dimensions that show two acceptable limits of size or location.
37. The origin symbol is a _____ drawn in place of one of the arrowheads on the dimension line.
38. Specific notes may be placed on the field of the drawing, but they are usually listed in a _____.

SHORT ANSWER

39. Define "unidirectional dimensions." *Always*
40. Define "aligned dimensions." *Parallel*
41. Explain the effect of using chain dimensions for all features along the length of a part.
42. Explain how tabulated dimensions are used to provide location and size requirements for holes.
43. If two parts in an assembly mount on the same surface, explain why the mounting surface should be used as a baseline for dimensioning each part.
44. Explain why larger dimensions are generally placed outside smaller ones.
45. Explain what is indicated by the origin symbol when it is applied to a drawing.
46. Explain what must be done to avoid dimensioning internal features where they appear as hidden lines.
47. Define "maximum material condition."

APPLICATION PROBLEMS

Each of the following problems shows a dimensioned pictorial drawing. Sketch the necessary orthographic views of each problem to permit application of all dimensions for the part. Apply dimensions to completely define size and location of all features. Dimensions are to be applied in compliance with guidelines covered in this and preceding chapters.

48.

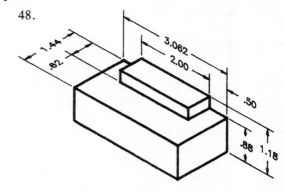

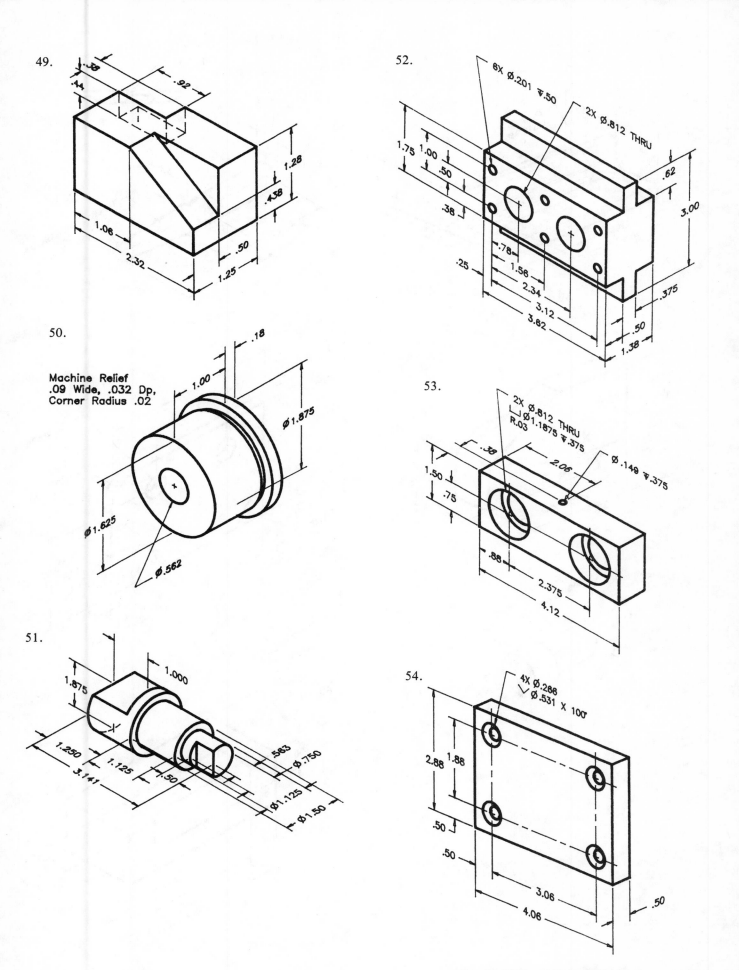

49.

.38
.44
.92
1.28
.438
1.06
2.32
.50
1.25

50.

Machine Relief
.09 Wide, .032 Dp,
Corner Radius .02

.18
1.00
⌀1.875
⌀1.625
⌀.562

51.

1.875
1.000
1.250
1.125
.50
3.141
.563
⌀.750
⌀1.125
⌀1.50

52.

6X ⌀.201 ⍯.50
2X ⌀.812 THRU
1.75
1.00
.50
.38
.62
3.00
.78
1.56
2.34
3.12
3.62
.25
.375
.50
1.38

53.

2X ⌀.812 THRU
⌀1.1875 ⍯.375
R.03
.38
2.06
⌀.149 ⍯.375
1.50
.75
.88
2.375
4.12

54.

4X ⌀.266
⌀.531 X 100°
2.88
1.88
.50
.50
3.06
4.06
.50

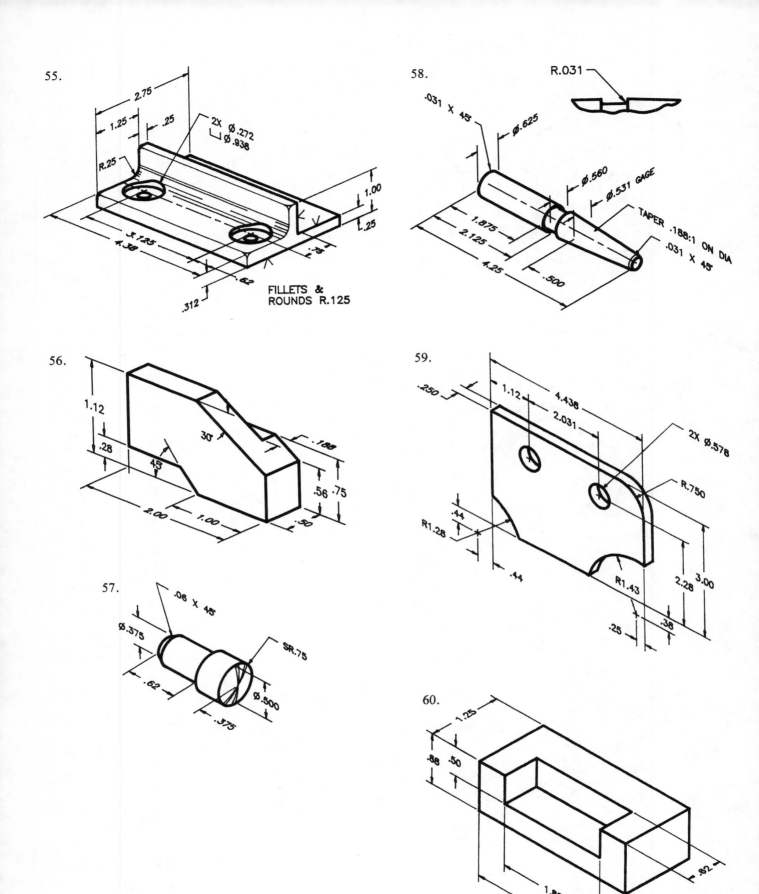

55.

2.75
1.25 .25
2X Ø.272
⌴ Ø.938
R.25
1.00
.25
3.125
4.38
.8
.62
.312
1.00
FILLETS &
ROUNDS R.125

56.

1.12
.28
30°
45°
.188
.56 .75
2.00
1.00
.50

57.

.06 X 45°
Ø.375
SR.75
.62
Ø.500
.375

58.

R.031
.031 X 45°
Ø.625
Ø.560
Ø.531 GAGE
TAPER .188:1 ON DIA
.031 X 45°
1.875
2.125
4.25
.500

59.

.250
1.12
4.438
2.031
2X Ø.578
R.750
R1.28
.44
.44
R1.43
2.28
3.00
.25
.38

60.

1.25
.88 .50
1.88
2.88
.82
.50

61.

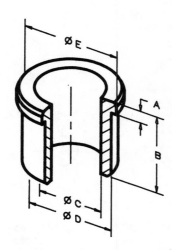

PROBLEM NUMBER	A	B	C	D	E
61A	.125	1.000	.7506 .7503	1.0018 1.0015	1.125
61B	.125	1.500	1.3760 1.3756	1.7523 1.7519	1.375

62.

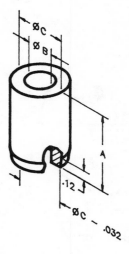

PROBLEM NUMBER	A	B		C
62A	.750	.6250	+.0001 −.0005	.8768 .8765
62B	1.000	.3281	+.0001 −.0005	.6267 .6264

63.

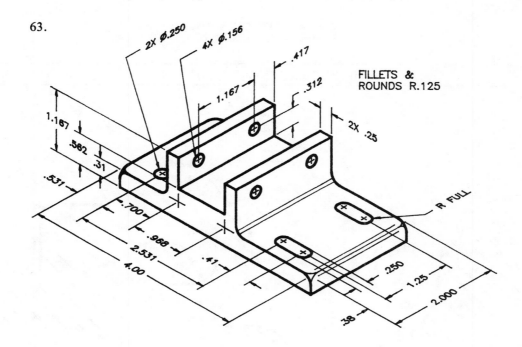

FILLETS & ROUNDS R.125

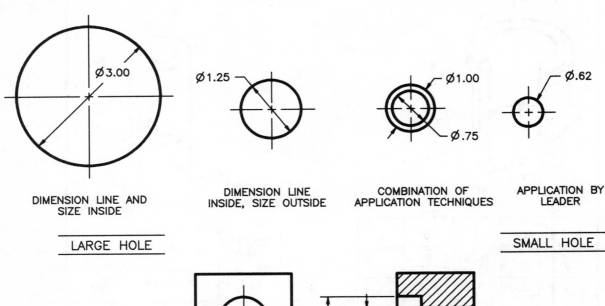

DIMENSION LINE AND
SIZE INSIDE

DIMENSION LINE
INSIDE, SIZE OUTSIDE

COMBINATION OF
APPLICATION TECHNIQUES

APPLICATION BY
LEADER

<u>LARGE HOLE</u>

<u>SMALL HOLE</u>

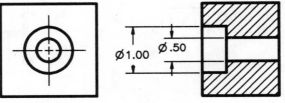

HOLE DIAMETER DIMENSIONED
ON THE NONCIRCULAR VIEW

DIMENSIONING HOLE DIAMETER

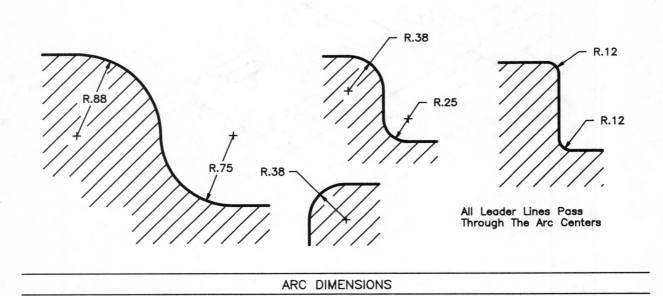

All Leader Lines Pass
Through The Arc Centers

ARC DIMENSIONS

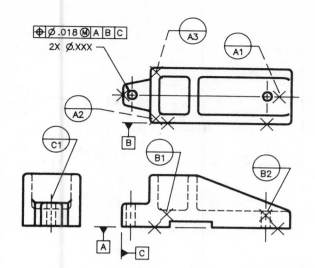

Chapter 4

Dimension Application and Limits of Size

CHAPTER TOPICS

☐ Guidelines for clear application of dimensions.
☐ Dimensioning requirements for geometric shapes and various features common to mechanical parts.
☐ Limits of size and methods for specifying limits of size.

☐ The rules contained in ASME Y14.5M.
☐ The effects that dimensions and the associated tolerances have on manufacturing.

INTRODUCTION

A general description of a part can be given through the standard views, section views, and auxiliary views of a multiview drawing. The size, location, and values affecting geometric form requirements are given through the application of dimensions.

Complete part definition requires that all aspects of each geometric feature on a part, including dimension requirements, be defined. Each part can be visualized as a composite of geometric shapes, and each of those shapes dimensioned. Methods for applying dimensions on various geometric shapes are contained in this chapter.

Each size dimension must have some allowable amount of acceptable variation since it is not possible to build multiple parts that are all the same size. The allowable variations for many applications are standardized. The means for showing the size limits on dimensions are also standardized. Allowable variations are calculated for parts on the basis of design function. The calculations are often completed using standard tolerance tables that provide assistance in determining limits of size.

DIMENSION APPLICATION

Specific dimension application techniques are used for common geometric shapes. The use of a specific application technique for each feature type provides a consistent and recognizable dimensioning system. Consistency in the application techniques prevents misinterpretations that might occur if random methods were used. A consistent set of application techniques ensures that dimensions are easy to interpret because the techniques are repeated every time the same geometric shape is repeated.

PRISMS AND PYRAMIDS

Prisms and pyramids have edges where the adjacent sides intersect. See Figure 4-1. The edges intersect to create outside the circle is also used when centerlines and concen-

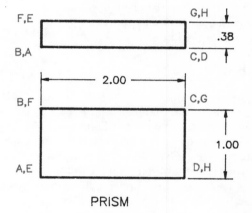

PRISM

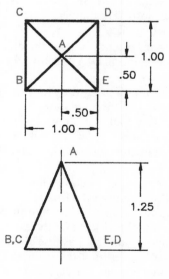

PYRAMID

Figure 4-1. Height, width, and depth dimensions must be defined for prisms and pyramids.

corners. Dimensions that show the locations for all the corners describe the size of the object.

The given right rectangular *parallelepiped* (prism) has eight corners. The corners are on the six flat surfaces that form the prism. Since opposite sides of this particular type of prism are known to be parallel, it is only necessary to show three dimensions. The three dimensions completely describe the size of the object by showing its height, width, and depth. The angle between surfaces is assumed to be 90° because the object appears as a parallelepiped. *It is not necessary to dimension 90° angles.*

The given pyramid is a *right square pyramid.* It has five points, all of which must be dimensioned. The apex (point A) is dimensioned from the base (surface B,C,D,E) at a height of 1.25″. Width and depth dimensions for the base are each given as 1.00″. Point A is shown centered above base B,C,D,E by locating dimensions of .50″.

Point A is not actually .50″ from edges B,E and D,E. The .50″ dimensions specify rectangular coordinates relative to the two edges. The 1.25″ height dimension must be considered if the true distance from the base edges to point A is desired. The dimensions shown are the correct dimensions for specifying the size of the given pyramid. All point locations are dimensioned, and the part can be produced when all point locations are known.

CYLINDERS

The *diameter* and *length* must be given for a cylinder. It is preferable for both the diameter and length to be dimensioned on the noncircular view. See Figure 4-2. The diameter dimension is normally placed between the noncircular view and the circular view. Cylinders are always dimensioned by specifying the diameter rather than the radius.

One view can be used to show a cylindrical object if all features can be clearly dimensioned, as shown in the second example. The diameter and length for each cylinder is specified; no additional dimensions are required. The current standard requires that a diameter symbol be placed in front of diameter dimensions. The diameter symbol is a circle with a diagonal line through it.

The 1973 standard did not include the diameter symbol. It required the abbreviation DIA to follow the size specification for the feature.

Some situations require that the diameter dimension for a cylinder be shown in a circular view. See Figure 4-3. The given cylinder has a flat cut down its length, and the noncircular view does not show the full diameter. Dimensioning the .75″ diameter in the noncircular view would be confusing.

On the circular view, an extension line is drawn using an arc that is equal in radius to the one for the cylinder. The extension line is drawn thin and dark like all extension lines. A gap is left between the object and the extension line. The extension line runs about .125″ past the dimension line. The dimension line passes through the center of the cylinder. One end terminates on the cylinder, and the other end extends outside the cylinder to a position where the dimension value is shown. A short horizontal line extends between the dimension line and the dimension value. The diameter sym-

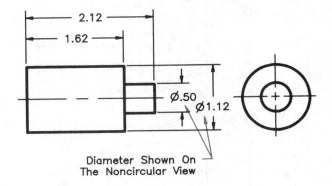

Diameter Shown On The Noncircular View

TWO VIEWS

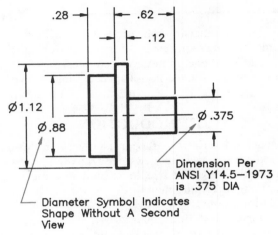

Diameter Symbol Indicates Shape Without A Second View

ONE VIEW

Figure 4-2. The diameter and length must be shown for a cylinder.

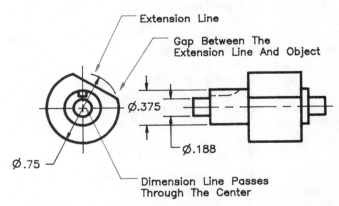

Figure 4-3. In some cases, it is necessary to dimension the diameter of a cylinder in the circular view.

bol is required to avoid the possibility of the dimension being interpreted as a radius. Arrowheads are drawn at the cylinder and extension line.

FLATS ON CYLINDERS

A flat is often cut into a cylinder to provide a surface suitable for clamping something in a fixed location on it.

The size of a flat on a cylinder is dimensioned as shown in Figure 4-4. The location of flat A,B,C,D can be specified relative to the centerline of the cylinder or relative to the far side of the cylinder. The flat's location can also be specified relative to other features on the cylinder when any other features exist.

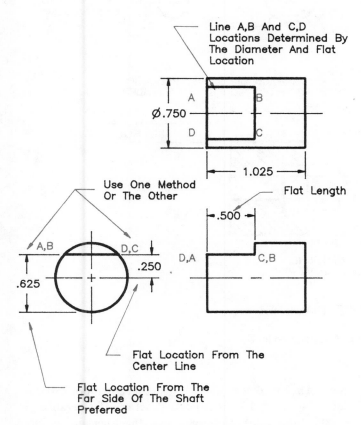

Figure 4-4. The location and length of a flat must be dimensioned on a cylinder.

The length of the flat is shown in a noncircular view. Locations of sides A,B and C,D are determined by the cylinder's diameter and the flat location dimension. Side A,B and C,D locations are not dimensioned directly.

Choice of which location dimension to use on a flat is based upon the part's function. If the flat is a clamping surface, a dimension from the far side is preferred. This dimension is easy to inspect and large errors relative to the cylinder centerline can be tolerated. If the flat is a locating surface for part of a mechanism, it's location to the cylinder's centerline may be specified. Inspection is made more difficult since the centerline must be established from the cylinder's diameter, but the function of the part sometimes takes precedence over the ease of inspection.

CONES

A right circular cone is normally dimensioned by giving the *base diameter* and the *cone height*. See Figure 4-5. It is also acceptable to show the base diameter and the cone angle. An oblique circular cone is dimensioned by giving the *base diameter, cone height,* and *apex offset*.

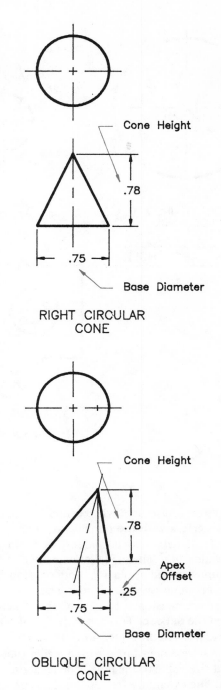

Figure 4-5. The diameter of the base, cone height, and any required apex offset are given when dimensioning a cone.

HOLES

Holes are normally dimensioned by giving the *diameter*. The dimensions may be shown in the views where the holes appear as circles. See Figure 4-6. Large diameter holes are dimensioned by placing the dimension line and size value inside the hole. The dimension line and value are placed inside the hole if it is large enough. If the diameter is too small for the value to be shown on the inside, the dimension line and arrowheads are placed inside and the value is placed outside. The dimension line must pass through the center of the hole. The technique of placing the dimension value

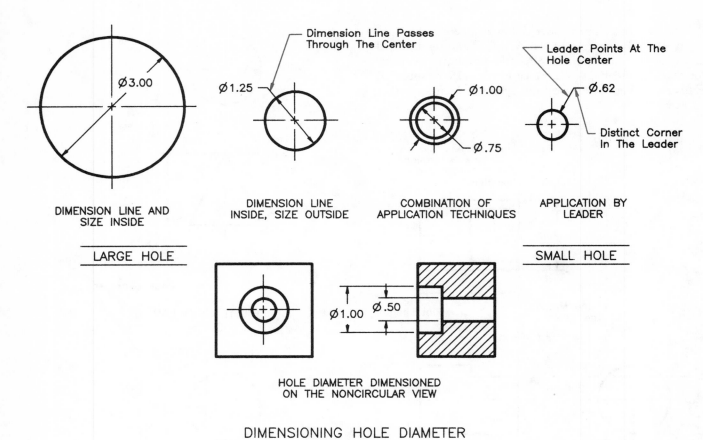

DIMENSIONING HOLE DIAMETER

Figure 4-6. The diameter must be specified for holes.

outside the circle is also used when centerlines and concentric circles interfere with the dimension value.

Concentric diameters can be dimensioned using a combination of techniques. The .75″ diameter hole in the given example is dimensioned by placing the dimension line on the inside and the value on the outside. The 1.00″ diameter counterbore is dimensioned by placing the dimension line and value on the outside. This combination of dimensions can be used to prevent the dimensions from being crowded.

Small holes are dimensioned with a leader that connects the specified size value to the hole. The leader may include a horizontal line extending from the dimension value. If a horizontal line is used, a distinct corner is made at the end of the horizontal line, and the leader extends from the corner toward the center of the hole. The leader is terminated with an arrowhead where the leader intersects the hole.

Hole diameters can be specified in a noncircular view if the hole is shown with object lines. This is only possible when the noncircular view appears in a section. Hidden noncircular views are not normally dimensioned.

The callout of a hole size does not require that a process for making the hole be specified. If a hole of .250 ± .005″ diameter is specified, the machinist or production planner must decide how to achieve the desired size. It is more important to control the size of the hole than to control how the hole is made.

The specified hole size and tolerance should allow the machinist a maximum amount of freedom in how the hole is made. The use of nominal hole sizes that correspond to standard drill diameters is a good practice. If tolerances of +.006″ and -.001″ are specified, the tolerance is large enough that most holes under .500″ diameter can be drilled, which is a one-step process.

When tolerances too small to permit drilling are used, it may be necessary to complete two steps to produce the holes. As an example, a small hole diameter tolerance can be achieved by first drilling the hole undersize, then finishing the hole by reaming it to the specified size.

As hole size tolerances are decreased, the amount of precision and time taken to produce them increases. When tolerances under .001″ are specified, the hole must be bored, ground, or lapped to meet the specification. These are expensive processes compared to drilling or reaming.

Hole diameters and tolerances must be specified to achieve the design goals, but specification of tighter-than-necessary tolerances must be avoided to keep manufacturing costs down. Hole size tolerances smaller than needed for the design function should be avoided since it forces more machine operations, and thereby increases product cost.

Hole locations are always dimensioned to the hole center. See Figure 4-7. This is done because the point of a standard drill first contacts the hole center. It is also done because hole-cutting tools rotate on a machine spindle, and the location of the machine spindle center can be accurately located.

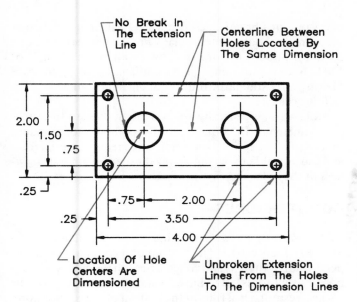

Figure 4-7. Hole locations must be dimensioned to the centers.

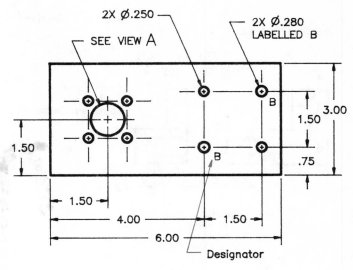

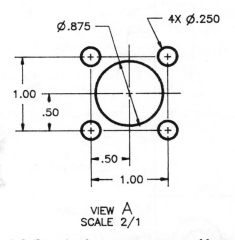

VIEW A
SCALE 2/1

Figure 4-8. Complex features or patterns of features may be dimensioned in removed views.

Extension lines project from the hole center to the dimension lines. No breaks are shown in an extension line once it is outside the hole, nor is there a break in the extension line at the hole. Centerlines are drawn between holes when a single dimension is used to locate multiple holes that are aligned with one another.

Hole sizes and locations must be dimensioned. The locating dimensions are placed on one side of the view, if possible, permitting hole size specifications to be placed on the other side of the view. This arrangement prevents leaders from crossing dimension lines and makes the drawing easier to read. As parts become complex, it is not possible to always stay in compliance with this general guideline.

A simple plate for which all holes are completely dimensioned is shown in Figure 4-8. A hole pattern is located on the plate, and dimensions for the hole pattern are shown in removed VIEW A. *Removed views* are used to reduce crossed dimensions on the drawing and to provide a clear definition of small features. The .875″ diameter hole is used as an origin in VIEW A. The location of the .875″ diameter hole is given in the main view. The four .250″ diameter holes are located from the center of the .875″ diameter hole.

The main view includes location dimensions for four holes in addition to those shown in View A. Two holes are .250″ diameter and two are .280″ diameter. A letter B is used to designate the two .280″ diameter holes. The designator is used since there is only a .030″ difference in diameter between the two hole sizes. The small difference in diameter makes it difficult to determine which holes are of what diameter without the designator. The designator is placed in the same position relative to each identified hole. This ensures association of the correct hole with its designator.

The hole size specification shows the number of holes to be made at each specified size. The number of holes can be written in one of two ways. One way is to write out the number of holes as shown in the following example. This method is no longer illustrated in the standard.

ϕ.250 2 Holes

The second way indicates the number of times a hole is repeated. This is the preferred practice since it is illustrated in the standard.
Example:

2X ϕ.250
(two times make hole diameter .250″)

Holes are assumed to go through the part on which they are shown unless a depth is specified. See Figure 4-9. If the drawing views do not make it clear that a hole goes through the part, the word THRU must follow the hole diameter. Depth must be specified for any hole that does not go through a part. Holes that do not go through are called *blind holes.*

The depth of a hole is specified after the diameter. The depth dimension is indicated using a symbol. Through and blind holes are shown in the given figure. The depth is the distance that the full diameter of the hole extends into the part. The drill point extends beyond the specified depth for a hole.

Depth of the full diameter is specified because it is the usable portion of the hole. The point protrusion past the full

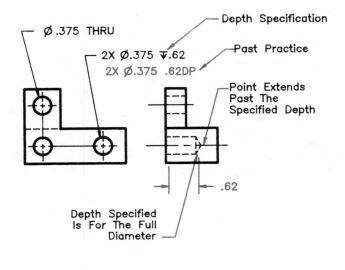

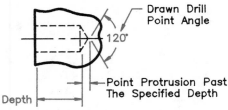

Figure 4-9. Hole depth must be specified if the hole does not go through the part.

diameter must be considered and steps taken to ensure that the point does not unintentionally break out the far side of a part. Drill points are drawn showing an included angle of 120°. Actual drill point angles and shapes vary, however 118° is fairly standard, thus a 120° included angle is sufficiently accurate.

Hole depth is generally not more than three to five times the diameter of the hole. This allows the use of standard tools. Greater depth-to-diameter ratios require special tools and cause problems in controlling the diameter and straightness of the hole.

COUNTERBORES

A *counterbore* is a stepped increase in the diameter of a hole. See Figure 4-10. The bottom of a counterbore is perpendicular to the axis and concentric with the hole. Counterbore cutting tools often have a small radius ground on the corner, resulting in a fillet at the bottom of the counterbore.

Counterbore holes are specified by giving the *hole diameter, counterbore diameter, counterbore depth,* and *corner radius.* Counterbores may be dimensioned by notation, applied dimensions, or a combination of the two. The symbol for counterbore is used when calling out counterbore sizes. Counterbore is abbreviated CBORE in notes lists.

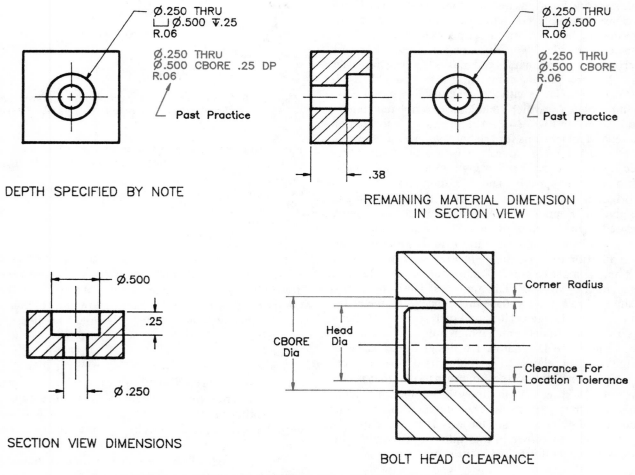

Figure 4-10. Counterbore dimensions provide a controlled depth.

The distance from the surface to the counterbore bottom can be noted in the counterbore callout or dimensioned in a section view. It is also permissible to dimension the thickness of remaining material to control counterbore depth. The remaining material is dimensioned only when a critical need exists for a specific material thickness. Controlling and inspecting the remaining material thickness is normally more expensive than controlling the counterbore depth from the surface.

The counterbore note is attached to the hole by a leader, and is usually applied to the circular view. The leader points toward the hole's center and terminates with an arrowhead on the first object line.

A common use for counterbores is to recess fastener heads, such as those found on screws and bolts. To recess a feature such as a screw head, the counterbore diameter must be equal to the head diameter, plus two times the corner radius, plus any location tolerance that affects the relative locations of the screw and counterbore. A counterbore diameter calculated in this way ensures that the head does not ride up on the counterbore corner radius. The minimum counterbore depth is equal to the maximum head height plus the minimum amount of recess desired.

COUNTERSINKS

A *countersink* is an angular increase in the size of a hole. It has the shape of a right circular cone and is coaxial with the hole. See Figure 4-11. Countersinks are used to recess fastener heads, such as those on flathead screws and to break (remove) the sharp edges of holes. The standard included angles for recessing flathead screws are 82° and 100°. The angle of the countersink must match the screw head angle. Countersink tools used to break sharp edges have a 90° included angle and are not meant to be used for recessing screws.

Countersunk holes are specified by giving the *hole diameter, countersink diameter,* and *countersink angle.* The included angle is always specified. The diameter of a countersink is the distance across the top. Irregular surfaces make measurement of a countersink diameter difficult. The countersink diameter on an irregular surface is equal to twice the minor radius of the countersink. The *minor radius* is the shortest distance between the axis of the hole and the intersection of the countersink with the surface.

Countersink requirements may be given through notation, applied dimensions, or a combination of the two. The countersink notation is attached to the hole by a leader. The leader points at the hole's center, but it terminates on the countersink.

Sizes for a countersink are determined by the function of the countersink. A countersink used to recess a screw head has an angle equal to the angle of the screw head. Flathead screws are made at angles of 82° and 100°. The minimum diameter of a countersink must be equal to the sharp diameter of the screw head or the screw will protrude above the

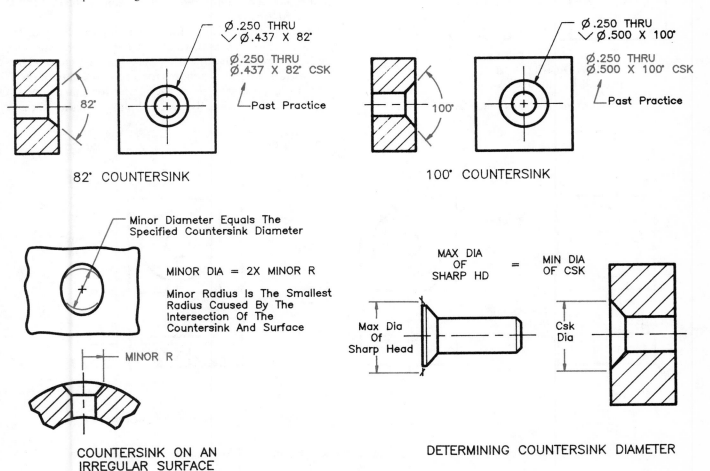

Figure 4-11. A countersink is used to recess a flathead machine screw or bolt head. A small countersink can also be used as a chamfer on a hole.

surface. The diameter of the screw head before the sharp edge is removed is known as the *sharp diameter*. A countersink diameter equal to the sharp head diameter causes the screw head to be flush with the surface. A countersink diameter larger than the sharp head diameter will ensure that the screw head is below the surface.

COUNTERDRILLS

A counterdrilled hole is somewhat of a cross between a counterbore and a countersink. See Figure 4-12. There is a stepped increase in the hole diameter. The change in hole diameter that is made by the counterdrill is conical. The counterdrilled diameter is coaxial with the through hole.

Counterdrilled holes are specified by giving the *hole diameter, counterdrill diameter, counterdrill angle,* and *counterdrill depth.* The counterdrill angle can be given in the note or it can be dimensioned on a section view. The counterdrill depth is the depth of the full diameter.

The note is attached to the counterdrilled hole by a leader. The leader extends toward the center of the hole and terminates on the first object line.

SPOTFACES

A spotface is similar to a counterbore, but does not go as far into the part. See Figure 4-13. The spotface is only intended to provide a flat surface against which a screw head or washer can rest; it is not intended to recess the screw head.

The minimum acceptable amount of information for a spotfaced hole is the *hole diameter* and the *spotface diameter.* If no depth is given for a spotface, the spotface depth is not more than the depth required to clean the surface to the specified diameter.

There are two methods for controlling the depth of the spotface surface; one method is to specify the spotface depth, the other is to dimension the remaining material in a section view. The corner radius for a spotface may be specified in the callout, given in a general note, or dimensioned in a section view. It is usually easiest to show it in the callout.

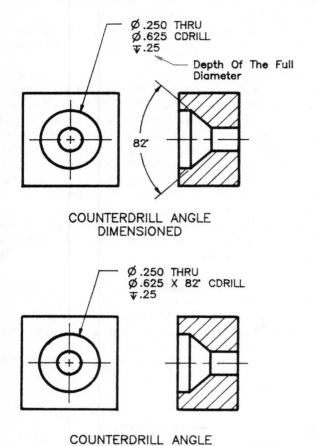

Figure 4-12. A counterdrill creates a recessed countersink.

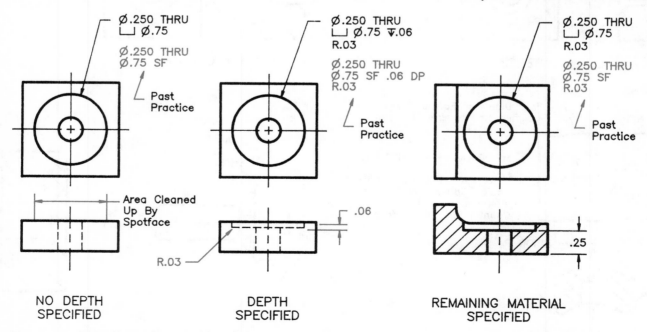

Figure 4-13. A spotface is similar to a counterbore, and its specification may or may not include a controlled depth requirement.

The diameter of a spotface is calculated by adding the maximum diameter of the fastener, two times the corner radius, and the associated location tolerance. Depth is determined by the amount of material that must be removed to ensure a flat surface.

ANGLES

Angles can be dimensioned by showing the *degrees between the sides* of an angle. See Figure 4-14. Angles are specified by using one of two methods. One method uses degrees, arc minutes, and arc seconds; the other uses degrees and decimal parts of a degree. The symbols used in dimensioning angles are:

degrees	30°
minutes	30° 15′
seconds	30° 15′ 30″
decimal	30.25°

Tolerances on an angle are specified using the same units as the nominal value.

30.50° ± .25°
30° 30′ ± 0° 15′

Sides of an angle are extended using extension lines, and a dimension line is drawn between the extension lines. The dimension line is an arc. The center point of the dimension line arc is located at the vertex of the angle. The extension lines, if extended, should intersect at the vertex of the dimensioned angle.

The dimension line arrowheads and dimension value can both be outside the extension lines, the arrowheads outside and the value inside, or the arrowheads and value inside. The arrangement of arrowheads and value depends on the angle size and the distance the dimension is located from the angle vertex.

One corner of an inclined surface must be located and the angle of inclination specified. The second corner of the inclined surface is located by the intersection of the inclined surface with another side of the part. If both ends of an inclined surface are located by coordinate dimensions, the angle does not need to be dimensioned.

Perpendicular surfaces are understood to be at a 90° angle without any dimension being shown. All other angles must be dimensioned, or endpoints must be defined by coordinate dimensions.

CHAMFERS

A chamfer is a small inclined surface cut on the edge of a part. It is used to eliminate sharp edges and to facilitate assembly of close-fitting parts. Chamfers of 45° can be dimensioned by note. All other chamfer angles must be dimensioned in one of two ways. One way is to show the *angle of inclination* and the *length of one side* of the chamfer. The other way is to show the length of both sides of the chamfer. See Figure 4-15.

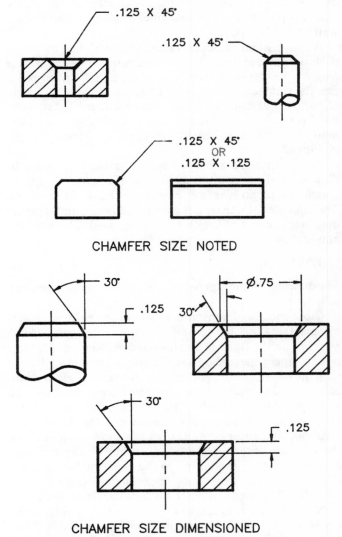

CHAMFER SIZE NOTED

CHAMFER SIZE DIMENSIONED

Figure 4-15. Chamfers eliminate sharp edges and facilitate assembly of close-fitting parts.

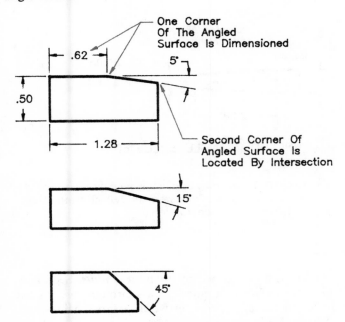

Figure 4-14. The arc used for a dimension line when dimensioning an angle is centered on the vertex of the angle.

A note can be used for 45° chamfers since both sides of a 45° triangle are equal in length. The sides of any other angle are unequal and therefore can't be specified by a note. A note can't specify which side of the angle is being given. The only way to be certain of control over chamfers other than 45° is to apply the chamfer dimensions to the part.

TAPERS

Tapers on a shaft or hole create a conical surface. Two purposes of tapers are to locate parts relative to one another and to hold parts together. A tapered pin sliding into a tapered hole provides a well-controlled location. The axial centerlines of the assembled tapered parts coincide. A taper of the correct angle provides a good clamping force between assembled parts. Some standard tapers provide enough clamping force to hold cutters in the spindles of milling machines.

Several methods are acceptable for dimensioning tapers. See Figure 4-16. The method used depends on the amount of control desired. Tapers on parts meant to mate existing machines and tools must be dimensioned using the appropriate standard machine taper. The machine taper to be mated can be determined by looking in the manufacturer's handbook that is supplied with any machine. Machine tapers are described in ANSI B5.10.

A standard machine taper can be dimensioned by noting the taper name and number. The *diameter at one end* and the *length of the taper* must be dimensioned. In the given example, the small end of the taper (.572″) and the taper length (2.562″) are dimensioned. American Standard Taper Number 2 is specified. The shaft made to these dimensions will mate a machine spindle with an American Standard Taper Number 2.

Tapers that are not made to the standard sizes can be dimensioned by giving the *taper on diameter per unit of length*. Taper on diameter means the change in diameter. The diameter of one end, the length of taper, and taper on diameter per unit of length are dimensioned. The given example shows:

TAPER .125:1

The diameter of the tapered part must change .125″ per inch of length.

Close control of nonstandard tapers can be dimensioned by specifying that the taper diameter be fitted to a gage. A position on the taper is dimensioned as the gage line location. A taper gage of the given diameter on one end will align with the gage location. Any error in the taper angle or diameter causes a substantial misalignment between the gage and gage line. Since some tolerance must be allowed on the taper, a relatively large tolerance is applied to the gage line location. The given example does not show a tolerance on the 1.80″ dimension, and a large title block tolerance is assumed to apply. In addition to the gage line location and diameter, the taper on diameter per unit of length must be specified. The length of taper must also be controlled.

Noncritical tapers can be dimensioned by giving the *included angle, length of taper,* and *diameter at one end*. It is also permissible to omit the angle if the diameter at both ends is specified.

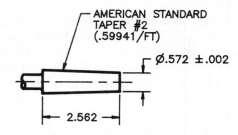

STANDARD TAPER SPECIFIED

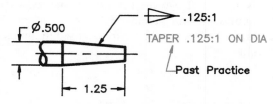

TAPER PER INCH SPECIFIED

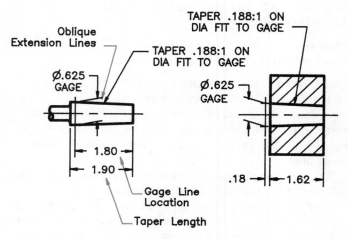

TAPER TO GAGE SPECIFIED

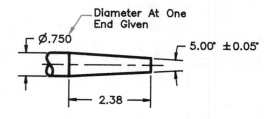

INCLUDED ANGLE SPECIFIED

Figure 4-16. Tapers may be dimensioned through several methods. The application determines which method is appropriate.

ARCS

An arc is dimensioned by giving its *radius*. See Figure 4-17. A leader line is oriented to pass through the arc's center. The arc center is shown by two short crossed lines.

When space permits, the leader line extends from the arc center to the arc. An arrowhead is placed on the end of the

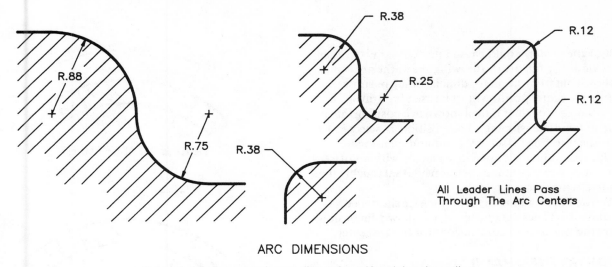

All Leader Lines Pass
Through The Arc Centers

ARC DIMENSIONS

Figure 4-17. Arcs are always dimensioned by giving the radius.

line that touches the arc. An arrowhead is never placed at the arc center. The dimension value is placed in a break in the leader line. The current national dimensioning standard requires that an **R** be placed in front of the dimension value to indicate that it is a radius. Prior to the 1982 standard, it was required that the **R** follow the dimension value.

An arc with a small radius is dimensioned with the leader line extended to the dimension value. If the arc center is inside the object, a leader line runs from the arc center to the outside of the object. If the arc center is outside the object, the leader line extends from the arc, through the center, and to the dimension value.

An arc with a very small radius is dimensioned with the leader line outside the object, regardless of whether the arc center is inside or outside the object. The leader lines are always oriented to pass through the arc center. It is not necessary to show arc centers for small radius arcs. Showing the arc center for small radii would interfere with the dimension arrowhead.

Arc locations are defined in one of two ways; either the *arc center* is dimensioned, or it is located by *arc tangents*. See Figure 4-18. Two short crossed lines are used to identify the arc center when the center is located by dimensions. The crossed lines are not required when an arc is located by tangents.

The positions of tangents are determined by the arc's location if the arc center point is dimensioned. If the arc center point is not dimensioned, then the tangent locations must be dimensioned. When tangents are dimensioned, the arc is located by the tangents. It is incorrect to provide dimensions for both the arc center and the arc tangents. Providing both is double dimensioning and is not acceptable.

The center for large radius arcs can be difficult to show in their true position. See Figure 4-19. In cases where the true

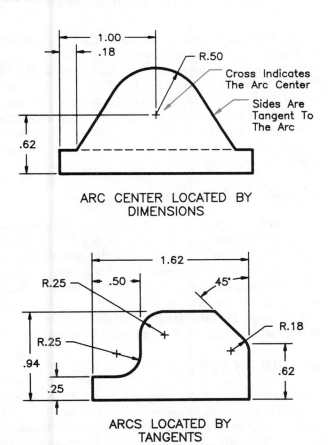

ARC CENTER LOCATED BY
DIMENSIONS

ARCS LOCATED BY
TANGENTS

Figure 4-18. Arc location must be defined.

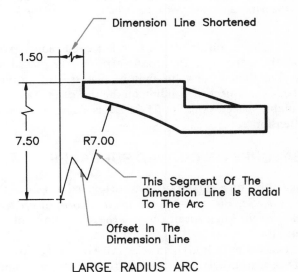

LARGE RADIUS ARC

Figure 4-19. The true dimensions defining arc center location must be shown even when the center point is not shown in its true location.

Dimension Application and Limits of Size   73

position of the center would cause interference with other features, views, or fall off the drawing sheet, the arc center can be shown out of position for dimensioning purposes.

The true arc center position must be used for drawing the arc. A simulated arc center is then shown in a position that is convenient for dimensioning. The position is selected to minimize the resulting offset in the radius dimension line.

The dimension line is drawn in segments with an offset included. The segment of the dimension line that touches the arc is radial to the arc.

Dimension lines locating the arc center are shortened. Although dimension lines are shortened, the shown dimension values are the true location dimensions for the arc center.

FORESHORTENED RADII

The radius of an arc is dimensioned in a true shape view whenever a true shape view is given. If no true shape view exists, the true radius can be dimensioned wherever a foreshortened view of the arc is seen. See Figure 4-20. A foreshortened view is seen when the line of sight is inclined to the surface on which an arc exists.

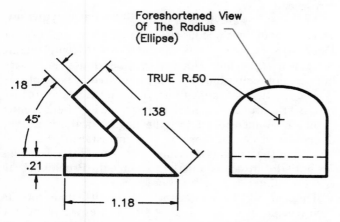

Figure 4-20. The true radius of an arc must be specified even if the dimension is shown on an elliptical view of the arc.

The dimension line for a foreshortened radius extends from the arc center to the arc. An arrowhead is placed at the arc. The dimension value may be placed either inside or outside the object, depending on the size of the radius and available space. A prefix of **TRUE R** is placed on the radius dimension.

FEATURES ON CURVED SURFACES

Features on a curved surface are located by one of three methods. See Figure 4-21. The *angle, chord,* or *arc length* can be given. Angle specification is the most commonly used method.

Location by a chord is an acceptable method of location, but fabrication accuracy can be difficult to check manually. This is especially true if the located features are holes normal to the curved surface. The reason for difficulty in checking chord measurements between radial holes is that the chord is

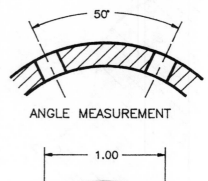

ANGLE MEASUREMENT

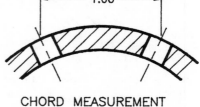

CHORD MEASUREMENT

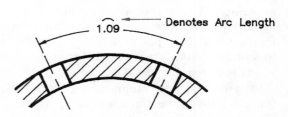

ARC LENGTH MEASUREMENT

Figure 4-21. The method used to locate features on a curved surface depends on the function of the part.

specified in relationship to the curved surface. This reference surface is cut away as the holes are drilled.

An arc distance dimension is an acceptable method for locating features on a curved surface, but it too can be difficult to check. An arc distance is the distance measured on a curved surface. The dimension line used for an arc distance dimension is concentric with the arc distance being shown. A small arc is placed above the dimension value to indicate that it is an arc length.

SPHERICAL RADII

A spherical radius must be specified by placing the letters **SR** in front of the dimension. See Figure 4-22. Failure to show the **SR** prefix could result in a simple arc being produced on the part. A spherical radius is easiest to produce and verify when the location dimension goes to the outside of the spherical surface.

Prior to the 1982 standard, it was required that the suffix **SPHER R** be applied to spherical radii dimensions.

IRREGULAR CURVES

Curved surfaces made of tangent arcs are dimensioned by showing the *locations* and *sizes* for the arcs. See Figure 4-23. In the given example, the position of each arc is determined by a combination of arc center location dimensions and tangent feature locations. Each radius is dimensioned.

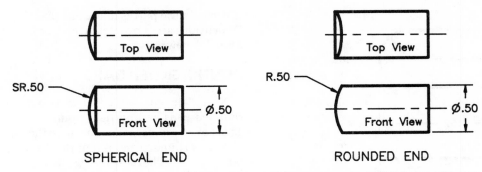

Figure 4-22. A spherical radius is noted by adding the spherical radius prefix to the dimension.

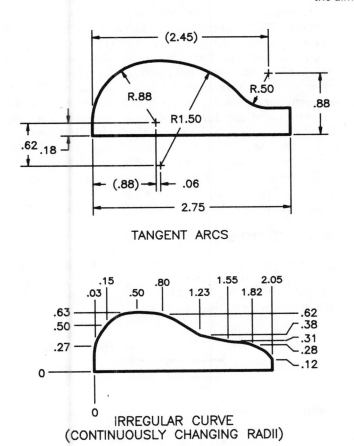

TANGENT ARCS

IRREGULAR CURVE
(CONTINUOUSLY CHANGING RADII)

Figure 4-23. Curved surfaces are sometimes dimensioned by showing point locations along the curve.

Coordinates for points along a curve or a *mathematical formula* defining the curve are given when the curve is not a common geometric shape. When dimensioning points on a curve, a sufficient number of points must be dimensioned to ensure production of a smooth curve.

SYMMETRICAL FEATURES

Only half of a symmetrical part must be dimensioned. See Figure 4-24. The dimensions shown must define the location and size of each feature relative to the line of symmetry. All locations and sizes are shown relative to the line of symmetry and are known to repeat on the opposite side of the line. The dimensions are known to repeat because sym-

metry is indicated on the drawing by two short lines drawn across each end of the centerline.

A symmetrical curved part is dimensioned in Figure 4-24. The centerline is indicated to be a line of symmetry. The distance from the centerline to the curve is dimensioned in four places. The locations for each of the four dimensions are given. The undimensioned side of the part is identical to the dimensioned side since a line of symmetry is identified.

Large symmetrical parts can be shown in partial or half views. The given view shows slightly more than half the part.

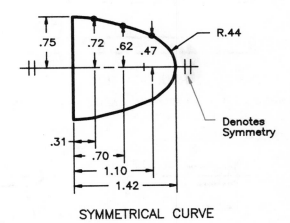

SYMMETRICAL CURVE

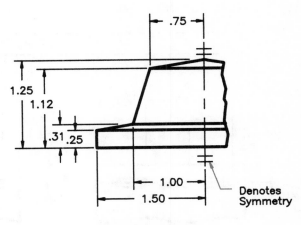

PARTIAL VIEW OF A
SYMMETRICAL PART

Figure 4-24. The dimensions for a symmetrical part are identical on both sides of the line of symmetry.

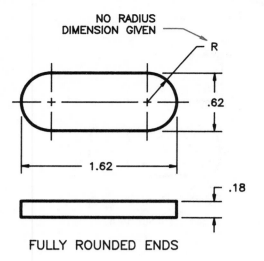

FULLY ROUNDED ENDS

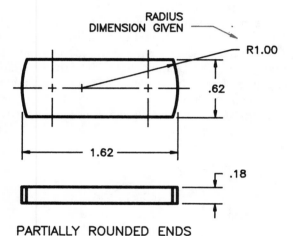

PARTIALLY ROUNDED ENDS

Figure 4-25. The overall length of round-ended bars is dimensioned.

A break in the part is made just past the line of symmetry. Dimensions are shown relative to the line of symmetry when a partial view is used.

ROUND-ENDED BARS

Round-ended bars come in two general shapes: fully rounded ends and partially rounded ends. Fully rounded ends have a radius that is tangent to the two sides. The radius of a fully rounded end is half the distance between the two tangent sides. The radius for partially rounded ends is always greater than half the distance between the two sides. A radius less than half the distance could not intersect both sides.

Examples of how to dimension round-ended bars are shown in Figure 4-25. The overall length, thickness, and the distance between sides are given. A radius value is not normally specified for fully rounded bars. The radius dimension line and the letter **R** are shown without giving a size value. No size is shown for the radius since it must be equal to half the distance between tangent sides. It is incorrect to show both a radius dimension and the distance between tangents on a fully rounded bar because showing both is double dimensioning. If the radius on a fully rounded end is critical, then the radius can be dimensioned and the distance between sides omitted.

The radius for partially rounded ends must always be specified. The distance between sides has no effect on the radius size except to determine the minimum possible radius.

SLOTTED HOLES

Slotted holes can be used when tolerances accumulate between parts to potentially prevent fasteners from passing through round holes. Slotted holes can also be used to allow adjustment in the position of a part.

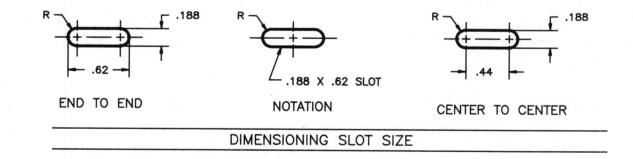

END TO END

NOTATION

CENTER TO CENTER

DIMENSIONING SLOT SIZE

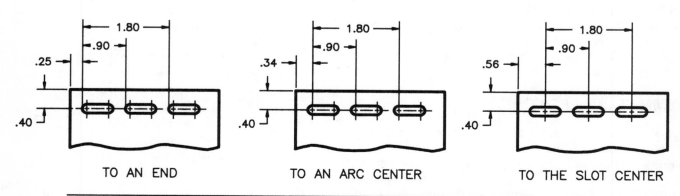

TO AN END

TO AN ARC CENTER

TO THE SLOT CENTER

DIMENSIONING SLOT LOCATION

Figure 4-26. Slotted holes are used to allow greater tolerance for the location of fastener holes and to permit adjustment.

Three dimensioning methods exist for dimensioning slotted hole size. See Figure 4-26. The method used depends on the design function and the machine process that will be used to produce the slot.

Slotted holes punched in sheet metal are normally dimensioned by giving the *overall length* and *width*. An R is shown on one end to indicate that both ends of the slot are made on a radius that is tangent to both sides of the slot. These dimensions may be applied directly to the slot or shown in a notation that is attached with a leader.

Slotted holes produced by machine cutting processes are dimensioned by showing the *slot width* and *distance between arc centers*. The slot width is equal to the cutter diameter, and the center distance equals the amount of cutter travel.

The method used to show slotted hole location depends on how the slot size is defined. When the overall length of the slot is given, location is specified either to a slot end or to its center. When slot size is defined by the center-to-center distance between arcs, the slot is located by giving the dimension defining a position for one of the arc centers. It is also possible to dimension the location of the slot center.

KEYSEATS

A *keyseat* is a recess cut into a shaft or hub. A key inserted into the keyseats of mating parts prevent a shaft from spinning inside the hub. Keyseat size is closely controlled to ensure a good fit between the keyseat and key. Keyseat *depth* and *width* must be dimensioned.

Depth of a keyseat is always dimensioned from the far side of the shaft or hole. See Figure 4-27. Dimensioning depth in this manner provides an existing feature from which dimensions can be checked. The width is dimensioned as shown. The keyseat is not assumed to be perfectly centered on the shaft. A position tolerance must be applied as defined in following chapters to define the allowable amount of location error.

NARROW SPACES

Small adjacent features require several dimensions in a small amount of space. To place all needed dimensions in a narrow space, dimensions are located at different distances from the part. See Figure 4-28. The shown extension lines are relatively close together and are broken where they cross arrowheads. The breaks help to clarify the point of application for each arrowhead. A combination of arrowhead and dimension value placements are used to maximize the clarity of dimension application.

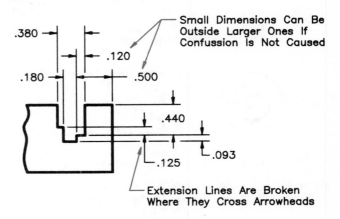

OFFSET DIMENSIONS

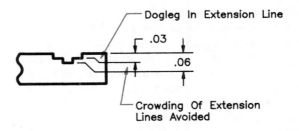

OFFSET EXTENSION LINES

Figure 4-28. Offsetting dimension values and using doglegs to offset extension lines makes it possible to dimension closely spaced features.

A *dogleg* (offset) is drawn in any extension line that is parallel within .06″ of another line. The dogleg is used to offset the extension line sufficiently to provide at least the .06″ distance. A dogleg allows clearance between adjacent lines without disconnecting the extension line from the point of application. A dogleg in an extension line is preferable to exaggerating the size of a feature.

When doglegs do not result in a clear application of dimensions, it is necessary to draw a larger scale view of the dimensioned features. Refer back to Figure 4-8.

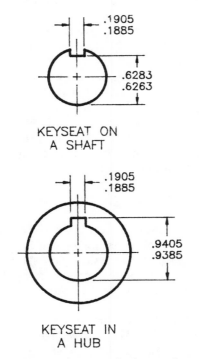

Figure 4-27. The depth of a keyseat is always dimensioned from the far side of the shaft or hole.

CENTERDRILLED HOLES

A *centerdrilled hole* is a hole and countersink cut into the end of a shaft or other cylindrical feature. One use for the centerdrilled hole is to mount the part in a lathe. A machine center on a lathe is inserted into the centerdrilled hole to ensure that the shaft turns on the lathe's axis of rotation when it is being turned.

A special tool is used for centerdrilling processes. The tool is identified by ANSI B94.11M as a combined drill and countersink, but is commonly known in the trade as a centerdrill. Common sizes for centerdrills are listed in ANSI B94.11. All centerdrills have a 60° angle countersink.

The centerdrill size is noted on a drawing by specifying the number of the centerdrill to be used. See Figure 4-29. The depth of the centerdrilled hole is controlled by the maximum diameter of the countersink.

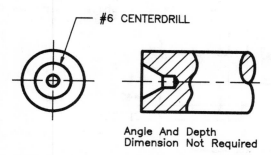

Figure 4-29. A machine center hole is called out by specifying the centerdrill tool number. Centerdrills are specified in ANSI B94.11.

MACHINING RELIEFS

A machine relief provides an area in which cutting tool travel can be stopped without damaging the tool or part. See Figure 4-30. As an example, a relief permits a single-point threading tool to emerge from the machined part and run freely. As another example, a grinding wheel can travel sufficiently past a properly located relief to ensure the cut surface is machined evenly and without fear of making contact between the grinding wheel and the shoulder. Side pressure on a grinding wheel can cause it to disintegrate.

Machine reliefs are specified in a note or dimensioned in a detail view. In either case, the *relief width, depth,* and *corner radius* must be given. Dimensions for the relief are determined by the size needed to clear the tools to be used in fabrication.

THREADS

The minimum amount of information that must be given for a thread is the *nominal size, threads per inch, thread form,* and *thread class.* See Figure 4-31. If the thread does not go all the way through the part, the thread depth must also be given. Thread depth is the distance that fully formed threads extend into the part or feature. Additional information pertaining to thread specification is found in ANSI Y14.6.

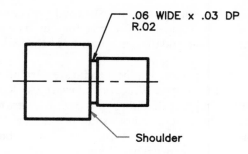

NOTED RELIEF DIMENSION

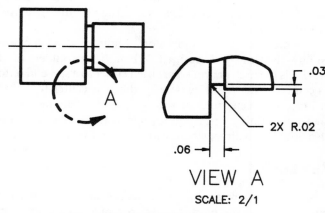

DETAIL RELIEF DIMENSIONS

Figure 4-30. Machine reliefs provide an area in which cutting tools are stopped without damaging finished surfaces.

KNURLS

Knurls are machined on cylindrical shapes to increase the friction between parts that are pressed together, improve the grip of a handle, or improve the appearance of parts. The purpose of the knurl determines the type of dimensions given. See Figure 4-32.

The diameter before and after knurling must be given on a shaft that fits into another part. The before and after knurling dimensions must include tolerances that show the acceptable amount of variation. Length of the full knurl, the diametral pitch (DP), and knurl form are all specified. Knurls used for appearance do not require specification of the after-knurling diameter.

SHEET METAL BENDS

Two mold lines are created when sheet metal is bent. See Figure 4-33. Mold lines are located by extending flat surfaces that lie on the same side of the sheet metal. Extending the surfaces inside a bend creates an inside mold line. Extending surfaces outside the bend creates an outside mold line.

Sizes of sheet metal parts are defined by extending dimensions to mold lines. The centerline of a bend is not normally dimensioned. The centerline of a bend is located by the tangent sides and the bend radius. Bend radii are specified by showing the inside radius. The bend radius used for a sheet

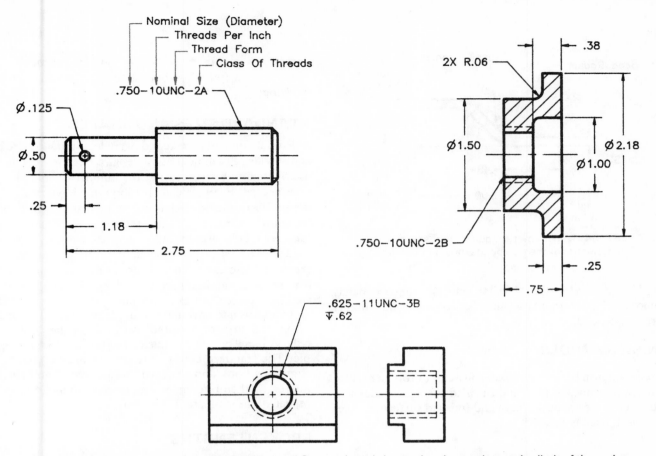

Figure 4-31. The nominal size for Unified National Course threads is equal to the maximum size limit of the major diameter on an external thread.

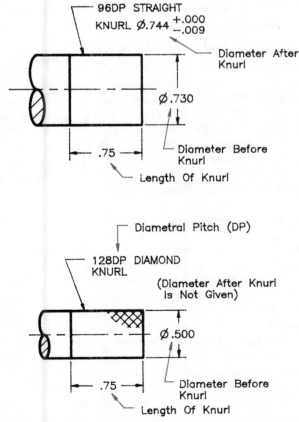

Figure 4-32. Knurls can perform functional and aesthetic purposes.

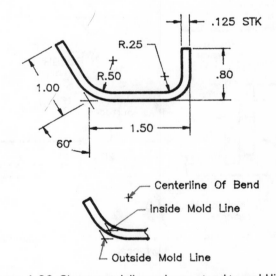

Figure 4-33. Sheet metal dimensions extend to mold lines.

metal part is primarily a function of the design application. However, a bend radius must not be smaller than can be tolerated by the material. A bend radius that is too small causes cracks in the material. The smallest allowable bend radius is a function of mechanical properties of the material.

JOGGLES

A *joggle* is an offset in a sheet metal part. See Figure 4-34. Joggles are dimensioned by giving the *location, height,*

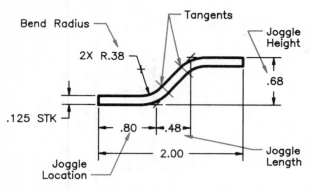

Figure 4-34. A sheet metal joggle requires two bends. The tangents of the two bends must not overlap.

length, and *bend radius.* The tangents for the two bends cannot overlap. A straight segment between the bend tangents is recommended.

FINISH SYMBOLS

It is common for some surfaces on a cast or forged part to be machine finished. Machine-finished surfaces are identified when there are both rough and finished surfaces on the same part. See Figure 4-35.

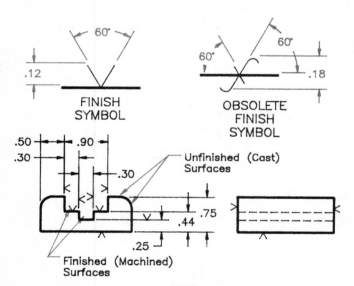

Figure 4-35. Finish marks are used to indicate machined surfaces.

Machined surfaces are identified with a finish symbol. The symbol looks like a letter V. The sides of the finish symbol form a 60° angle. The vertex is placed on the finished surface where it appears as an object line. When the finish symbol cannot be placed on the surface, it is placed on an extension line from the surface. The symbol size is .12".

The finish symbol is shown on all views where the surface appears as an edge view. The symbol is placed on the object line or extension line so that it points toward the machined surface.

A symbol still seen on some drawings, but no longer applied to new drawings, looks somewhat like a lowercase letter F. Its use may be required when revising an old drawing. The two finish symbols should not be mixed on one drawing.

STANDARD SIZES AND SHAPES

Production costs are kept down when standard-size parts and stock sizes can be used in designs. See Figure 4-36. The use of standard parts reduces the number of special operations needed. Every special operation that is eliminated from the production of a design means increased profit for the company and reduced cost to the consumer. Standard sizes for screws, bolts, washers, pins, nuts, etc., already exist. Standard sizes for wires, sheet metal, metal plates, bar stock, and extrusions also exist. There is no need to dimension all the features on a standard part, since dimensions and tolerances are already specified in a standard part drawing.

In addition to national standards, individual companies often standardize parts and features that they use repeatedly. A panel cutout for an electrical connector is an example. The standard cutouts are documented in a company standards book and assigned an identifying number. The number is then referenced on drawings to avoid having to repeat all the standard dimensions.

BROKEN LENGTHS

Long parts can be broken and a portion removed to reduce the drawing space required for the part. See Figure 4-37. The dimension used to show the length is not broken. The dimension value shown is the size to which the part is to be made.

LIMITS OF SIZE

All dimensions, except basic dimensions, include a tolerance. *Tolerance* is the acceptable amount of variation that is allowed on a dimension. A general title block tolerance applies to each dimension that does not show a specific tolerance applied to it. The tolerance on size dimensions define the acceptable limits of size.

There are several methods for applying tolerances to size dimensions. These include limit dimensions, plus and minus tolerances, and single limit dimensions.

LIMIT DIMENSIONS

Limit dimensions specify the minimum and maximum acceptable sizes for a feature. See Figure 4-38. The difference between the minimum and maximum limits is known as the tolerance. The maximum limit is written above the minimum limit when both values are shown in the dimension. A line is not drawn between the two values.

If the limits are specified in a note, then the smaller limit precedes the larger value. The values are separated by a diagonal line when shown in a note.

PLUS AND MINUS TOLERANCES

Plus and minus tolerances specify the acceptable amount of variation in both directions from a nominal dimension

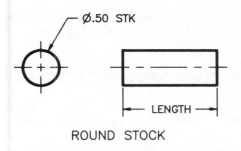

Ø.50 STK

ROUND STOCK

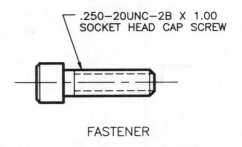

.250—20UNC—2B X 1.00
SOCKET HEAD CAP SCREW

FASTENER

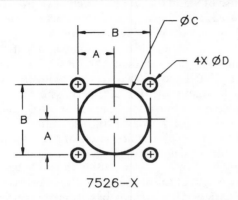

7526—X

DASH NO (X)	A	B	C	D
—1	.50	1.00	.969	.188
—2	.62	1.25	1.188	.219
—3	.75	1.50	1.438	.250

STANDARDS BOOK
SPECIFICATION

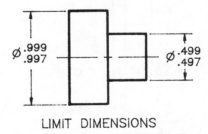

7526—3 CUTOUT

DRAWING CALLOUT
Standard Feature Dimensions
May Be Located In A Company
Standards Book

Figure 4-36. It is not necessary to specify all the dimensions for standard stock shapes or for features specified in company standards books.

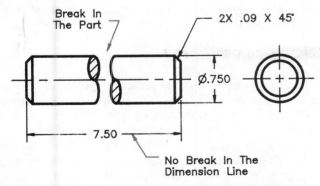

Break In The Part

2X .09 X 45°

Ø.750

7.50

No Break In The Dimension Line

Figure 4-37. Part of an object can be removed to shorten a view if the removed portion is a continuation of the shown features.

Ø $\frac{.999}{.997}$

Ø $\frac{.499}{.497}$

LIMIT DIMENSIONS

Figure 4-38. The maximum and minimum acceptable size of a feature can be shown by using limit dimensions.

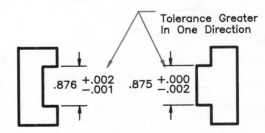

Tolerance Greater In One Direction

.876 +.002 −.001   .875 +.000 −.002

Figure 4-39. Plus and minus tolerances applied to a dimension may show an unequal amount of variation from the nominal value.

value. See Figure 4-39. The amount of acceptable variation is shown adjacent to the nominal dimension. It is acceptable for the tolerance in one direction to be zero. Even if the tolerance in one direction is zero, both tolerance values must be shown.

One tolerance value can be used to permit an equal amount of variation above and below the nominal dimension

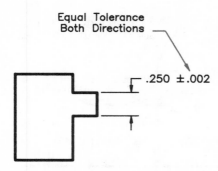

Figure 4-40. Plus and minus tolerances applied to a dimension may show an equal amount of variation in both directions from the nominal value.

value. See Figure 4-40. The tolerance value is preceded by a plus or minus sign (±). If the dimension is a diameter, the diameter symbol is placed before the nominal size. A diameter symbol is not shown in front of the tolerance value.

SINGLE LIMIT DIMENSIONS

Single limit dimensions are acceptable only when the second limit is controlled by the part geometry. Calling out a **R.03 MAX** indicates that a minimum radius of .00″ is acceptable. See Figure 4-41. Calling out a minimum knurl length of .50″ indicates that a maximum length knurl can be as long as the diameter on which the knurl is rolled.

TOLERANCES ON ANGLES

Tolerances on angles can be specified using limit dimensions or plus and minus tolerances. Angles can also be con-

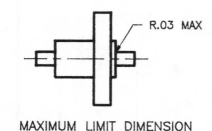

MAXIMUM LIMIT DIMENSION

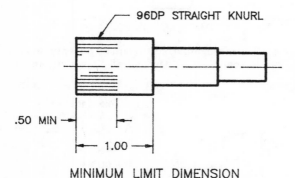

MINIMUM LIMIT DIMENSION

SINGLE LIMIT DIMENSIONS

Figure 4-41. A single limit dimension establishes one acceptable size. The part geometry controls the other limit.

trolled with orientation tolerances as explained in a later chapter. Single limit dimensions are not practical for use on angles.

CALCULATION OF SIZE TOLERANCES

Standard tables for different classes of tolerances per ANSI B4.1 can be used to determine the limits of size for mating features. ANSI B4.1 is based on the inch measurement system. Limits of size can also be calculated without the use of standard tables. Whether or not standard tables are used, the allowance and tolerances applied to the parts are based on the parts' function(s).

When using metric dimensions, limits of size can be calculated using the tables in ANSI B4.2. The limits of size can be calculated and applied to the drawing in a similar manner to inch dimensions. It is also permissible to show the nominal metric size and note the applicable tolerance code shown in the standard. However, the use of the code is not recommended since this requires users of the drawing to calculate the limits of size.

Size limits are calculated using either the basic hole or basic shaft system of tolerancing. The choice between the basic hole and basic shaft system is based on the design application. Either system can be utilized if all parts in the design are produced by the designing company. If one of the parts is purchased from a vendor, the choice between systems of tolerancing is dictated by the tolerances on the purchased parts.

The word **BASIC** *when used in the context of limits of fit means something entirely different than when it is used in the context of geometric dimensioning and tolerancing as defined in Chapter 3.* When used in the context of limits of fit, a basic value is one from which the limits of size are calculated. A basic size is often equal to the nominal size, but may not be. An example is the outside diameter for a pipe. Neither the minimum nor maximum diameter for a 3/4″ pipe is actually .750″. Therefore, the nominal size is only a size name applied to a part.

BASIC HOLE SYSTEM

The *basic hole system* is the preferred system for calculating limits of size. See Figure 4-42. In this system, the limits of size for the hole are calculated first. Then, the limits of size for the shaft are calculated to fit the hole.

Usually, the basic size is used as the lower limit of size for the hole. The upper limit of size for the hole is determined by adding a tolerance to the lower limit. When using standard tables, such as ANSI B4.1, the amount of tolerance added to the basic size is obtained from the tables.

The limits of size for a mating shaft are calculated after the limits of size are established for the hole. The limits of size for the shaft are determined by the minimum and maximum amounts of clearance or interference that is acceptable for the given part.

The basic hole system can be used, and is preferred, when two mating parts are simultaneously designed within one company. It must be used when there is an existing hole, and a shaft is being produced to mate the predefined hole size.

Nominal Size Range, Inches		Class RC6			Class RC7		
Over To	Clearance	Standard Tolerance Limits					
		Hole H8	Shaft e7				
0.40 — 0.71	1.2 3.8	+1.6 0	−1.2 −2.2				
0.71 — 1.19	1.6 4.8	+2.0 0	−1.6 −2.8				

Partial Table From ANSI B4.1 Tolerance Values Are In Thousandths Of An Inch

Ø.8750   Nominal Size
RC6   Class Fit
.71 — 1.19   Nominal Size Range In Tolerance Table
.0016   Minimum Clearance
.0048   Maximum Clearance
+.0020 −.0000   Tolerance On Hole
−.0016 −.0028   Tolerance On Shaft

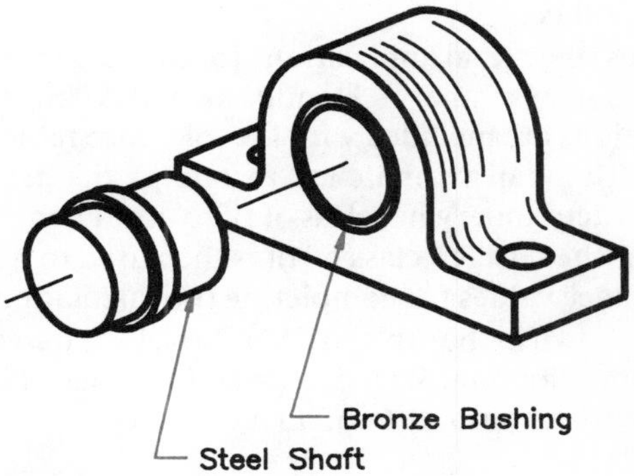

Bronze Bushing
Steel Shaft

BASIC HOLE CALCULATIONS

HOLE

Basic Size .8750
Tolerance +.0020
Maximum Hole .8770

Basic Size .8750
Tolerance −.0000
Minimum Hole .8750

Limits Of Size   Ø .8770 .8750

SHAFT

Basic Size .8750
Tolerance −.0016
Maximum Shaft .8734

Basic Size .8750
Tolerance −.0028
Minimum Shaft .8722

Limits Of Size   Ø .8734 .8722

Figure 4-42. Calculation of the limits of size for a specific class of fit requires that standard tolerance tables be used.

It is a good practice to assign standard tool dimensions as the basic size for holes. This permits the utilization of standard tools to produce the holes. If the limits of size for a hole are .125″ minimum and .130″ maximum, then a standard .125″ diameter drill can be used to make the hole. The .125″ drill will not produce an undersized hole unless the drill is worn. It will produce a slightly oversize hole because of errors in tool sharpening, poor tool alignment in the machine, and shavings scraping the sides of the hole as they travel up the flutes of the drill.

BASIC SHAFT SYSTEM

The *basic shaft system* for calculating limits of size results in assignment of the basic size as one limit dimension for the shaft. A disadvantage of this system is that the basic size of the shaft may result in calculated limits of size for the hole that do not permit utilization of standard tools. A nonstandard hole size can require the use of special processes or tools to produce the hole.

The basic shaft system for calculating limits of size is typically used when the limits for the shaft must be determined first, or when the size of the shaft is predetermined. One situation requiring utilization of the basic shaft system is when the shaft is purchased, and a mating hole must be specified.

An example of when the basic shaft system must be used is when a bearing is purchased. Limits of size for the outside diameter of the bearing are assigned by the bearing manufacturer. The bearing bore (hole) must be designed to fit the outside diameter of the bearing. The designer must determine the limits of size for the bearing bore according to the predefined bearing size.

CALCULATING SIZE LIMITS USING STANDARD TABLES

Tolerances and allowances for many design applications and a wide range of nominal sizes have been standardized. These tolerances and allowances are organized into tables and categorized according to the class (type) of fit. These tables can be found in the Machinery's Handbook and in ANSI B4.1. The classes of fit are:

- Running and sliding clearance (RC).
- Location clearance (LC).
- Location transition (LT).
- Location interference (LN).
- Force fits (FN).

The function of the parts must be considered and a class of fit selected before the tolerance tables can be used.

Each of the five general classes of fit are divided into multiple categories. Each category provides a different amount of tolerance.

Descriptions of general applications for the categories of fit are provided in Machinery's Handbook and ANSI B4.1. These descriptions are included with the tolerance tables. A design application can be matched to one of the general descriptions to determine what class of fit to use. The appropriate table for the selected class of fit is then used to determine the tolerance values for completing the calculations.

The values shown in the tables of ANSI B4.1 are based on the inch system. Nominal sizes are given in inches. Tolerances are shown in thousandths of an inch.

Running and Sliding Clearance Fits

There are nine running and sliding clearance classes of fit. Running or sliding clearance fits always provide a clearance between the hole and shaft. The parts move freely, but to varying degrees according to the class of Running or Sliding Clearance (RC) fit used. Smaller RC classes of fit (RC1, RC2, etc.) provide less clearance than do larger ones (RC7, RC8, RC9). Enough clearance is provided by the RC classes to allow lubrication.

The amount of tolerance permitted by RC classes of fit is proportional to the size of the toleranced feature. Small features are allowed smaller tolerances than large features. This is logical since a .001″ clearance causes a relatively large amount of freedom on a .125″ diameter shaft, but is hardly noticeable on a 2.000″ diameter shaft.

Location Clearance Fits

There are eleven Location Clearance (LC) classes of fit. They are used for locating parts that must assemble without any interference between them. The LC1 class results in a very small amount of clearance, while the LC11 class results in a large amount of clearance. The smaller the clearance, the more accurate the location. Location clearance fits provide freedom in assembly of parts, but are not intended to be used for parts that move after assembly.

Location Transition Fits

There are six Location Transition (LT) classes of fit. All classes can result in either a clearance or interference. They are used for location of parts when a slight clearance or interference is acceptable. The LT1 class fit is more likely to result in a clearance than an interference. The LT3 class fit has an approximately equal chance for either a clearance or interference. The LT6 class fit is more likely to result in an interference than a clearance.

Location Interference Fits

There are three Location iNterference (LN) classes of fit. They are used for locating parts when a clearance fit is unacceptable. A press fit ensures a more exact location than can be guaranteed if any clearance is permitted, since a clearance allows some movement. An interference fit does not allow any movement. The LN1 class fit provides the least interference, and the LN3 provides the most interference. None of the LN classes of fit are used for transmitting loads; they are only intended to be used for location.

Force Fits

There are five Force and shriNk (FN) classes of fit. All classes result in interference conditions. Each class fit provides a constant bore pressure throughout the range of nominal sizes. This means that the amount of interference between the shaft and hole varies proportionally to the nominal size. The tolerances are relatively small to prevent large variations in the bore pressure.

Limit Calculations Using Tables

The class of fit to be used on a pair of parts depends on the function of the parts. When the function is known, a selection of the class of fit can be made. The selected class of fit determines which tolerance table must be used. The nominal size of the parts is also affected by the function of the parts. After a tolerance table and nominal size are selected, the limits of size for the parts can be calculated.

The example in Figure 4-42 shows how to calculate the limits of size for a shaft and hole. The nominal size for the shaft is .8750″. The inside diameter of the bushing has the same nominal size as the shaft. A class RC6 fit has been selected for use. The figure shows the appropriate portion of the table for running and sliding fits.

The nominal size is located in the column of the table labeled "Nominal Size Range, Inches." The nominal size of .8750 is in the 0.71-1.19 nominal size range shown by the table. Reading across the table from the 0.71-1.19 nominal size, the applicable tolerances are located under the columns headed by "Class RC6." There are three subcolumns under the Class RC6 heading.

All values shown in the tolerance table are given in thousandths of an inch. The first subcolumn under Class RC6 shows the amount of clearance that will result from an application of the given tolerances. A nominal size of .8750″ requires a minimum clearance of .0016″ and a maximum clearance of .0048″.

The second subcolumn shows the tolerances that are to be applied to the nominal size of the hole. A .8750″ diameter hole has an RC6 tolerance of +.0020″ and +.0000″. Applying these tolerances to the nominal size gives limits for the hole of .8770″ and .8750″. One size limit for the hole is equal to the basic size since the tables are based on the basic hole system.

The RC6 subcolumn for the shaft shows a -.0016″ and a -.0028″ tolerance that must be applied to the nominal size. Applying these tolerances to the nominal size results in limits of size for the shaft that are .8734″ and .8722″.

The calculations can be checked to ensure that no mistakes have been made. The difference between the minimum hole size and the maximum shaft size must equal the minimum clearance (allowance) value given in the clearance column.

Minimum hole diameter	.8750
Maximum shaft diameter	-.8734
Minimum clearance	.0016

The calculated minimum clearance value equals the one shown in the table, so the maximum shaft diameter and the minimum hole diameter are correct.

The difference between the values for the minimum shaft diameter and maximum hole diameter are determined. This difference must equal the maximum clearance value shown in the table.

Maximum hole diameter	.8770
Minimum shaft diameter	-.8722
Maximum clearance	.0048

Both the minimum and maximum clearance values equal the values in the table, therefore the calculated limits of size are correct.

Limit Calculations When One Design Exists

It may be necessary to calculate the limits of size for one part when the limits of size for the mating part are already established. This situation occurs when one of the parts is purchased from a vendor.

When calculating the limits of size for features that mate with a vendor's part, the limits of size for the vendor's part must be determined. This is done by requesting a drawing from the vendor, or by checking dimensions in the vendor's catalog.

The maximum and minimum clearance (or interference) values are obtained from the tolerance tables and applied to the limits of size for the purchased part. This determines the limits of size for the part being designed. The following example shows how such a calculation is made.

A bearing with an outside diameter of 1.2500″ is to be pressed into a housing using an FN1 class fit. The limits of size for the outside diameter of the bearing are 1.2500″ and 1.2495″. The tolerance table shows a minimum acceptable interference of .0003″ and a maximum interference of .0013″. The acceptable hole sizes are found as follows:

Maximum shaft	1.2500
Maximum interference	-.0013
Minimum hole	1.2487

Minimum shaft	1.2495
Minimum interference	-.0003
Maximum hole	1.2492

LIMIT CALCULATIONS WITHOUT TABLES

Size tolerance calculations are often made without the use of a tolerance table. See Figure 4-43. In such cases, it is necessary to determine the desired allowance value and the maximum desired clearance or minimum interference. In the case of an interference fit, the acceptable bore pressure is one of the factors that should determine the maximum interference. The maximum amount of relative movement between parts is one factor used to determine the maximum clearance.

Figure 4-43 shows a dowel pin that presses into one part and slips into the other. All features other than the pin holes are omitted to simplify the problem. The dowel pin limits of size are known to be .1877″ and .1875″. Limits of size for the two holes must be determined.

Calculation of size limits require that the definition of allowance be understood. *Allowance is the difference between the maximum shaft size and the minimum hole size.* The allowance for a clearance fit is a positive number, and it indicates the minimum clearance between the shaft and hole. The allowance value for an interference fit is a negative number, and it indicates the maximum interference. The maximum interference can be thought of as the least clearance, or the furthest condition from a clearance.

Clearance hole limits of size can only be determined after an allowance value and maximum clearance value are

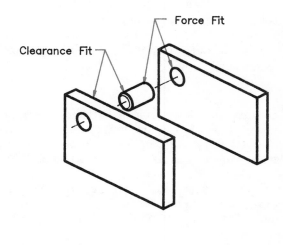

.1875 Nominal Size

CLEARANCE

Max Dowel	.1877
Allowance	+.0010
Min Hole	.1887

Min Dowel	.1875
Max Clearance	+.0030
Max Hole	.1905

CLEARANCE HOLE
Limits Of Size

ø .1905
.1887

INTERFERENCE

Max Dowel	.1877
Allowance	-.0013
Min Hole	.1864

Min Dowel	.1875
Min Interference	-.0004
Max Hole	.1871

INTERFERENCE HOLE
Limits Of Size

ø .1871
.1864

Figure 4-43. If the nominal size, allowance, and either a maximum clearance or minimum interference are known, the limits of size for mating parts can be calculated.

selected. For this example, an allowance of .0010″ and a maximum clearance of .0030″ is selected.

The limits of size are calculated as follows. The allowance of .0010″ is added to the maximum dowel pin diameter of .1877″ to determine the minimum hole diameter (.1887″). The .0010″ allowance is the minimum clearance between the pin and hole. The maximum clearance value (.0030″) is added to the minimum diameter of the dowel pin (.1875″). The result is a maximum hole diameter of .1905″. The acceptable limits of size for the clearance hole are .1877″ and .1905″ which ensures a clearance of not less than .0010″ or more than .0030″.

The limits of size for the interference hole in the given figure can only be determined after an allowance value (maximum interference) and a minimum interference value are selected. For this example, an allowance of .0013″ and a minimum interference of .0004″ are selected.

The limits are calculated as follows. The allowance (maximum interference) is subtracted from the maximum dowel pin diameter to determine the minimum hole diameter. The allowance value of .0013″ is subtracted from the maximum dowel pin diameter of .1877″ to calculate a minimum hole diameter of .1864″.

The minimum interference value of .0004″ is subtracted from the minimum dowel pin diameter of .1875″ to determine the maximum hole diameter of .1871″.

Limits of size for the interference fit hole are .1871″ and .1864″. These limits of size create a minimum interference of .0004″ and a maximum interference of .0013″ when assembled with the dowel pin.

RULES IN ASME Y14.5M

There are many methods for applying dimensions as defined in the national standard. Only two of the guidelines in the standard are identified as rules. The fact that only two guidelines are identified as rules gives a strong signal that these rules need to be understood and remembered as dimensions and tolerances are applied to drawings.

There should be no misunderstanding about the importance of other requirements in the standard. None of the requirements should be violated simply because they are not identified as rules. The identification of rules only indicates the fundamental nature of these guidelines.

RULE #1

The most fundamental rule that affects dimension interpretation is Rule #1. To paraphrase Rule #1, it states that *any dimensioned feature of size must have perfect form when the feature is at its Maximum Material Condition (MMC)*. The *maximum material condition* of a feature is the size at which the most material is in the part. See Figure 4-44. The MMC of an external feature is equal to the maximum size limit. The MMC of an internal feature is equal to the smallest size limit.

Rule #1 requires that a feature of size have perfect form at MMC. This requires that a round hole be a perfect cylinder when it is at its smallest permissible diameter, and that a shaft be a perfect cylinder when it is at its largest diameter. See Figure 4-45. As a size feature departs from the MMC,

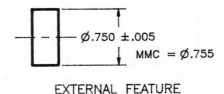

EXTERNAL FEATURE

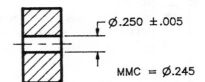

INTERNAL FEATURE

Figure 4-44. The maximum amount of material exists in the part when a dimension is at Maximum Material Condition.

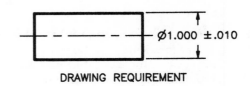

DRAWING REQUIREMENT

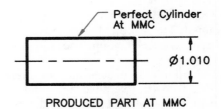

PRODUCED PART AT MMC

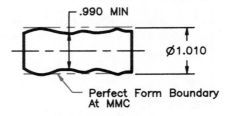

PRODUCED PART WITH SIZE VARIATIONS

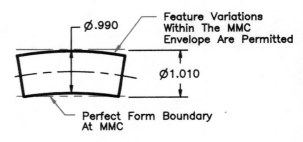

PRODUCED PART AT LMC

Figure 4-45. A perfect form boundary exists at the maximum material condition for a feature of size.

the feature is permitted to have form variations. The form variations are not permitted to extend beyond the perfect form boundary.

There is no requirement for perfect form at the *Least Material Condition (LMC)*. The *least material condition* of a feature is the size at which the least material is contained in the part. A feature produced at LMC is permitted to have form variations within the MMC boundary of perfect form. See Figure 4-45.

Exceptions To Rule #1.

There are some exceptions to Rule #1. The exceptions apply to parts that cannot be expected to have a perfect form. Parts that cannot be expected to have a perfect form are those subject to *free state variations*. This means the part is flexible to some extent. A rubber gasket may need to have a thickness dimension that is controlled by a relatively small size tolerance, but the gasket cannot be expected to stay flat when it is picked up. Another type of free state variation is the distortion of a thin-walled part that is caused by stresses built into the part. When the thin-walled part is removed from clamps that hold it during machining, the form of the part is likely to change. Rule #1 does not apply to parts that are subject to free state variation.

The perfect form at MMC requirement does not apply to standard stock material such as bar, sheet, tube, extrusions, or other similar items. These items have standard form tolerances that permit variation of form outside the perfect form boundary at MMC.

Invoking Exceptions To Rule #1

Sometimes, the functional application for a part requires that a feature have a relatively small size tolerance. The functional application on the same feature may permit a relatively large form tolerance. See Figure 4-46. It is not permissible to specify a form tolerance larger than the size tolerance unless a note is placed on the part for which Rule #1 is not to apply. When exception to Rule #1 is noted, a form tolerance must be shown on the feature.

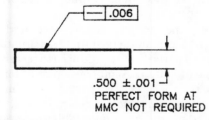

Figure 4-46. Exception to Rule #1 can be noted adjacent to a dimension if a form tolerance exceeding the size tolerance is specified.

Interrelationships

Rule #1 only establishes a perfect form boundary for individual features. It does not require a perfect relationship between multiple features. See Figure 4-47. The given rectangle includes two features of size. They both have size tolerances of ± .010″. The 90° angle between them has a tolerance of ± .5°. When both features are produced at

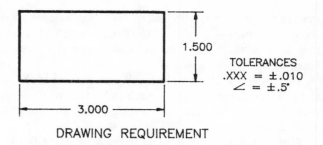

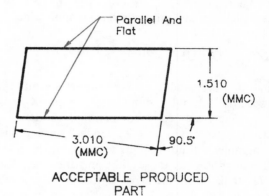

Figure 4-47. The perfect form boundary required by Rule #1 does not require perfect interrelationships between features.

MMC, they each must have perfect form. However, the 90° angles still have a ± .5° tolerance.

There are methods for forcing perfect interrelationships at MMC. *They must not be used except when part function requires this level of control.* Using unnecessary part accuracy requirements unnecessarily forces part cost to increase. One method for forcing perfect interrelationships at MMC is to include a specific note that references dimensions to a datum reference frame. Datum reference frames are defined in the chapter on datums.

RULE #2

Rule #2 states that regardless of features size (RFS) is assumed for all tolerances. Material condition modifiers must be shown when a maximum material condition or least

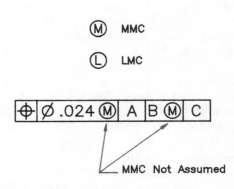

Figure 4-48. RFS is assumed on all tolerance specifications.

material condition is applicable. If no modifier is shown, RFS is assumed to apply. Modifiers are applicable on the tolerance or referenced datums only when the associated features are features of size.

This rule is a change from previous issues of the standard. The change brings the national standard into agreement with international practice of assumed RFS applicability.

Which standard applies to the drawing being drawn or reviewed? There should be a general note on the drawing that states which dimensioning standard is applicable.

ALTERNATE TO RULE #2

Although it is permissible to show the RFS symbol in a tolerance specification, it is redundant. See Figure 4-49. The symbol is no longer defined by the national standard, so it must be drawn as defined in the 1982 standard when it is used.

Most companies will not find it necessary to continue using the RFS symbol. Only those still using drawings created to the 1973 standard should have a need for the symbol. Drawings created to the 1973 standard are assumed to include the MMC modifiers when no modifier is shown. Companies still using the 1973 drawings may want to use the RFS symbol to avoid confusion between 1973 drawings, which have assumed MMC modifiers, and 1994 drawings, which have assumed RFS modifiers.

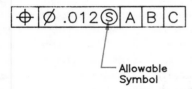

Figure 4-49. Although the RFS symbol may be used, it is redundant.

DIMENSIONING FOR INDUSTRY

Specification of size and location for each feature is the primary requirement that must be met when dimensioning, but many other factors need to be considered. An object can be dimensioned in any of several ways without violating dimensioning standards. However, in order to produce a functional part, one way of dimensioning a particular part is generally preferable. Unless the sizes and locations are specified correctly, the produced part may not assemble with its associated parts, or it may not function as intended.

Dimensions must be applied in industry with consideration given to the part's use, associated parts, interchangeability, and fabrication capabilities. These considerations affect dimension and tolerance application. Consideration of dimension interpretations must be made for the dimension and tolerance application to be correct. The person reading the drawing has to interpret its meaning. If the interpretation is different from the intent, the part will not be produced as expected.

Dimensioning of a single part without knowledge of the part's function requires that size and locations be given without complete consideration of the functional relationships with associated parts. This is not an acceptable practice in industry.

After a drawing is released for production, changes to the drawing are likely to be required. Changes made after a drawing is released are known as *revisions*. Revisions to dimensions are common in industry. However, keeping the number of revisions to a minimum helps to keep production costs down. The number of drawing changes can be kept to a minimum through careful application of dimensions on the original drawing.

INTERCHANGEABILITY

Parts in mass-produced assemblies are normally made to be interchangeable. A design is interchangeable only if parts produced to the design documentation can be installed in place of any other part produced to the same drawing. As an example, the windshield wiper blade from a specific model car will fit any other car of the same model. The concept of interchangeable parts has existed for a long time; Eli Whitney is credited with making the first interchangeable parts in 1798.

Interchangeability can be ensured by careful analysis of the dimension tolerances to determine possible size, location, and geometry variations. For 100% interchangeability, the analysis must show that the worst case conditions of all parts and their features will permit assembly of the parts. An explanation of tolerance calculation is given in following chapters.

Designing interchangeable parts is not difficult unless the assembly is a high-precision assembly. On high-precision assemblies, tolerances may become very small, thereby causing production of parts to become more difficult. Part of the designer's job is to find ways to maximize tolerances to keep part production as simple and inexpensive as possible, and thus reduce overall costs.

It is nearly always preferred that parts in a design be interchangeable. It reduces assembly time, spare parts cost, and maintenance costs.

Sometimes, the design accuracy required for interchangeable parts is not practical. In these situations, it may be best to design parts that are produced together (mated). A common type of mated assembly is a pair of parts through which a hole is drilled. The two parts are held in the desired assembly position, and a hole is drilled through both parts. This is referred to as *mate drilling*. This process does ensure that the two holes are in line with one another.

A problem with mate drilled holes is that a mated assembly is created. There is no assurance that parts from one assembly will fit onto any other assembly; therefore, interchangeability is lost.

When only a few assemblies are to be made, mated (or matched) parts may be preferred rather than designing interchangeable parts. Mated parts cannot be guaranteed to be interchangeable. For this reason, the parts must be marked to indicate their correct association. This is normally done by marking mated parts with numbers indicating matched sets. The first mated assembly might have all parts marked with a number one, the second assembly with a number two, etc.

Mated assemblies reduce the number of tolerances that must be analyzed. *Mate with dimensions* do not require a tolerance analysis since the features are machined as matched sets. However, it is not a good practice to use mated assemblies for the sole purpose of avoiding tolerance calculations.

Mated parts are initially less expensive than interchangeable parts when only a few assemblies are produced. The initial cost is lower because design time and production tooling requirements are reduced, and machinists can work to larger tolerances.

The life cycle cost of mated parts is not always less than interchangeable parts. Life cycle cost of mated parts must take the cost of maintenance into account. If a part breaks in a mated assembly, the entire assembly must be replaced or a special replacement part must be made by transferring dimensions from the mated parts. Both alternatives for fixing the broken assembly are expensive.

Another factor in the mated part cost is the effect of a machining error made after the mate machining operations are completed. If a machining error is made during or after the mate machining operations, the entire assembly might be ruined. This requires producing all parts in the assembly for a second time. If one part in an interchangeable assembly is ruined, only that one part must be produced a second time.

Clear requirements must be shown on a drawing when a mated assembly is to be produced. If a hole is to be mate drilled on two parts, the hole may be specified on the assembly drawing. The assembly drawing hole specification includes a note that states **MATE DRILL**.

It is also possible to dimension a mate drilling operation on separate detail drawings. If a hole in part 001 is to be mated with a hole in part 002, the hole specification in part 001 has a note added to state **MATE WITH PART 002**. The hole specification shown for part 002 has a note added to state **MATE WITH PART 001**.

DIMENSION INTERPRETATIONS

Correct interpretation of tolerances is necessary to make effective decisions regarding dimension application. The following paragraphs show interpretations of tolerances applied directly to angles and arcs. The accumulation of tolerances as applied directly to dimensions is defined in the previous chapter. Explanations of how to apply geometric tolerances and also define the resulting tolerance zones are given in following chapters.

The tolerance zone for an angle is affected by the manner is which it is dimensioned. See Figure 4-50. An angle defined with a dimension measured in degrees, and having a tolerance expressed in degrees, has a wedge-shaped tolerance

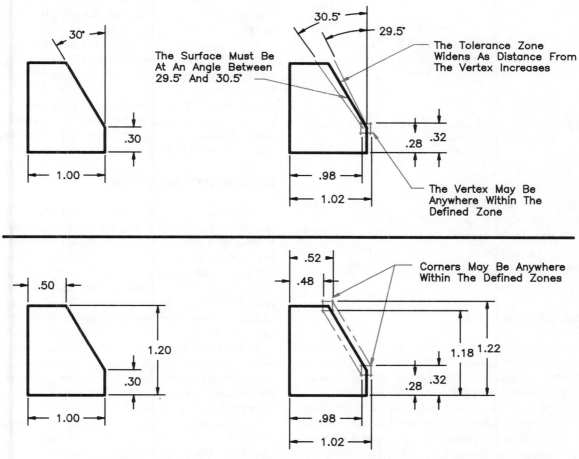

Figure 4-50. The method used to dimension an angle affects the tolerance on the angle.

zone. The wedge has a point at the location of the angle vertex. The vertex of the angle may be located anywhere within the tolerance on the vertex location dimensions.

Location of the two ends of an angle create a tolerance zone for the location of each corner. Unfortunately, this method of dimensioning an angle doesn't provide a clear definition of the tolerance zone that exists between the two corners. A well-defined means for specifying angularity tolerances is explained in the chapter on orientation tolerance. The methods of controlling orientation as described in that chapter provide a clear definition of the requirements for surface variation.

Radius dimensions have a tolerance that affects the dimensioned surface. See Figure 4-51. A radius dimension of .07″ ± .02″ can be any radius between .05″ and .09″. According to ASME Y14.5M-1994, the surface may lie anywhere within the boundaries. If an arc without reversals is needed, a controlled radius (CR) is specified.

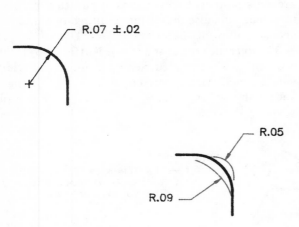

Figure 4-51. A curved surface must fall within the size limits defined by the radius dimension.

The interpretation of radius (R) dimension tolerances has been the subject of many standardization discussions. The interpretation permitting reversals of the surface is a 1994 requirement. This requirement varies in some previous version of dimensioning standards.

MACHINE CAPABILITY

Each manufacturing machine is capable of a certain range of accuracy. The tolerances applied on a drawing are going to impact the machine type and machine process selected for production of the part. See Figure 4-52.

Machines capable of holding small tolerances are more expensive than similar machines that are less accurate. If tolerances of ± .005″ are the smallest required in a factory, the equipment is considerably less expensive than for a factory that must produce tolerances of ± .00005″.

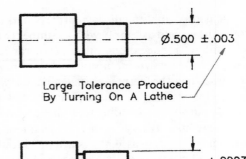
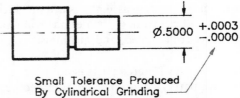

Figure 4-52. Machine capabilities must be considered when assigning tolerances.

If the machines in a factory are only capable of holding a tolerance of ± .001″, then a specification of ± .0001″ on a drawing forces the company to make a decision. A new machine must be purchased, or the part must be subcontracted for production by an outside company.

Accuracies that can be held by a machine depend to some extent upon the size of part being made. A machine that can bore a .500″ diameter hole to a tolerance of .0002″ in diameter cannot hold the same tolerance on a 6.000″ diameter hole. The larger diameter hole requires larger tooling that can have more play in it, and the machine play can be affected by the diameter cut made.

It shouldn't be universally assumed that the smaller the hole, the smaller the achievable tolerance. Very small diameter holes can be difficult to produce to very small tolerances.

Machining handbooks provide information about machine process accuracies. The data in these books can provide general guidance as to the size tolerance that should be worked toward as design calculations are completed.

DIMENSION REVISIONS

Any change to a drawing original after it has been released for production must be completed through what is known as a *drawing revision*. Revisions may be required because of errors made on the original drawing, design changes, or changes for improving manufacturing ease.

Revisions made to a dimension value also require the affected object size or location to be revised. Changing only the dimension value without changing the feature is not usually acceptable.

In CAD models, it is extremely important to change the object geometry so that it is always to scale with the dimension. The geometry is used for creating Numerical Control (NC) tapes to run machines. If the geometry is out of scale, then parts will be made incorrectly.

Dimension revisions on a drawing are made as shown in Figure 4-53. The width of the top step on the given part is changed. The revised drawing shows the corrected front view, and the dimension is changed to show the new size. A

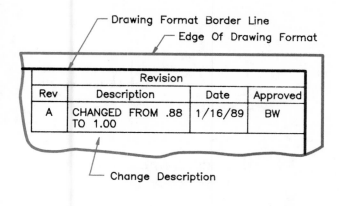

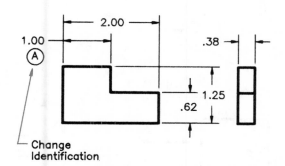

Figure 4-53. A drawing revision is normally identified where it is made and then explained in the revisions block.

revision letter (change identification) is placed inside a circle near the appropriate dimension to indicate that the drawing has been changed. A description of the change is entered in the revisions block on the drawing format.

Changes of less than .030″ can be made to a dimension on a manually produced drawing without changing the object view. (Do not make this type change to a CAD drawing.) Only the dimension is changed and a revision letter placed next to it. If a dimension change greater than .030″ is made without changing the view, the dimension is out-of-scale. Such a dimension must be underlined with a straight solid line. See Figure 4-54.

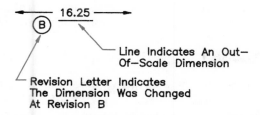

Figure 4-54. Out-of-scale dimensions should always be avoided and are only permitted when making a drawing revision.

It is not a good practice to change dimensions without changing the affected features. Out-of-scale drawings become difficult to work with as additional changes are made. Out-of-scale dimensions are generally reserved for use in situations where a small change would cause a major revision to a drawing. An example would be changing the location of one hole by .125″ when the hole is the origin for several other features. Moving the hole would require moving all the other features.

SURFACE CONTROL DIMENSIONS

Surface quality requirements for a part are determined according to the part's function. Examples of surface finish requirements determined by function are bearing surfaces and the outside surfaces of a gear box. Bearings require a smooth surface finish to reduce friction. The outer surface of a gear box housing can be relatively rough.

Surface finish is a function of the fabrication process. A sand cast part has a surface finish that is affected by the size of sand used in the mold. The finish on a lathe-turned part is affected by the cutting tool, machine speed, and feed. The finish on a ground part is smooth because of the small grain size in the grinding wheel, the high speed of the wheel, and a relatively slow feed.

Specification of tolerances on dimensions have an indirect effect on surface finish. As tolerances become small, the fabrication processes become more exacting. More exacting fabrication processes result in better quality surface finishes. A tolerance of $\pm .0002″$ on a shaft diameter forces the part to be ground. Grinding to achieve a tolerance this small will probably result in a surface finish of 32 microinches or better. (A microinch is $.000001″$.) In effect, the $\pm .0002″$ size tolerance forced a good-quality surface finish.

Surface finish can be specified by showing surface control symbols applied to the profile view of the controlled surface. This is done if the size tolerance is not certain to result in the desired finish. The surface control symbol is applied where the surface appears as a line. The minimum information that is specified using the symbol is surface roughness. Additional information is shown if the additional control is required. Surface roughness is interpreted in accordance with ANSI B46.1, and applied per ANSI Y14.36.

Manufacturing costs increase as surface control increases. Manufacturing methods can be dictated by the degree of surface control specified. In order to keep costs down, surface control must not be overspecified. Care in applying surface control must be exercised to prevent specifying a machine process that is more restricting than is required by the design function.

Surface conditions include variations known as roughness, waviness, and lay. Surface *roughness* is the height of small peaks and valleys on a surface. See Figure 4-55. *Waviness* is a larger variation in surface finish than roughness. Roughness variations are superimposed on the waviness variations. *Lay* is the direction of tool marks, scratches, or the grain that affects surface finish.

Surface control symbols are drawn as shown in the given figure. The horizontal bar can be omitted from the symbol when only surface roughness is specified. The symbol is always placed horizontally; it is not to be inclined or upside down.

SURFACE ROUGHNESS

Surface roughness is often the only surface control specified. The surface control symbol, for roughness control

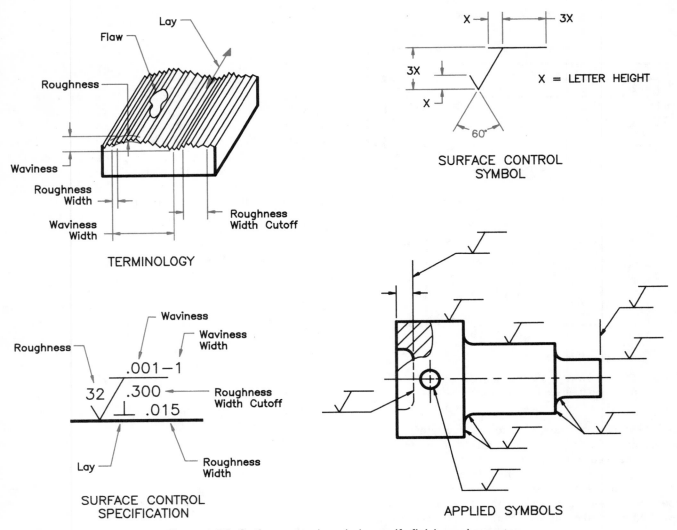

Figure 4-55. Surface control symbols specify finish requirements.

only, is drawn without the horizontal line. See Figure 4-56. The surface roughness value is shown in the "V". Microinches are specified if the inch system is being used. Micrometers are specified if the metric system is in use. Standard values for surface roughness are listed in the given figure. If only one roughness value is shown in the symbol, it is the maximum acceptable roughness. Two roughness values indicate a maximum and minimum acceptable roughness. The roughness value represents the arithmetic average deviation from a mean line on a surface profile.

A general note can be used to indicate the surface roughness requirement for all surfaces except those with a symbol applied to them.

Surface roughness can be measured by electronic inspection equipment or by comparison to sample surfaces of known roughness. The comparison method is less accurate, but it is also less expensive. If roughness requirements are not very demanding, the comparison method may be preferred.

The distance across peaks and valleys that cause surface roughness is known as the *roughness width*. If a maximum distance is acceptable, the width is shown inside the surface control symbol near the bottom.

The standard distance across which roughness values are measured is .080″. This distance is called the *roughness width cutoff*. Any roughness width cutoff distance other than .080″ must be specified. The value is placed directly under the horizontal bar of the surface control symbol. Commonly used cutoff values are shown in the given figure.

WAVINESS

Waviness is the large surface variations on which surface roughness is superimposed. See Figure 4-57. These variations are expressed in decimal parts of an inch or millimeters. Common values used for waviness are shown in the given figure.

Waviness is not meant to be a replacement for the flatness tolerance explained in the chapter on form tolerances. If a waviness tolerance value is shown, it should always be less than any flatness tolerance that is on the same surface.

Waviness values are shown above the horizontal line on a surface control symbol. A maximum width over which waviness may be measured is shown beside the waviness value if control of the waviness width is necessary.

A minimum surface contact area can be specified in place

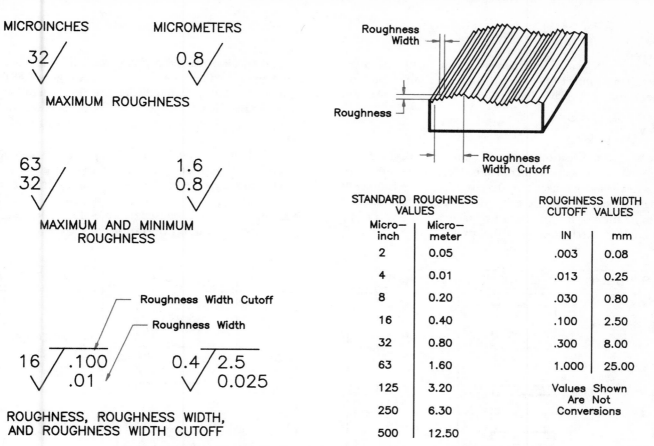

STANDARD ROUGHNESS VALUES

Micro-inch	Micro-meter
2	0.05
4	0.01
8	0.20
16	0.40
32	0.80
63	1.60
125	3.20
250	6.30
500	12.50

ROUGHNESS WIDTH CUTOFF VALUES

IN	mm
.003	0.08
.013	0.25
.030	0.80
.100	2.50
.300	8.00
1.000	25.00

Values Shown Are Not Conversions

Figure 4-56. Only the required level of control for surface finish is specified.

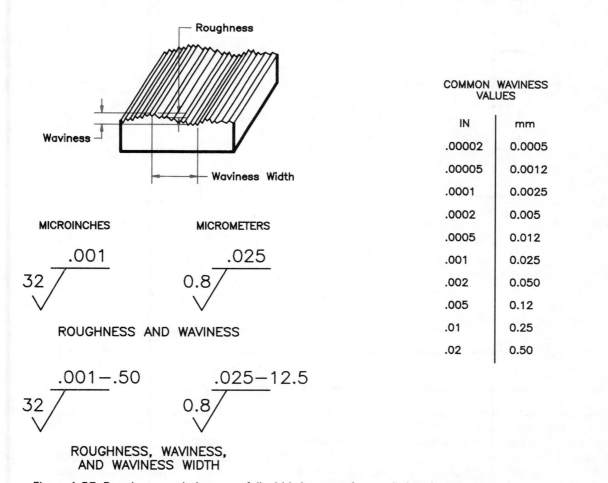

COMMON WAVINESS VALUES

IN	mm
.00002	0.0005
.00005	0.0012
.0001	0.0025
.0002	0.005
.0005	0.012
.001	0.025
.002	0.050
.005	0.12
.01	0.25
.02	0.50

Figure 4-57. Roughness variations may fall within larger surface variations known as waviness.

of surface waviness. The percentage of surface area that must be in contact is shown on the surface control symbol in place of the surface waviness value.

LAY

Lay is the direction of surface lines caused by cutting tools. The lay direction can have an effect on the function of a part. A bearing is used to reduce friction in a machine; the lay on a bearing surface has a definite effect on friction. Specification of lay on a bearing surface ensures control of the amount of friction that will be present.

Lay is specified on a drawing by showing a lay direction symbol in the surface control symbol. See Figure 4-58 for an example. The lay symbol indicates the direction of lay relative to the profile view on which the symbol is shown. Lay must always be indicated on a profile view of the affected surface.

Parallel and perpendicular lay paths are caused by straight cutter motions. Angular paths are generated by straight and large radius cutter motion. Circular lay is caused by operations such as facing parts on a lathe. Multidirectional lay can be caused by an end mill. Radial lay is caused by some grinding operations.

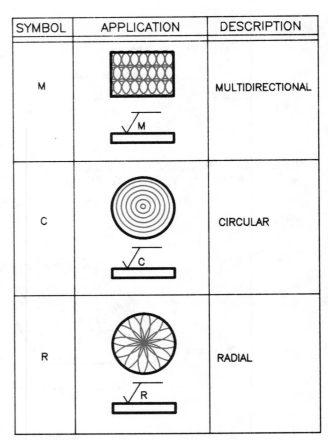

SYMBOL	APPLICATION	DESCRIPTION
⊥		PERPENDICULAR
//		PARALLEL
X		ANGULAR IN TWO DIRECTIONS
M		MULTIDIRECTIONAL
C		CIRCULAR
R		RADIAL

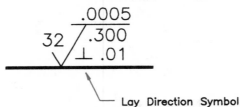

Lay Direction Symbol

Figure 4-58. Lay is the direction of surface lines caused by the path of the cutting tools.

CHAPTER SUMMARY

Generally, a feature of size must have perfect form when at the maximum material condition.

Every feature of size must have some amount of permissible variation in its size.

Size tolerances are calculated using either the basic hole or basic shaft system.

Standard tables provide the allowances and tolerances required for classes of fit that meet many common design applications.

There are two rules in the current dimensioning and tolerancing standard. These rules define the requirement for perfect form at MMC and the requirements for applicable material condition modifiers in feature control frames.

Interchangeability is an important reason for properly applying dimensions and tolerances.

Surface control specifications are used only to the extent necessary for the part function.

Surface roughness on a drawing is specified in microinches or micrometers.

REVIEW QUESTIONS

Answer the following questions on a separate sheet of paper. Do not write in this book.

MULTIPLE CHOICE

1. The current dimensioning standard requires that diameter dimensions be _____.
   A. preceded by a diameter symbol
   B. followed by a diameter symbol
   C. preceded by the diameter abbreviation
   D. followed by the diameter abbreviation

2. A flat on a shaft must be dimensioned by specifying the distance from the _____.
   A. centerline to the flat
   B. far side of the shaft to the flat
   C. top of the removed portion of the shaft to the flat
   D. Either A or B.

3. A dimension showing the location of a round hole must be applied to _____.
   A. the center of the hole
   B. the edge of the hole
   C. provide two methods for determining location
   D. Either A or B.

4. The depth of a hole is the distance into the part that the _____ extends.
   A. drill point
   B. full diameter
   C. centerline
   D. None of the above.

5. A counterbore depth is controlled by specifying _____.
   A. remaining material
   B. depth from the surface
   C. noted depth from the surface
   D. Any of the above.

6. The specified angle of a countersink is the _____.
   A. included angle
   B. angle between the centerline and countersink surface
   C. angle between the part surface and the countersink surface
   D. None of the above.

7. An angle dimension includes a dimension line that is drawn with an arc. The center for the arc is located at the _____.
   A. edge on the part
   B. vertex of the angle
   C. location that makes the arc appear correct
   D. midway between the part and the dimension line location

8. The leader used to show a radius dimension must point at or pass through _____.
   A. one tangent point
   B. a location dimension
   C. the arc center point
   D. Either A or C.

9. The arc length can be dimensioned by using an arc to draw the dimension line and placing a(n) _____.
   A. straight line above the dimension
   B. arc above the dimension
   C. straight line below the dimension
   D. arc below the dimension

10. A dimension showing the location of a sheet metal bend applies to the _____ location.
    A. mold line
    B. neutral line
    C. bend centerline
    D. bend tangent point

11. A symbol that looks like the letter V can be applied to show _____.
    A. critical surfaces
    B. cast surfaces
    C. critical dimensions
    D. finished surfaces

12. Rule #1 requires that a part have perfect form when at _____.
    A. MMC
    B. LMC
    C. RFS
    D. Both A and B.

13. Two features shown at a 90° angle on a drawing are both produced at the size limit at which perfect form is required. The 90° angle _____.
    A. must be a perfect 90°
    B. can vary by the angular tolerance for the 90° dimension
    C. can vary only within the perfect form boundary
    D. Inadequate information given.

14. On all tolerances, _____.
    A. LMC is assumed to apply unless specified otherwise
    B. RFS is assumed to apply unless specified otherwise
    C. MMC is assumed to apply unless specified otherwise
    D. material condition modifiers must be shown

15. The limits for a radius dimension permit the arc to vary in radius, _____.
    A. but no flats are permitted on the surface
    B. but no reversals are permitted on the surface
    C. and the form of the surface may include reversals
    D. Both A and B.

16. The tolerances applied to a feature affect _____.
   A. which machine processes can be used to produce the part
   B. the cost of the part
   C. how well the part meets its functional requirements
   D. All the above.
17. The _____ indicates the required pattern for the surface roughness.
   A. roughness value
   B. waviness value
   C. roughness width
   D. lay direction

TRUE/FALSE

18. The height, width, and depth must be given when dimensioning a prism. (A)True or (B)False?
19. A diameter may only be specified when the feature is a complete circle. (A)True or (B)False?
20. A counterdrill specification should include the angle at the bottom of the counterdrill. (A)True or (B)False?
21. Any chamfer can be dimensioned with a note. (A)True or (B)False?
22. A tapered shaft or hole can be dimensioned by specifying the amount of change in the diameter over a given length of the part. (A)True or (B)False?
23. The center point for an arc must never be moved from its true location for dimensioning purposes. (A)True or (B)False?
24. A spherical radius must be dimensioned differently than a nonspherical radius. (A)True or (B)False?
25. The true size of a feature must be dimensioned even if a segment of the feature is removed to reduce the size of a view. (A)True or (B)False?
26. The tolerance zone for an angle does not control the surface conditions of the feature oriented by the angle. The tolerance only limits the angle. (A)True or (B)False?
27. Dimensions define the part requirements, and also include instructions as to which machine processes are required. (A)True or (B)False?
28. Dimension text in a CAD model should never be edited without changing the model geometry to accurately reflect the dimension numbers. (A)True or (B)False?

FILL IN THE BLANK

29. When dimensioning a cylinder, the _____ and length must be specified.
30. An arc is dimensioned by specifying its _____ and location.
31. The prefix used for a spherical radius is _____.
32. A _____ symbol can be used to specify surface roughness, waviness, and other surface irregularities.

SHORT ANSWER

33. List the information that must be provided in a counterbore specification.

34. List two methods for providing the necessary definition for the location of an arc.
35. Explain how the line of symmetry is shown on a drawing.
36. Under what conditions might it be necessary to draw offsets in extension lines?
37. List three methods of giving the acceptable size limits for a dimension.
38. How is exception to Rule #1 taken when a form tolerance greater than the size tolerance is to be specified?
39. List the two measurement units that are used to measure surface roughness.
40. Complete a specification for a counterbored hole that meets the following requirements. Use symbology in the specification.

Hole diameter	.250″
Counterbore diameter	.375″
Counterbore depth	.250″
Corner radius	.030″

41. Complete a specification for a countersunk hole that meets the following requirements. Use symbology in the specification.

Hole diameter	.375″
Countersink diameter	.625″
Countersink Angle	100°

Pg 68-69

APPLICATION PROBLEMS

Each of the following problems requires that a sketch be completed. All sketches should be neat and accurate. Dimensions must be applied to the sketch. Unspecified dimensions can be approximated using the shown grid. Dimensions for approximated sizes and locations will be evaluated on the basis of how the dimension is applied, not on the accuracy of the approximation. Some of the following problems require calculations. Dimensions applied on the basis of calculations must be the correct number values.

42. The views in the following problem are incomplete. Make a sketch of the required views and apply all dimensions.

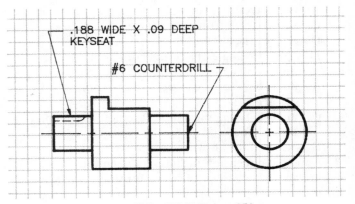

.188 WIDE X .09 DEEP KEYSEAT

#6 COUNTERDRILL

GRID SPACING = .250

43. Sketch the shown object and apply all dimensions. Hole sizes provided in the table are to be applied using leader lines and notations. Either the inch or metric part may be completed.

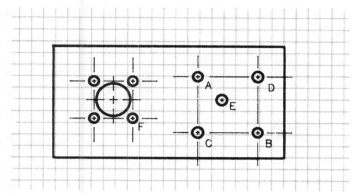

| HOLE | DIA | |
	INCH	mm
A	.250	9.0
B	.250	9.0
C	.280	10.0
D	.280	10.0
E	.280	10.0
F	.188	8.0

GRID SPACING = .250
OR
GRID SPACING = 10.0mm

INCH AND METRIC PARTS ARE DIFFERENT SIZES. VALUES GIVEN ARE NOT CONVERSIONS.

44. Apply the radius dimensions to the given figure. Either the inch or metric part may be completed.

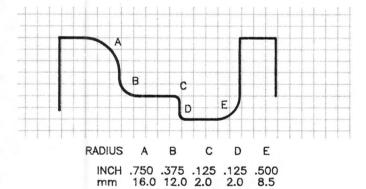

RADIUS	A	B	C	D	E
INCH	.750	.375	.125	.125	.500
mm	16.0	12.0	2.0	2.0	8.5

GRID SPACING = .250
OR
GRID SPACING = 10mm

45. Complete a drawing of the following part at a scale of 1/1. Do not increase the scale. Apply all dimensions. Doglegs may be placed in the extension lines, if needed. Do not use ordinate dimensions.

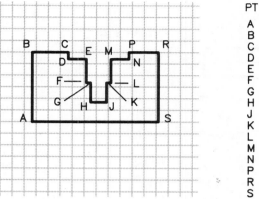

PT	X	Y
A	0	0
B	0	1.00
C	.50	1.00
D	.50	.90
E	.75	.90
F	.75	.56
G	.81	.56
H	.81	.28
J	1.03	.28
K	1.03	.56
L	1.09	.56
M	1.09	.90
N	1.34	.90
P	1.34	1.00
R	1.75	1.00
S	1.75	0

46. Completely dimension the two parts. The 1.250″ nominal diameter features are to be dimensioned for an RC7 class fit. Tolerance values for an RC7 class fit can be obtained from ANSI B4.1 or a Machinery's Handbook.

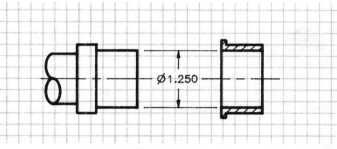

GRID SPACING = .250

47. Dimension the hole to result in a .0025″ maximum and .0010″ minimum interference with the shown rest button.

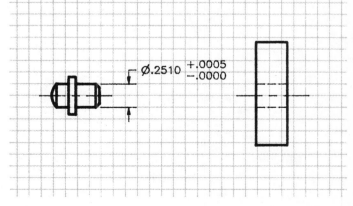

$\varnothing$.2510 $^{+.0005}_{-.0000}$

48. Show one acceptable distorted shape for the given part when all diameter measurements on a produced part are at 2.48″ (LMC). Show whether the part is within the MMC boundary.

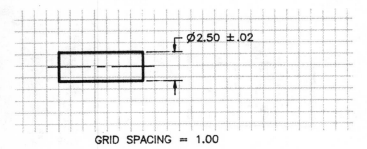

GRID SPACING = 1.00

49. Show two acceptable extremes for the angled surface with the vertex produced at a location of .485″ from the bottom surface.

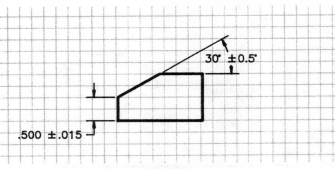

GRID SPACING = .250

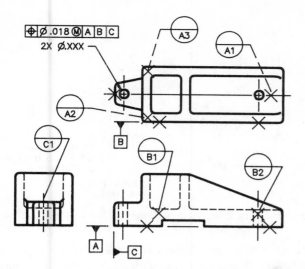

Chapter 5

Form Tolerances

CHAPTER TOPICS

- ☐ Symbols used to indicate form tolerances.
- ☐ Feature control frames and material condition modifiers for form tolerances.
- ☐ Form control through limits of size.
- ☐ Straightness tolerances on surface elements and on an axis.
- ☐ Flatness tolerance applied to surfaces.
- ☐ Control of center plane form.

- ☐ Circularity to control the form of circular cross sections.
- ☐ Cylindricity tolerances applied to cylindrical features.
- ☐ Virtual condition when a form tolerance is applied to a feature of size.

INTRODUCTION

All parts include shape and surface variations. The amount of variation on individual features is controlled by size tolerances, and can be further controlled through the application of *form tolerances.*

Tolerances associated with size dimensions permit form variations equal to the size tolerance. If the variations permitted by the size tolerances are acceptable, then no form tolerance specification is needed. When form variations must be less than is permitted by the size tolerances, a form tolerance is specified. This allows a large size tolerance to be utilized while controlling the form variations to a smaller value.

Form tolerances only control variations on individual features. Controlling the flatness of a single surface only controls that particular surface. Form tolerances do not control the relationship between features. A flatness form tolerance, for example, does not imply a parallelism or perpendicularity requirement relative to other features.

Since smaller size tolerances do reduce permitted form variations, it might seem that form tolerances could be eliminated by using smaller size tolerances to achieve the desired form control. However, controlling form variations through small size tolerances unnecessarily increases part cost.

The use of form tolerances is desirable since it is easier to control form of a surface, such as flatness, than it is to control location. As an example, controlling flatness of a surface to .003″ is easier than controlling its location to within ±.0015″ (.003″ total).

A part that has a flatness tolerance specification of .003″ and a thickness (size) tolerance of ±.010″ can be produced at a relatively low cost. If, for some reason, a flatness toler-

ance is not specified but the surface flatness is still desired, then the surface flatness must be controlled by reducing the size tolerance. Decreasing the size tolerance will increase part cost. To achieve the .003″ flatness through size tolerance specification, a size tolerance of ±.0015″ is required. This small size tolerance is much more difficult to achieve than the combination of a ±.010″ size tolerance and a .003″ form tolerance specification.

Size tolerances should be used to control the size of a part. When form variations need to be controlled to a value less than the size tolerance, form tolerances need to be specified. Part cost can be minimized by properly utilizing size and form tolerances.

FORM TOLERANCE CATEGORIES

The form tolerance types are straightness, flatness, circularity, and cylindricity. Each type of form tolerance has a specific definition. Application of the proper tolerance specification requires understanding the possible uses for the form tolerances.

Material condition modifiers are only used with straightness tolerance specifications, and then only used when straightness is applied to features of size such as a cylinder. Material condition modifiers were introduced in Chapter 2 and defined to a greater extent in Chapters 3 and 4. How they impact straightness tolerances will be explained in this chapter.

FORM TOLERANCE SYMBOLS

Symbols used for form tolerances are easy to remember since their shape is symbolic of the indicated control. See Figure 5-1. A straight line indicates straightness. A parallelogram

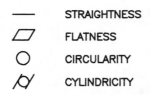

— STRAIGHTNESS

▱ FLATNESS

○ CIRCULARITY

⌀ CYLINDRICITY

Figure 5-1. There are four form tolerance symbols.

represents flatness. A circle stands for circularity, and a circle with tangent lines means cylindricity.

FEATURE CONTROL FRAMES

Proper format and application of the tolerance specification is necessary to convey the correct information. See Figure 5-2. The proper specification format requires the tolerance be shown in a *feature control frame*. Feature control frames must be organized as shown in the given figure. The manner in which the feature control frame is shown on the drawing affects the meaning of the tolerance.

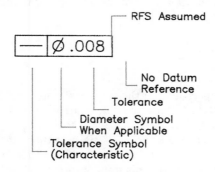

Figure 5-2. A feature control frame for a form tolerance always contains a tolerance symbol and a tolerance value.

The feature control frame for a form tolerance only has two cells. The first cell always contains the *tolerance symbol*. The second cell contains the *tolerance value* that indicates the amount of acceptable tolerance. Depending on the tolerance being specified, there may be a diameter symbol in front of the tolerance value. A modifier follows the tolerance value when applicable. The modifier may be either the MMC or RFS symbol. However, the RFS symbol need not be shown since it is assumed. The statistical symbol and the free state symbol are shown only when applicable. When shown, these symbols are placed after any applicable material condition modifier.

A datum reference is never included in a form tolerance specification. Datum references would be incorrect since each form tolerance is a control of an individual feature and does not include any requirement for control of relationships between features.

Material Condition Modifiers

Straightness tolerances applied to a *feature of size* requires that a material condition modifier be associated with the specified tolerance value. The maximum material condition (MMC) modifier must be shown in the feature control frame if it is to apply. See Figure 5-3. If a material condition modifier is not shown in the feature control frame, then the

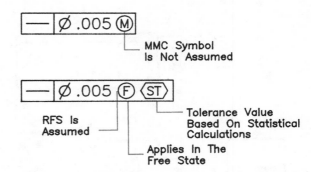

Figure 5-3. The material condition modifier on a form tolerance is assumed to be RFS unless shown otherwise.

regardless of feature size (RFS) modifier is assumed to apply. This is in compliance with *Rule #2* of the current national standard.

Care must be taken to show the MMC modifier in all situations where the tolerance needs to apply at MMC. The MMC modifier must not be shown if the tolerance is to apply regardless of feature size (RFS). Which material condition modifier is applied to a tolerance causes significant differences in the resulting part requirements.

Material condition modifiers are only used when the tolerance is applied to a feature of size. Application of a form tolerance to surface elements does not require the application of a material condition modifier. In fact, it is incorrect to show a material condition modifier on a form tolerance applied to surface elements. (*not a feature of size*)

Control Of Single Features

Each form tolerance only affects the single feature to which it is applied. See Figure 5-4. A flatness tolerance applied to the top surface of a part only expresses a requirement for that surface. If control of another surface is desired, another feature control frame must be drawn. It is also possible that two leaders can extend from one feature control frame to two surfaces. If this is done, it is the same as two separate feature control frames on the drawing.

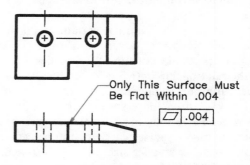

Figure 5-4. A form tolerance only applies to the single feature to which it is attached.

SURFACES AND FEATURES OF SIZE

It is extremely important that the difference between surfaces and *features of size* be understood. The method of application of a feature control frame to a part determines whether the tolerance is on a surface or on a feature of size. A feature control frame can have a significantly different meaning according to which feature type it is applied.

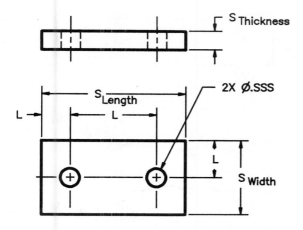

6 FLAT SURFACES

2 HOLES ARE FEATURES OF
   SIZE

3 FEATURES OF SIZE DEFINED
   BY THICKNESS, WIDTH, AND
   LENGTH

Figure 5-5. The difference between surfaces and features of size must be understood to properly apply form tolerances.

The difference between the two feature types are shown in Figure 5-5. Features of size are those defined by size dimensions. The given rectangular part includes two holes. The hole diameters are defined by size dimensions. The holes are therefore features of size.

There are six flat surfaces on the part. Each of these, when considered independently of all other features, is a surface.

There are three features of size defined by the thickness, width, and length of the part. Each of these features of size are made of two surfaces separated by a size dimension.

FORM CONTROL THROUGH SIZE LIMITS

Every feature of size must have a *size dimension,* and every size dimension has a tolerance. The tolerance on a size dimension creates a maximum and minimum size limit. The maximum amount of variation that may exist on a feature is controlled by the limits of size.

Variations in size have an effect on the amount of material in a part. See Figure 5-6. The *maximum material condition (MMC)* is when the maximum permissible amount of material exists. The *least material condition (LMC)* is when the least amount of material exists.

An external feature, such as a shaft, is at MMC when the feature is at its maximum size limit. LMC is when the external feature is at its minimum size limit. An internal feature, such as a hole or slot, is at MMC when the feature is at its minimum size limit. LMC is when the internal feature is at its maximum size limit.

RULE #1, PERFECT FORM BOUNDARY AT MMC

The most fundamental of the guidelines given in the dimensioning and tolerancing standard is *Rule #1.* It defines

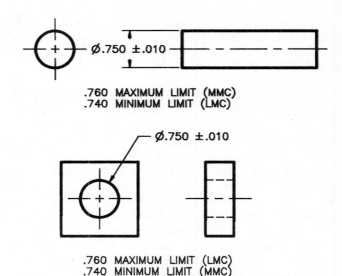

.760 MAXIMUM LIMIT (MMC)
.740 MINIMUM LIMIT (LMC)

.760 MAXIMUM LIMIT (LMC)
.740 MINIMUM LIMIT (MMC)

Figure 5-6. Maximum Material Condition (MMC) is when the most material exists in the part. It is the largest shaft and the smallest hole that is permissible.

what is often referred to as the *envelope principle.* The standard actually states:

> "Where only a tolerance of size is specified, the limits of size of an individual feature prescribe the extent to which variations in its geometric form, as well as size, are allowed."

This means that when a feature is produced at MMC, the feature must have perfect form. See Figure 5-7. MMC size establishes a perfect form boundary and no part of the feature may extend outside the boundary.

The example in Figure 5-7 shows the effect of Rule #1 on a shaft, which is a cylindrical feature of size. The diameter dimension has an MMC value of .508″. A theoretically perfect .508″ diameter cylinder is the perfect form boundary for this feature.

As the size of a feature departs from MMC, its form is permitted to vary. Any amount of variation is permitted, provided the perfect form boundary is not violated. Of course, the size limits must not be violated either.

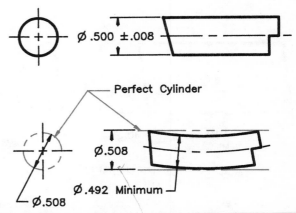

Figure 5-7. Perfect form is required at MMC. The part can vary in form as it departs from the MMC size.

It is possible for the given cylinder to be produced with form variations equalling .016″ if the produced diameter of the part is only .492″. It is not likely that this extreme condition will exist, but it is permissible according to the requirements of Rule #1.

From the given figure, it can be seen that perfect form of the feature is not required when the feature is at LMC. *Perfect form at the LMC is not required even when all cross-sectional measurements along the length of the part are at LMC.* This fact is probably overlooked or misunderstood as much as any other item related to dimension interpretation. Although a requirement for perfect form at LMC can be forced through special methods, the size tolerance alone does not establish a requirement for perfect form at LMC.

The interrelationship between features is not affected by the perfect form boundary. See Figure 5-8. When a feature of size is at MMC, it must have perfect form. This does not require that it have perfect orientation to any other feature.

When each of the three features of size in Figure 5-8 are at MMC, there is still an allowable tolerance on the 90° angle between the features. In this example, the given angularity tolerance of ± .5° is permitted on each angle.

Rule #1 establishes a perfect form boundary at MMC for individual features of size. It does not force any control of interrelationship between features of size.

Perfect Form Boundary Effect On Flat Surfaces

A size dimension applied to the distance separating two flat surfaces must have a size tolerance. See Figure 5-9. The

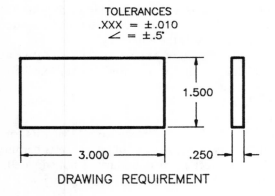

TOLERANCES
.XXX = ±.010
∠ = ±.5°

DRAWING REQUIREMENT

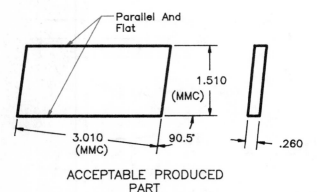

ACCEPTABLE PRODUCED PART

Figure 5-8. Perfect form at MMC is only a requirement on individual features. Features at MMC do not have a required perfect orientation to other features.

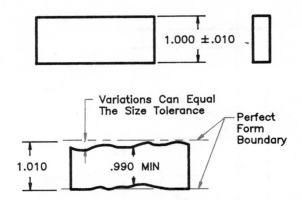

Figure 5-9. Form variation within the perfect form boundary can equal the size tolerance.

distance measured between the two surfaces at any location must be within the acceptable limits of size. There is no requirement for all measurements to be of a single value.

If the limits of size for a feature are known, it is possible to determine the amount of form error that can occur on the feature. The given figure shows a size dimension of 1.000″ ± .010″ applied to an external feature bounded by two parallel planes. The feature has size limits of .990″ and 1.010″, therefore a perfect form boundary exists at the 1.010″ limit.

If the feature is produced at MMC (1.010″), then the two surfaces must be perfectly flat and parallel. As the actual produced size varies from MMC, the surface form and orientation may vary, provided the perfect form envelope is not violated. This means that the acceptable surface variations are equal to the amount of size departure from the MMC value. The maximum possible form variation is equal to the size tolerance. In the given figure, the permitted form variations on one surface could be as much as .020″.

The acceptable amount of form variation on a feature is dependent on the produced (measured) size of the feature. The permitted form variation is equal to the amount of size departure from the MMC size.

Perfect Form Boundary Effect On Cylinders

The perfect form boundary established by a diameter dimension is a cylinder. The perfect form boundary for a shaft is a cylinder equal to the maximum size limit of the shaft. The perfect form boundary for a hole is a cylinder equal in diameter to the minimum size limit of the hole.

Figure 5-10 shows a shaft with a 1.000″ ± .010″ diameter dimension. This results in a perfect form cylinder of 1.010″ diameter. At any point along the length of the cylinder, the measured diameter may be any size between .990″ and 1.010″. At any point where the diameter is less than 1.010″, form variations on the shaft are permitted to equal the amount of departure from MMC.

If the shaft is produced so that all cross-sectional measurements are .990″ (LMC), the shaft could have surface errors equal to .020″. If surface errors exist on the shaft, the shaft axis may not be straight.

When form errors as large as the limits of size are not acceptable, form tolerances or other controls can be applied to the drawing.

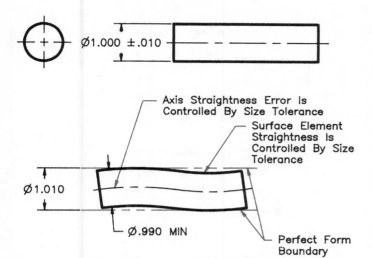

Figure 5-10. Both the surface error and axis straightness errors of a cylinder are controlled by the diameter dimension.

Exceptions To Rule #1

The requirements of Rule #1 do not apply to everything. Exceptions include parts that are subject to free state variation. Some stock materials are also excluded from Rule #1.

Free state variation is not well defined. Some common sense must be used to determine when a part is subject to free state variation. Examples of free state variation are given here to provide guidance. A thin, metal gasket made from a piece of shim stock will have a small size tolerance. It is not practical to expect the part produced from shim stock to be within a perfect form envelope at MMC since the thin material bends so easily. A shaft with a long length-to-diameter ratio may tend to flex. This makes compliance with the perfect form envelope impractical.

A thin-walled tube when machined in a lathe chuck may meet the perfect form envelope requirement while clamped in the lathe chuck. Since the part is produced while under the stresses exerted by the chuck, it will change shape when it is released. This is one type of free state variation.

Parts subject to free state variation may need to be measured prior to removal from machining clamps. It is also possible to use drawing notes that require the part to be restrained for inspection. Circular parts subject to free state variation are sometimes inspected by making multiple measurements to establish an average diameter.

Some stock material shapes have standardized form tolerances that are larger than the size tolerances for the shapes. Generally, Rule #1 does not apply to standard *stock shapes. Sheet, plate, rod,* and *bar stock* all have form tolerances that are larger than the required size tolerance. *Extrusions* and *structural shapes* also have form tolerances that exceed size tolerances.

STRAIGHTNESS

A *straightness tolerance* can be used to control the straightness of surface elements or an axis. The tolerance specifies how close to perfectly straight a feature must be made. The form of any straight surface can be controlled with this type tolerance. The form of the axis for a feature of size, such as a shaft, can also be controlled with a straightness tolerance.

The way in which a straightness tolerance is shown on the drawing indicates whether surface elements or an axis is to be controlled. Application on an extension line from a surface has one meaning, while application adjacent to a dimension has a completely different meaning. The meaning of a straightness tolerance specification varies according to the type of feature to which the tolerance is applied.

Each straightness tolerance specification is shown in a feature control frame. See Figure 5-11. The feature control frame includes the straightness tolerance symbol and the tolerance value. The symbol indicates the required tolerance zone shape, and the tolerance value shows the allowable size of the tolerance zone. A datum reference is never shown in a straightness tolerance specification.

APPLIED TO SURFACE ELEMENTS

Previous paragraphs in this chapter explained that form variations on a surface must be contained within the size tolerance. It is necessary to apply a feature control frame containing a form tolerance when the form variations must be controlled to a smaller value than is permitted by the size tolerance. See Figure 5-11.

The given part has a size tolerance of ± .005″ between the top and bottom surfaces. The allowable form variation due to this tolerance is as follows. If the bottom surface is perfect, the maximum form variations on the top surface are equal to .010″. If .010″ variations on the top surface of the given part are acceptable, then no other specification is needed.

A form tolerance is only applied to a surface if the required control is smaller than the size tolerance. The shown form tolerance is .005″, but could be any value less than .010″. Application of a .010″ form tolerance would not be wrong, but it would be redundant since the size tolerance already controls the surface to within .010″.

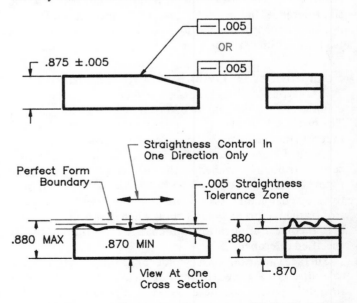

Figure 5-11. Straightness tolerances on a flat surface control line elements in one direction.

A form tolerance larger than the size tolerance is not normally specified on a surface. Showing a form tolerance larger than the size tolerance creates a conflict between the size and form tolerance requirements. A form tolerance larger than the size tolerance is not permitted on a surface unless a note specifies that perfect form at MMC is not required on the designated dimension.

Material condition modifiers are not used in form tolerance specifications when they are applied to surfaces. A surface does not have a material condition, so application of a material condition modifier is incorrect.

Straightness Of Flat Surfaces

The straightness requirements for line elements on a flat surface are specified as shown in Figure 5-11. The feature control frame is drawn to show the straightness symbol and the allowable tolerance value. The feature control frame is applied to the surface by a leader, or it can be placed on an extension line from the surface.

It is important to select the correct view for application of a straightness specification. Straightness tolerance specifications only control the straightness of line elements that are parallel to the line on which the specification is shown. Straightness control should not be confused with flatness.

Figure 5-11 shows the interpretation of a size tolerance and a straightness tolerance. The size dimension between the top and bottom surfaces has a maximum limit of .880″. The part has a perfect form boundary at the dimension of .880″, and no part variations may extend outside this boundary. The size dimension has a minimum limit of .870″, and no measurement across the part may be less than this value.

The straightness tolerance is applied to the top line in the front view, therefore all surface elements parallel to this line must be straight within the given .005″ tolerance. Each line element on the produced part is verified separate from all others since the specification is straightness, not flatness. Line elements perpendicular to the required direction of control may be out-of-straight by an amount equal to the size tolerance. In the given figure, the straightness tolerance applied to the top surface does not imply any requirements for parallelism to the bottom surface.

Straightness Of Cylinder Surface Elements

Two types of straightness control can be specified on a cylinder. One is the control of surface element straightness. The other is control of the axis straightness. Control of cylinder surface elements is explained in this section.

Surface element control is indicated when the feature control frame is attached to the surface of a cylinder. See Figure 5-12. This is done either by a leader or by attachment to an extension line. Proper application of the feature control frame is extremely important when controlling a cylinder. Incorrect application will change the meaning of the specified tolerance.

Surface variations on a cylinder are controlled by the size tolerance just as they are for a flat surface. Application of a form tolerance is only necessary when the desired control is a smaller value than the size tolerance.

The given example shows a shaft with a diameter of .875″

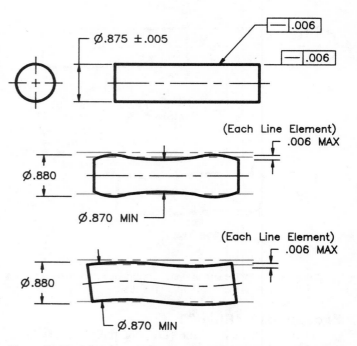

Figure 5-12. Straightness tolerances on a cylindrical surface only control line elements on the cylinder.

± .005″. This results in a size tolerance of .010″. A straightness tolerance of .006″ is applied to the surface.

Two of the many possible part configurations that could result from the given size and straightness specifications are shown. In the first example, the .880″ diameter perfect form boundary is shown. None of the part variations fall outside this boundary. The smallest diameter at any point along the shaft is not less than the smallest size limit, .870″. The variations of surface straightness are within the .006″ tolerance. Surface variations along the length of the part are caused by a change in diameter. The changing diameter has caused surface straightness variations, but the axis has remained perfectly straight. This example proves that a straightness tolerance applied to the surface of a cylinder does not directly control the axis straightness. It also shows that axis straightness does not necessarily control the surface.

The second example shows the same .880″ diameter perfect form boundary. The smallest measured diameter is .870″. Each surface element on the shaft is straight within the allowable .006″ tolerance. On this part, the surface errors are such that the shaft axis is not straight.

A straightness tolerance applied to the surface of a cylinder controls the surface elements. It does not control the axis straightness. The straightness errors on each element are independent of errors on any other element. Regardless of the specified surface form tolerance, the axis straightness errors can be as much as is permitted by the size tolerance. There is no orientation requirement included in the straightness tolerance.

Straightness Of Cone Elements

Straightness tolerances may be applied to any feature that has straight line elements. See Figure 5-13. A cone is an example. Straightness of the elements that lie in a plane with

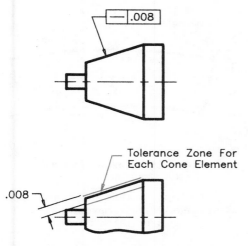

Figure 5-13. A straightness tolerance can be applied to any feature that contains straight line elements.

the cone axis can be controlled. Each cone element is controlled independently. Although the straightness tolerance zone must be in a common plane with the cone axis, there is no requirement to maintain a specific angle to the axis. The cone angle is controlled either by the size and angle tolerances, or by a profile tolerance as defined in another chapter.

Verification Of Surface Straightness

The part in Figure 5-14 has a surface straightness requirement of .009″ on the inclined surface. The surface elements parallel to the edge of the part must be straight within the specified .009″ tolerance. There is no other requirement implied by the straightness tolerance. Surface straightness inspection is completed independent of the size requirement.

One method for inspecting the straightness of the surface elements is shown. The part is set on an inclined *sine bar*.

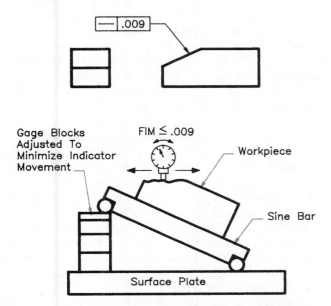

Figure 5-14. Straightness only controls form. The orientation of a part can be adjusted to minimize measured errors since the tolerance does not control orientation.

The sine bar is inclined on top of a flat plate or surface table by placing gage blocks under one of its feet. The angle of the sine bar can be adjusted to minimize the dial indicator readings since the straightness requirement does not include an orientation requirement.

A dial indicator is moved in a straight line along the surface. This is repeated at several locations across the surface. The errors read on each pass of the dial indicator are considered separately. The error on any one pass is determined by the full movement of the dial indicator needle. This is referred to as the full indicator movement (FIM) or total indicator reading (TIR).

Noted Exceptions To Rule #1

The requirement for perfect form at MMC can be very restrictive when the size tolerance is small. At times, this requirement can make production of the part expensive, if not impossible. If the size limits for a part make production within the perfect form boundary difficult, a decision must be made as to whether or not an exception to Rule #1 will be taken. The requirement for perfect form at MMC must be considered when designing a part, and the functional design should be the final determining factor in whether or not to take exception to Rule #1.

Variance outside the perfect form boundary may need to be permitted for a large, thin part produced with a small thickness tolerance. See Figure 5-15. A part can be permitted to vary outside the perfect form boundary by specifying a form tolerance that is larger than the size tolerance. In addition to the form tolerance, there must be a note applied in a manner that indicates Rule #1 is not applicable to a specific dimension. The note may be applied on the drawing, or a *note flag* can be placed beside either the feature control frame or the applicable dimension. If a note flag is used, the note is placed in the notes list.

Exception to Rule #1 should only be taken when required for production and when the functional design is such that the specified variation is acceptable. When taking exception to Rule #1, it is necessary to show both a form tolerance and the note. A noted exception to Rule #1 can be applied to a dimension by using the method shown in Figure 5-15. Application of a form tolerance larger than the size tolerance is an error if the note is not present. Application of the note without a form tolerance would be an error. The form tolerance must be specified or the amount of permissible variation is not known.

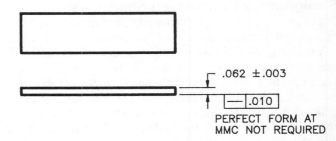

Figure 5-15. If a form tolerance larger than the size tolerance is applied to a surface, a note must be applied to the size dimension to take exception to the requirements of Rule #1.

APPLIED TO FEATURES OF SIZE

Axis straightness for a cylinder can be controlled through a form tolerance applied to the feature of size. See Figure 5-16. The tolerance is applied to a cylindrical feature of size by placing a feature control frame on the diameter of a shaft or hole. The feature control frame may be placed adjacent to the size dimension, or it may be attached to the dimension line. This indicates that the axis straightness is being controlled.

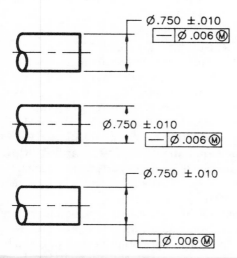

Figure 5-16. The axis of a feature can be controlled by placing a form tolerance adjacent to the size dimension or on the dimension line.

Care must be used not to make contact between the feature control frame and any extension lines. If an extension line is contacted, then surface control is established instead of axis control.

Axis straightness is not restricted by the limits of size when an axis straightness tolerance is specified. It is permissible to apply an axis straightness tolerance that is larger than the size tolerance. This does not require that exception to Rule #1 be noted.

A straightness tolerance applied to a feature of size is assumed to apply RFS unless shown otherwise. The given example includes MMC modifiers. When the tolerance specification is applied to control axis straightness, a diameter symbol is placed in front of the tolerance value.

Straightness To Control A Shaft Axis

An axis straightness specification applied to a shaft creates a cylindrical tolerance zone that extends the full length of the shaft. See Figure 5-17. The figure shows the effect of the size tolerance and axis straightness tolerance specified in Figure 5-16. There is a .006″ diameter cylindrical zone that extends the length of the shaft. This tolerance zone does not control the surface elements, but instead controls the axis form. The variations in the surface of the shaft may be of any value permitted by the size tolerance, provided the effect on the axis does not cause any violation of the .006″ zone.

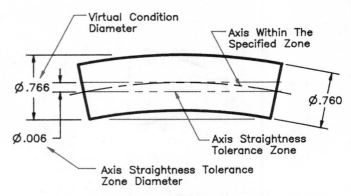

Figure 5-17. A straightness tolerance applied to control axis straightness permits the part to have an axis straightness error when the part is at MMC.

Diameter measurements made at any cross section along the shaft must be within the limits of size–between .740″ and .760″ diameter. The straightness tolerance specification permits a .006″ diameter axis straightness error when the part is at MMC. For this particular part, if all cross-sectional measurements are .760″, the axis straightness error is permitted to be .006″.

The given figure shows the effect of a .006″ axis straightness error with the shaft produced at MMC. The MMC value for the shaft is .760″. An axis error of .006″ with the shaft at MMC (.760″) causes the shaft to have an effective diameter of .766″. The apparent increase in diameter because of the combined size and form tolerances is known as the *virtual condition.*

Virtual Condition Of A Shaft

An axis straightness error in addition to the shaft diameter at MMC creates a condition known as the *virtual condition.* When a tolerance is specified with the MMC modifier, virtual condition is the MMC size and the effects imposed by the applied tolerances. Another way to think of the virtual condition is to visualize the envelope required to contain the combined effects of the MMC size and the tolerance zone.

The virtual condition for a shaft can be determined by adding the MMC size of the shaft and the axis straightness tolerance. See Figure 5-18. The virtual condition for the given shaft is 1.267″. This was determined by adding the straightness tolerance of .007″ to the MMC size of 1.260″.

The concept of a virtual condition is important to understand. Calculations for parts that must slide together are completed based on virtual conditions. If the virtual conditions of two parts in an assembly will fit together, then any part that is within the virtual condition envelope will also fit in the assembly. This is a primary justification for using MMC modifiers on tolerance specifications.

An MMC modifier applied to an axis straightness tolerance permits the tolerance zone to increase as the feature size departs from MMC. Although the straightness tolerance is being allowed to increase, the envelope that encloses the shaft does not change in size.

A part produced at MMC has a form tolerance equal to the value shown in the feature control frame. For the given figure, this would be a 1.260″ diameter shaft with an axis

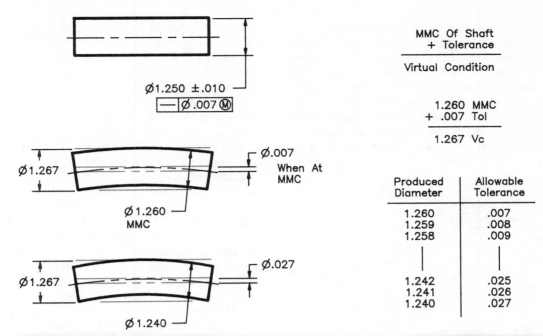

Figure 5-18. The permitted axis error applied to a cylinder has an effect on the apparent diameter of the cylinder. The combined effect of the MMC size and the permitted axis error is known as the virtual condition.

straightness tolerance of .007". This results in a virtual condition, or envelope, of 1.267" diameter.

The tolerance of .007" diameter shown in the given figure is only required while the shaft is at MMC (1.260"). As the diameter of the shaft gets smaller, the straightness tolerance is permitted to increase. The tolerance zone increases at a value equal to the departure of the shaft from its MMC size.

If a part is produced with a diameter smaller than 1.260", that part can have a larger straightness error and still fit inside the same virtual condition envelope. The amount of departure from MMC is the amount of additional tolerance that can be added to the value shown in the feature control frame.

The given table shows various produced shaft diameters and the allowable straightness tolerances. Each diameter and corresponding tolerance, when added together, equals the virtual condition.

Since the straightness tolerance increases by an amount equal to the departure from MMC, the example shaft is permitted to have a straightness error of .027" when the shaft diameter is 1.240". Increasing the tolerance zone as the shaft size decreases results in an envelope that remains the same size as the virtual condition.

An MMC modifier should be used when the function of the part only requires clearance for assembly. If axis straightness tolerance must remain unchanged regardless of the produced size, then the MMC modifier must be omitted, and RFS used. The impact of the RFS modifier is important to understand. If the tolerance in the given figure is specified at RFS, then .007" maximum error is required regardless of the produced diameter of the shaft.

The permitted amount of error can easily be calculated for any produced diameter when the MMC modifier is used. The allowable straightness tolerance is determined by adding the tolerance in the feature control frame to what is

commonly referred to as the *bonus tolerance*. The bonus tolerance for a particular part is determined by finding the difference between the produced size and the specified MMC size.

The following example shows how calculations are made to determine the allowable tolerance for a produced part. The example is based on the specified size and form tolerance in Figure 5-18.

Specified MMC	1.260
Produced Shaft DIA	-1.248
Bonus Tolerance	.012
Specified Form Tolerance	.007
Bonus Tolerance	+ .012
TOTAL ALLOWABLE FORM ERROR	.019

To Control Axis Straightness Of A Hole

A straightness tolerance to control the axis for a hole may be applied in either of two ways. See Figure 5-19. The feature control frame may be applied to the dimension line for the hole diameter, or it may be placed adjacent to the diameter dimension value. It must not be placed on the extension lines. Placement on the extension lines would result in control of the surface elements rather than the axis.

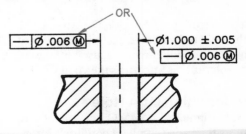

Figure 5-19. Form tolerance can be applied to internal features such as holes.

An axis straightness specification applied to a hole creates a cylindrical tolerance zone that extends the full length of the hole. See Figure 5-20. The figure shows the effect of the size tolerance and axis straightness tolerance. The straightness tolerance creates a .005″ diameter cylindrical zone that extends the length of the hole. This tolerance zone does not control the surface elements, but instead controls the axis. The variations on the surface of the hole may be of any value within the size limits, provided the effect on the axis does not cause any violation of the .005″ zone.

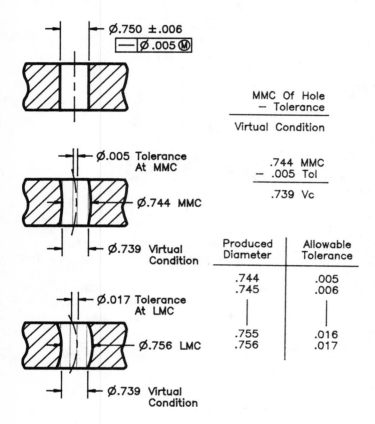

Figure 5-20. The virtual condition of a hole has a diameter smaller than the MMC diameter permitted by the size dimension.

Diameter measurements made at any cross section along the hole must be within the limits of size–between .744″ and .756″ diameter. The straightness tolerance specification permits a .005″ diameter axis straightness error when the part is at MMC. For this particular part, if all cross-sectional measurements are .744″, the axis straightness error is permitted to be .005″.

An axis error of .005″ with the hole at MMC (.744″) causes the hole to have an effective diameter of .739″. The apparent decrease in diameter caused by the combined size and form tolerance is known as the virtual condition.

Virtual Condition Of A Hole

The hole diameter at MMC and the effect of the axis straightness tolerance creates a condition known as the virtual condition. When a tolerance is specified with the MMC

modifier, virtual condition is the MMC size and the effect of the applied tolerances. Another way to think of the virtual condition for a hole is to visualize the maximum mating envelope that fits inside the hole when the hole is at MMC and the straightness tolerance is applied.

The virtual condition for a hole can be determined by subtracting the form tolerance from the MMC size. See Figure 5-20. The virtual condition for the given hole is .739″.

A MMC modifier applied on an axis straightness tolerance permits the tolerance zone to increase as the feature diameter departs from MMC. Although the straightness tolerance is being allowed to increase, the mating envelope inside the hole does not change in size.

The tolerance of .005″ diameter shown in the given figure is only required while the hole is at MMC (.744″). As the diameter of the hole gets larger, the straightness tolerance is permitted to increase. The tolerance zone increases at a value equal to the departure of the hole from its MMC size.

Since the straightness tolerance increases by an amount equal to the departure from MMC, the example hole is permitted to have a straightness error of .017″ when the hole diameter is .756″. Increasing the tolerance zone by the same amount as the increase in hole size results in a mating envelope that remains the same size as the virtual condition.

The tolerance should be specified at MMC when the function of the part only requires clearance for assembly. If axis straightness tolerance must remain unchanged by the produced size, then the MMC modifier must be omitted, and RFS assumed. If the tolerance is specified at RFS, then the .005″ axis straightness requirement applies regardless of the produced diameter of the hole.

Verification Of Axis Straightness

Tolerances specified at MMC can be inspected using *functional gages*. A functional gage makes inspection faster and easier than using free-standing manual inspection methods. Design and fabrication costs for gages are recovered when the gages are used to inspect large numbers of parts. If production quantities are not large enough to justify the cost of gages, then free-standing inspection methods can be used.

A shaft with an axis straightness tolerance can be inspected with a functional gage that contains a hole equal in diameter to the virtual condition of the shaft. See Figure 5-21. The gage hole diameter of 1.015″ for the shown shaft is determined by adding the shaft MMC of 1.010″ to the axis straightness tolerance of .005″. This is the theoretical size requirement for the hole in the gage.

The gage only verifies the straightness tolerance. The size limits of .990″ and 1.010″ must be verified by *snap gages* or diameter measurements made with a *caliper* or *micrometer*.

The functional gage used to check axis straightness of a hole is a perfect form shaft that has a diameter equal to the virtual condition of the hole. See Figure 5-21. The gage pin diameter of .985″ is determined by subtracting the .005″ axis straightness tolerance from the .990″ MMC of the hole. The .985″ diameter is the theoretical size required for the gage pin.

No attempt has been made to explain how to assign gage tolerances to the theoretical gage dimensions previously

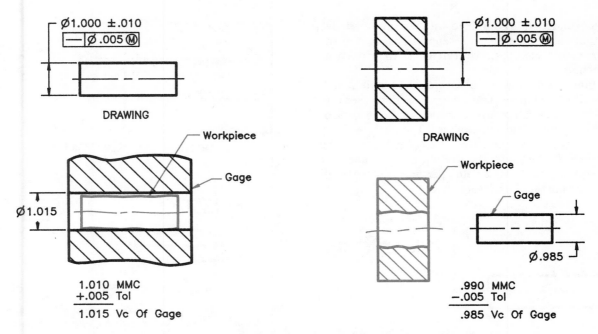

Figure 5-21. Functional gages sized at the virtual condition of a feature can be used for inspection if the form tolerance is specified with an MMC modifier.

defined. Dimension and tolerance calculations for gages is an extensive subject and is not required to understand dimensioning and tolerancing of production parts.

The use of fixed size functional gages is generally limited to inspection of tolerances that are specified with the MMC modifier. If the RFS modifier is used (or assumed), then parts can be verified using free-standing manual inspection methods or gages with adjustable features. See Figure 5-22. Computer-driven *Coordinate Measurement Machines (CMM)* can be used to verify tolerances specified at MMC or RFS.

Verification of axis straightness with a free-standing setup can be accomplished by measuring surface errors on opposite sides of the cylinder. Care must be taken to observe locations at which measurements are being taken, since the axis error is calculated from the surface measurements. In Figure 5-22, the surface errors are such that the axis error is calculated to be one-half the total surface error.

It is necessary to take multiple sets of measurements to determine the full amount of axis error. Attempting to determine axis location from surface measurements can be frustrating, and the accuracy of the calculated axis error is subject to being questioned.

OVERALL AND UNIT LENGTH CONTROL

Situations sometimes occur making it necessary to control axis straightness within a small tolerance zone. Care must be taken to make the tolerance zones no smaller than is needed since small tolerance zones increase part cost. If the tolerance zones become too restrictive, the parts can't be produced.

Achievable tolerances are affected by the proportions of the part. A relatively small straightness tolerance can be achieved on a shaft that has a small length-to-diameter

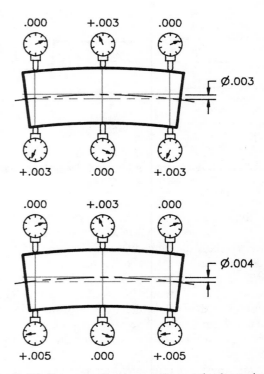

Figure 5-22. Free-standing inspection methods can be used when functional gages are not available.

ratio. A relatively large straightness tolerance is needed for the full length of a shaft that has a large length-to-diameter ratio.

Sometimes it is necessary to confine a segment of a long shaft to a small straightness tolerance. In these situations, the part can be made more producible by using a combined overall and unit length straightness tolerance. Figure 5-23 shows a liberal straightness tolerance applied to the full length of a cylinder. It is combined with a more restrictive tolerance that only applies to a unit length. This permits the tolerance zone that effects the overall length to be made as large as possible.

A combined specification of overall and unit length straightness is placed in one feature control frame. See Figure 5-23. Only one straightness tolerance symbol is shown. The feature control frame has two lines. The top line shows the overall straightness tolerance. The second line of the feature control frame shows the straightness tolerance per unit of length. This tolerance value per unit of length will always be less than the overall tolerance.

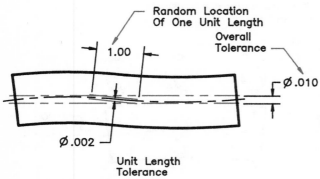

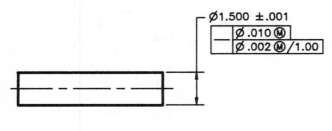

Figure 5-23. Overall straightness and straightness per unit length can be combined.

The given example shows an overall straightness tolerance of .010″ diameter at MMC. The tolerance per 1.00″ of unit length is .002″ diameter at MMC. This means that along any 1.00″ of length on the shaft, it must be straight within .002″ diameter at MMC. If the MMC modifiers are not shown, the tolerances would be assumed to apply RFS.

The unit length across which the straightness tolerance is applied may be any value. The given example shows 1.00″. It could be 1.50″ or any other dimension dictated by the function of the part.

A straightness per unit length can be specified without any overall straightness tolerance, but the practice has few applications. It would appear in a feature control frame with one line. If straightness per unit length is given without an overall straightness, the unit length straightness is permitted to accumulate across the full length of the part. Design applications usually require that an overall tolerance be specified to control the accumulation of the unit length tolerance.

Verification Of Unit Length Control

A method for verifying overall straightness and unit length straightness tolerance requirements is shown in Fig-

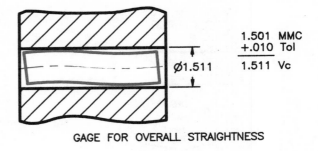

GAGE FOR OVERALL STRAIGHTNESS

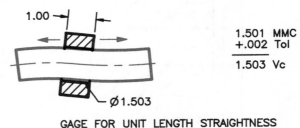

GAGE FOR UNIT LENGTH STRAIGHTNESS

Figure 5-24. Two functional gages can be used to check the requirements of overall and unit length straightness.

ure 5-24. The given example is based on the dimensions in Figure 5-23. The gages and calculations shown in Figure 5-24 are only applicable to tolerances specified at MMC. Overall axis straightness is verified by designing a gage that encloses the entire shaft length. The gage diameter is equal to the virtual condition of the shaft.

The axis straightness per unit length is verified in a similar manner to overall straightness. The gage diameter is calculated to equal the virtual condition across the unit length distance. In the given example, the gage diameter of 1.503″ is determined by adding the 1.501″ MMC of the shaft and its .002″ axis straightness tolerance.

The gage length must equal the specified unit dimension that is shown in the feature control frame. For the given figure, the length is 1.00″. The full length of the shaft must slip through the gage without binding or getting stuck. This verifies the .002″ tolerance zone at all 1.00″ lengths along the shaft.

FLATNESS

Flatness tolerances are only used to specify a form control for flat surfaces. The flatness symbol is not used to control the flatness of the center plane for a feature of size. According to the standard, a straightness symbol is used for this purpose. To be in compliance with the standard, a straightness symbol is used to control a center plane.

No material condition modifier is applied to flatness since it is a surface control. No datum references are shown in a flatness tolerance specification.

APPLIED TO A SURFACE

Flatness tolerance specifications only include the flatness tolerance symbol and the tolerance value. See Figure 5-25. Since the controlled feature is a flat surface, no material condition modifier is shown. The feature control frame is attached to the surface with a leader, or it is placed on an extension line from the surface.

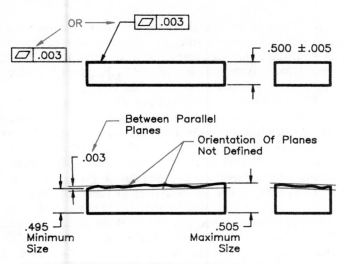

Figure 5-25. Flatness tolerances create a tolerance zone bounded by two parallel planes.

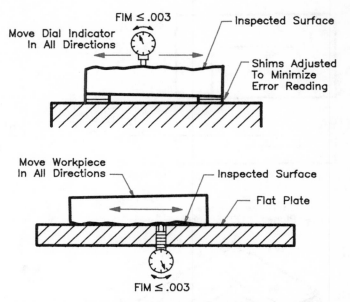

Figure 5-26. The orientation of the flatness tolerance zone is not controlled except that size tolerances must be met.

The tolerance zone is bounded by two parallel planes separated by a distance equal to the tolerance value. It is not necessary for the tolerance zone to be in a fixed orientation relative to the part. The tolerance zone can be at any angle provided the part surface does not violate the size limits.

The requirement for perfect form at MMC controls the surface flatness within the size limits. This means that a flatness tolerance should only be applied if the surface form must be better than is required by the size limits.

The given part has a total size tolerance of .010″. Because of the requirement for a perfect form boundary at MMC, the flatness errors on the top surface cannot be more than .010″. A flatness tolerance specification of .003″ is attached to the top surface to refine the required level of control on that surface.

As shown in the given figure, one end of the part may be produced at .495″, and the other end at .505″. These variations in size are within the tolerance limits. The top surface is also shown to be flat within the .003″ tolerance zone. This part configuration is acceptable. Any other configuration that meets both the size limit requirements and the flatness tolerance is also acceptable.

The flatness tolerance specification applies to the entire surface. It simultaneously controls the surface in all directions. This is different from straightness, which controls the feature in only one direction.

Verification Of Surface Flatness

Flatness of a surface can be verified by running a dial indicator across the surface. See Figure 5-26. The part is rested on an inspection tool such as a surface plate. The dial indicator can be held in a height gage or dial indicator stand and moved across the surface. The height of the dial indicator is not important provided it is kept at one height. It is only important that the indicator stay in one plane as it is moved across the workpiece.

If a flatness tolerance of .003″ is specified, then the full indicator movement must be .003″ or less. Should the first attempt to verify flatness show too much error, the part

orientation can be changed and another check of the surface made. The orientation should be adjusted to minimize the indicator movement. Part orientation can be adjusted since flatness does not control orientation. Changing the part orientation and repeating the surface measurement is an iterative process.

Adjusting the part to obtain the optimum flatness reading can be difficult. To avoid this problem, another method may be used to check flatness. See Figure 5-26. The inspection tool is a dial indicator installed in a hole on a flat plate. The dial indicator probe is positioned to extend above the surface of the flat plate. The workpiece is placed over the dial indicator probe, and the workpiece is moved. The full indicator movement is equal to the amount of flatness error on the surface and must not exceed the specified flatness tolerance.

OVERALL AND UNIT AREA CONTROL

A flatness tolerance applied to a surface extends across the entire surface. It is possible to apply an overall flatness tolerance in combination with a flatness control per unit area. See Figure 5-27. The combined overall and unit area specification is completed in a similar manner to the one used for straightness per unit length.

A flatness requirement of .010″ is specified across the top surface of the given part. The requirement only affects the top surface. The second line of the combined flatness specification requires that flatness be controlled within .003″ across any 1.00″ square area on the surface. The feature must meet the size requirements in addition to the two flatness requirements.

CONTROL OF A CENTER PLANE

A straightness tolerance can be applied to a feature of size defined by parallel flat surfaces. This controls the center plane of the feature. See Figure 5-28. The feature control

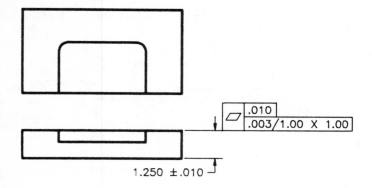

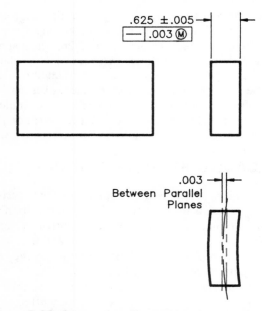

Figure 5-27. Flatness per unit area can be combined with overall surface flatness.

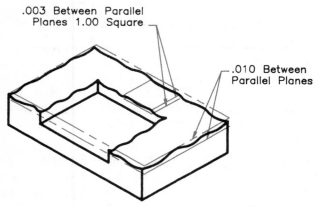

Figure 5-28. Center plane flatness is toleranced by placing the feature control frame, showing a straightness tolerance, adjacent to the dimension value.

frame is placed adjacent to the dimension value or on the dimension line. Showing the feature control frame in one of these positions indicates that the center plane must be straight in all directions within the given tolerance value.

Application of a straightness tolerance to a feature of size does not directly control the surface conditions. The surfaces can vary to any size within the given size limits if the

center plane stays within the specified straightness tolerance.

When a straightness tolerance is applied to a flat feature of size, the tolerance is assumed to apply at RFS. If it is desired to apply the tolerance at MMC, the modifier must be applied as it is shown in the given figure.

The given figure has a maximum size limit of .630″. The straightness tolerance applied to the feature of size permits .003″ error when the feature is at MMC. The combined effect of the form tolerance and MMC is a virtual condition of .633″.

Care must be taken to apply straightness tolerances in a manner that achieves the desired control. Placement of the feature control frame on the dimension results in control of the center plane. This type of application has a virtual condition equal to the MMC size and the effect of the tolerance. Placement of the feature control frame on an extension line from a surface only controls the form of one surface. This type of application has a perfect form boundary at the MMC size of the part, and the form errors must be contained within that boundary.

Verification Of Center Plane Flatness

Flatness of a center plane when a straightness tolerance is specified at MMC may be inspected with a functional gage. The gage must simulate the virtual condition of the feature being inspected. See Figure 5-29.

The given figure has a straightness tolerance of .003″ at MMC. This value plus the MMC size of .630″ results in a virtual condition of .633″. The gage to check the center plane flatness for the given part must have a slot size of .633″. The gage slot must be large enough to totally enclose all of the feature being inspected.

If the straightness tolerance requirement is specified at RFS, then a functional gage of a fixed size is not acceptable. When RFS is applicable, individual readings can be taken on

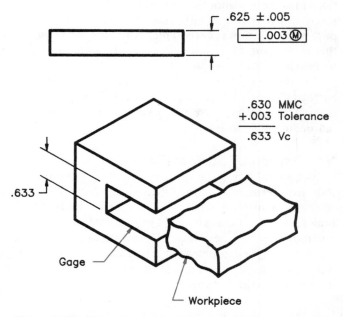

Figure 5-29. Flatness applied to control the center plane at MMC can be verified with a functional gage.

both sides of the part and the center plane location calculated as was done for axis straightness in Figure 5-22.

CIRCULARITY

Circularity tolerances control the roundness of any feature that has a circular cross section, including cylinders and cones. Although a cone does not have a constant diameter, it does have a circular cross section. Circularity tolerances are often referred to as roundness tolerances.

APPLIED TO A CYLINDRICAL SURFACE

Circularity is only applied as a control of surface form. See Figure 5-30. The feature control frame is always attached to the surface of the part and not associated with the size dimension. Since the tolerance is not attached to a size feature, no material condition modifier is shown. The tolerance is most clearly specified when a leader is used to attach the feature control frame to a circular view.

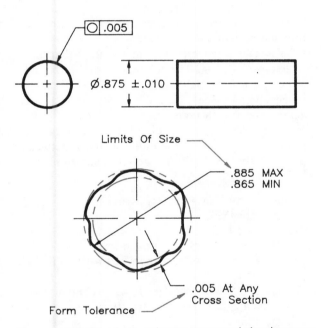

Figure 5-30. Circularity tolerances control circular cross sections.

A diameter dimension for a cylinder has a perfect form boundary at MMC. This means that circularity errors cannot be greater than is permitted by the size tolerance. The given figure has a perfect form boundary at .885″, the MMC size. Because of the .865″ minimum size limit, it is possible for surface errors as large as .020″ to occur inside the perfect form boundary. Circularity tolerance should not be specified unless it requires better control of the surface conditions than is required by the size tolerance.

The tolerance zone for a circularity tolerance is bounded by two concentric circles. The tolerance value is the *radial distance* between the circles. It is not the diameter difference. The radial difference between the circles is half the diameter difference. This means that a circularity tolerance

specification of .005″ is bounded by two circles that have a diameter difference of .010″.

The given part has a perfect form boundary of .885″ diameter. The surface variations must not violate this boundary. All diameter measurements across the part must be within the size limits of .865″ and .885″.

The surface errors must be contained within the circularity tolerance boundary created by two concentric circles separated .005″ radially. The circles forming the boundary are concentric, but are not required to be centered on the axis of the shaft. The tolerance zone boundaries may be any diameter, provided they are separated by .005″ and the part surface does not violate the size dimension limits.

The circularity tolerance controls the form of the surface at each cross section. The errors at each cross section are considered separately from all others. It is possible that two cross sections meet the circularity tolerance, and also have the tolerance zone at those cross sections offset in different directions from the axis.

APPLIED TO A CONE

The form of the circular cross section of a cone can be controlled with a circularity tolerance. The tolerance zone at each cross section is bounded by two concentric circles. However, the diameter of the tolerance zone boundaries change according to the location along the cone axis. Circularity has no affect on the control of the cone angle or the cone size limits. It only controls the circular cross section. The feature control frame can be attached to the surface using either a leader or by placement on an extension line.

CYLINDRICITY

The form of a cylindrical surface can be controlled by a *cylindricity* tolerance. It can be applied to a shaft or hole. See Figure 5-31. This tolerance type is, in effect, a combination of circularity and straightness tolerances. It also controls

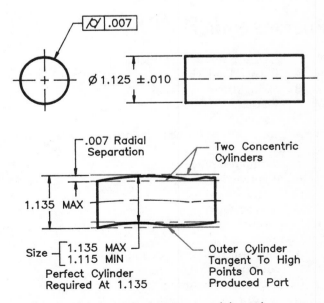

Figure 5-31. Cylindricity tolerances result in a tolerance zone bounded by two concentric cylinders.

parallelism of the sides of the cylinder to prevent the part from being tapered. Material condition modifiers are not included in the feature control frame since cylindricity is a surface form tolerance.

Two concentric cylinders create the cylindricity tolerance boundary. The concentric cylinders are radially separated by a distance equal to the tolerance value.

The given figure shows a shaft with a 1.135″ maximum size, which is the diameter of the perfect form boundary. The minimum size limit of 1.115″ establishes the smallest permitted cross-sectional measurement. Since there is a cylindricity tolerance of .007″, there cannot be cross-sectional measurements at both limits of size.

If the largest cross-sectional measurement is 1.132″, then the diameter of the cylinder forming the outside of the cyl-indricity tolerance must be 1.132″. The diameter of the inside cylinder of the tolerance zone is .014″ less than the large zone (.007″ radially). This means the inside cylinder has a 1.118″ diameter. No part of the surface may extend inside this cylinder.

The requirements of the size limits and the form tolerance must both be met. It is not possible to have cross-sectional measurements of 1.135″ and 1.115″ for the given part, since these measurements would mean the form tolerance (cyl-indricity) has been violated. Any measurements greater than 1.135″ or less than 1.115″ are unacceptable regardless of how perfect the cylindrical shape.

Cylindricity tolerances may only be applied to cylindrical features such as holes and shafts. It is not applicable to any other features.

CHAPTER SUMMARY

The four form tolerances are straightness, flatness, circularity, and cylindricity.

Straightness is the only form tolerance that may be applied to control an axis or center plane.

Limits of size control part form. There is a perfect form boundary at the maximum material condition.

The perfect form boundary at MMC does not control the orientation or relationships between multiple features.

The feature control frame for a form tolerance never includes a datum reference.

Form tolerances are assumed to apply at RFS when applied to features of size.

One form tolerance may be applied to either a surface or a feature of size. The way in which the feature control frame is applied to the part indicates the type of feature being controlled.

Straightness, when applied to a surface, controls single line elements in one direction.

Flatness applied to a surface creates a tolerance zone bounded by two parallel planes.

Circularity can be specified on any feature that has a circular cross section. The tolerance zone is bounded by two concentric circles.

The circularity tolerance value is the radial distance between the concentric circles that form the tolerance zone.

Cylindricity can only be specified on cylindrical features. The tolerance value is the radial distance between concentric cylinders that form the tolerance zone.

REVIEW QUESTIONS

Answer the following questions on a separate sheet of paper. Do not write in this book. Accurately complete any required sketches.

MULTIPLE CHOICE

1. _____ is not a form tolerance.
   A. Perpendicularity
   B. Straightness
   C. Flatness
   D. Cylindricity

2. A feature control frame for a form tolerance must contain a _____.
   A. datum reference
   B. diameter symbol
   C. tolerance value
   D. All the above.

3. Each form tolerance specification applied on an extension line _____.
   A. only controls surface conditions
   B. controls an axis or center plane
   C. may control two surfaces
   D. is a drawing error

4. A feature must have perfect form when _____.
   A. it is at MMC
   B. it is at LMC
   C. form tolerances are not specified
   D. the form tolerance is specified without an MMC modifier

5. A _____ tolerance specification controls the form of line elements in only one direction.
   A. circularity
   B. cylindricity
   C. flatness
   D. straightness

6. A flatness tolerance results in a boundary defined by two _____.
   A. concentric circles
   B. irregular curves
   C. parallel lines
   D. parallel planes

7. Axis straightness for a cylinder can be specified on a drawing by placing the feature control frame _____.
   A. adjacent to an extension line
   B. on the centerline
   C. adjacent to the diameter dimension
   D. None of the above.

8. _____ can be used to control the cross-sectional shape of a cone.
   A. Circularity
   B. Straightness
   C. Cylindricity
   D. Flatness

9. A tolerance zone bounded by two concentric cylinders is established when _____ is/are specified.
   A. cylindricity
   B. circularity and straightness
   C. circularity
   D. Both A and B.

10. A form tolerance only need be applied to a surface if the needed amount of surface control is _____ the amount of the size tolerance.
    A. more than
    B. equal to
    C. less than
    D. unrelated to

11. Verification of a form tolerance is _____ if the tolerance is applied to a feature of size and has an MMC modifier.
    A. ignored
    B. checked only with coordinate measuring machines
    C. not necessary at MMC
    D. possible to be checked with functional gages

12. Application of a straightness tolerance on a cylindrical feature of size results in a/an _____.
    A. impossible to check part
    B. virtual condition
    C. drawing error
    D. envelope equal to the MMC size

13. A cylindricity tolerance specification defines the _____ distance between two concentric cylinders that define the tolerance boundary.
    A. radial
    B. diameter
    C. cone
    D. axial

TRUE/FALSE

14. Cylindricity tolerances are always required since the size tolerance on a diameter dimension does not control the shape of the cylinder. (A)True or (B)False?

15. Bar stock is not required to have perfect form at MMC, but thin-walled tubing and sheet stock must have perfect form at MMC. (A)True or (B)False?

16. Surface variations on the top surface of a flat plate must not be greater than the size tolerance. (A)True or (B)False?

17. An axis straightness tolerance on a hole results in a virtual condition that is smaller in diameter than the MMC of the hole. (A)True or (B)False?

18. An MMC modifier is required on a form tolerance only if the tolerance is applied to control the axis or center plane of a feature of size. (A)True or (B)False?

19. A form tolerance must be smaller than the size tolerance only if the feature is a cylinder. (A)True or (B)False?

20. Two flatness tolerance requirements, in a single feature control frame, can be placed on one surface if one tolerance is applied to the entire surface and a smaller tolerance is applied to unit areas on the surface. (A)True or (B)False?

21. The concentric circles that form a circularity tolerance zone have a diameter difference equal to the tolerance value. (A)True or (B)False?

FILL IN THE BLANK

22. Size dimensions control the size and _____ of the dimensioned feature.

23. Only a _____ form tolerance may be applied to a feature of size.

24. The MMC size of a hole is the _____ limit of size for the hole.

25. _____ may be specified to control the allowable straightness error for a specific unit length on a shaft.

26. Verifying that parallel line elements are straight in one direction on a surface does not verify flatness, since flatness requires measurements in more than one _____.

27. The _____ symbol must be placed in front of the tolerance value when straightness is applied to control straightness of an axis on a shaft or hole.

28. The abbreviation FIM stands for _____ _____ _____.

29. A form tolerance applied to a flat surface does not require a specific _____ to any other feature on the part.

30. A feature control frame is placed _____ to indicate a surface requirement.

SHORT ANSWER

31. Describe how the virtual condition for a shaft is calculated when the shaft has an axis straightness tolerance applied and the tolerance specification includes an MMC modifier.

32. Name the four form tolerances.

33. Explain why form tolerances are used instead of reducing size tolerances to control surface variations.

34. Show the note that must be applied to a size dimension when a form tolerance greater than the size tolerance is being applied to a surface.

35. An MMC modifier may only be used on a form tolerance if it is applied to which kind of feature?

36. Explain what a bonus tolerance is.

37. Sketch the feature control frame to show a straightness tolerance that requires a .008″ diameter zone when the part is produced at MMC.

38. Sketch a feature control frame that requires a surface flatness of .011″.

APPLICATION PROBLEMS

Some of the following problems require that a sketch be made. All sketches should be neat and accurate. Each problem description requires the addition of some dimensions for completion of the problem. Apply all required dimensions in compliance with dimensioning and tolerancing requirements.

39. Apply a feature control frame to control the flatness of the top surface of the given part to .006″.

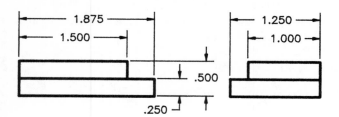

40. Apply a straightness tolerance of .012″ diameter to control the axis straightness of the .563″ diameter. Also apply a straightness tolerance of .008″ to control the axis straightness of the .785″ diameter. Show the MMC modifier on both tolerances.

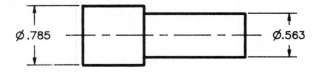

Use the following drawing for questions 41, 42, and 43.

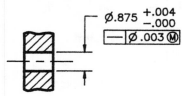

41. Using the given hole specification, calculate the virtual condition of the hole. Show your calculations.
42. The given hole specification permits .003″ axis straightness error at MMC. Calculate the allowable straightness tolerance for a hole produced at .877″ diameter. Show your calculations.
43. What is the MMC size for the given hole?

Use the following drawing for questions 44, 45, and 46.

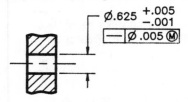

44. Using the given hole specification, calculate the virtual condition of the hole. Show your calculations.
45. The given hole specification permits .005″ axis straightness error at MMC. Calculate the allowable straightness tolerance for a hole produced at .629″ diameter. Show your calculations.
46. What is the MMC size for the given hole?

47. Sketch the given part to show the possible surface conditions for the top surface. Show the size tolerance and form tolerance zones.

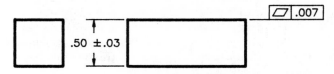

48. Sketch one possible effect of the given tolerances. Show the virtual condition boundary and show the axis straightness tolerance zone.

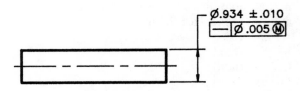

49. Show a straightness tolerance to control the cone elements to within .009″.

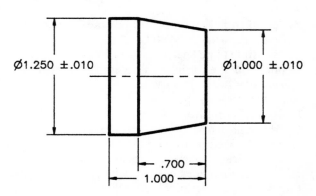

50. The first attempt to verify the required .004″ flatness tolerance on the given part resulted in readings of .005″. The orientation of the part was adjusted, and the second attempt to check the part resulted in a reading of .003″. Explain whether or not the part is good.

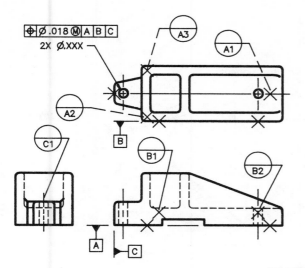

Chapter 6

Datums and Datum References

CHAPTER TOPICS

- ☐ Differences between datums and datum features.
- ☐ Characteristics of a datum reference frame.
- ☐ Methods for identifying datum features.
- ☐ How datum references are made in a feature control frame.
- ☐ Target points, lines, and areas for establishing datum planes and axes.

- ☐ Primary, secondary, and tertiary datum references and how they are simulated.
- ☐ Rules regarding the application of material condition modifiers on datum references.
- ☐ The effect of material condition modifiers on datum references.

INTRODUCTION

Tolerance specifications can be used to define relationships between features. Examples include position, orientation, and profile tolerances.

The feature control frames that define these types of tolerances include *datum references*. See Figure 6-1. The datum references indicate which *datum features* establish the theoretical *datums* from which measurements are made.

Three similar sounding but significantly different terms appear in the preceding paragraph. These terms are datum references, datum features, and datums. It is important to understand the difference between them.

Datum references are the letters shown in a feature control frame. The tolerance requirement in the feature control frame is actually referenced to the theoretical datums.

Datum features are physical features on a part. Datum features are identified on a drawing by the use of datum

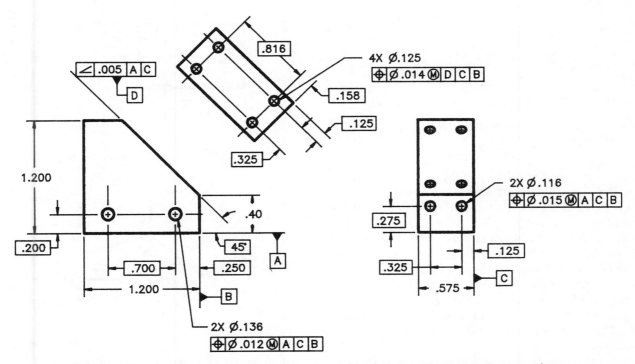

Figure 6-1. Datums and datum references make it possible to define relationships between features.

feature symbols combined with a datum feature triangle. A hole can be a datum feature since it is something that can be seen or touched. Centerlines and theoretical axes are not identified as datum features since they do not physically exist.

A datum is different than a datum feature. A *datum* is a theoretically perfect point, line, or plane. These theoretical geometric entities are located by the physical datum features that are identified on the drawing.

Although datums cannot be touched or seen, they are an important part of tolerance application and verification. Tolerance specifications reference theoretical datums. Datum features are identified on the part to define the features which are used in the location of the datums from which to make measurements.

Since measurements are made with respect to the theoretical datums, the datums must be simulated with tools or inspection equipment. The *datum simulators* provide something real from which measurements can be made.

In Figure 6-1, the bottom surface of the part is identified as datum feature A. Three feature control frames in this figure reference datum A. These feature control frames require that tolerance zones be established relative to datum plane A.

DATUM REFERENCE FRAME

Position, orientation, runout, and concentricity tolerances almost always include references to datums. An occasional exception is possible. See Figure 6-2. Profile tolerances may or may not include datum references, depending on the required level of control.

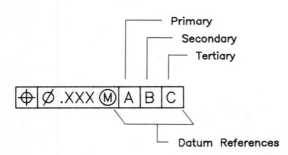

Figure 6-2. Datum references are included on the right end of the feature control frame. They are read from left to right.

Datum references are shown on the right end of the feature control frames. Depending on the tolerance specification and the needed level of control, there may be one, two, or three datum references.

The shown order of letters in the feature control frame signifies datum precedence. The letter of the alphabet used to identify a datum does not affect precedence.

Datum references shown in a feature control frame establish a datum reference frame. See Figure 6-3. *A datum reference frame* is made up of three mutually perpendicular planes. These planes locate the three axes that form a X, Y, and Z coordinate system.

Planes in a datum reference frame are always mutually perpendicular. Although a perfect datum reference frame is

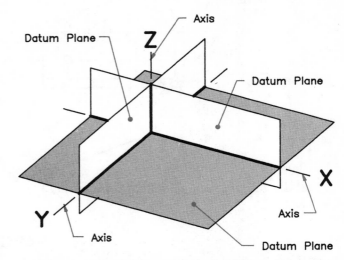

Figure 6-3. A datum reference frame is theoretically perfect and is made of three mutually perpendicular planes.

located by imperfect datum features, the feature imperfections are not permitted to affect the datum reference frame. The planes remain mutually perpendicular regardless of the conditions of the datum features.

Datum references shown in a feature control frame are chosen according to which part features are used to establish the datum reference frame. See Figure 6-4. Selection of the features to use is normally driven by the functional requirement of the part, but the impact of datum feature selection on fabrication and inspection methods must also be considered.

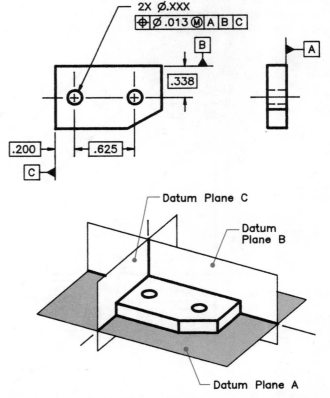

Figure 6-4. Datum references made in a feature control frame determine how a part is located in the datum reference frame.

Three surfaces on the given part are selected to act as datum features A, B, and C. A feature control frame shows the position tolerance that is applied to the two holes. Datums A, B, and C are referenced in the feature control frame. The tolerance specification, as shown, requires that the three datum features be used to locate a theoretically perfect datum reference frame.

There are advantages to having a perfect datum reference frame from which to make measurements. One advantage is that surface variations on the part will not impact the locations of holes. Another advantage is that measurements can always be made relative to a known coordinate system. This is in contrast to making measurements from surfaces that may not be perfectly perpendicular.

DATUM IDENTIFICATION

Datum features must be identified if datum references are to be made in tolerance specifications. Although standards prior to the 1982 issue permitted datums to be implied, it is no longer acceptable. Even when working to an old standard, it is not advisable to use implied datums because of the possible misunderstandings that can occur.

SYMBOLS

A *datum feature symbol* is combined with a datum feature triangle to identify surfaces and features of size as datum features. See Figure 6-5. The datum feature symbol is a square containing a letter to identify the datum. It is attached to a triangle which is applied in a manner that indicates a datum feature. The symbols are only applied to physical features and are not attached to centerlines, axes, or other theoretical entities.

Datum feature symbols are applied to actual features because of the clear definition of requirements that this provides. They are not applied to centerlines on a drawing

because of the confusion that can be caused. An ambiguous (and wrong) application of a datum identifier is to place it on the centerline of a counterbored hole. Such placement does not indicate which feature–the hole or counterbore–establishes the datum. It is important to know which feature locates the datum axis since the axes of the counterbore and hole may not coincide on the produced part. If the counterbore is not perfectly centered on the hole, then the two features will have separate axes.

A *datum target symbol* is used to identify datum targets. See Figure 6-6. Datum targets are used when it is not desirable or possible to use an entire feature to establish the datum.

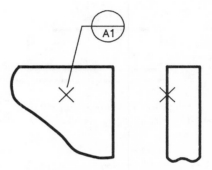

Figure 6-6. The leader on a datum target symbol does not include an arrowhead.

The datum target symbol is a circle with a horizontal line across it. The bottom half of the circle is used to identify the datum target. The top half is left empty except when specifying the diameter of a datum target area.

FLAT SURFACES (PLANES)

Flat surfaces are commonly used as datum features. See Figure 6-7. A datum feature symbol placed on an extension line from a flat surface identifies that surface as a datum feature. The datum feature symbol may also be attached directly to the surface.

A flat surface identified as a datum feature establishes a datum plane. If the surface of the produced part isn't perfect, then the high points on the surface establish the location of the theoretical datum. The theoretical datum can be simulated by placing the part on a very accurate surface, such as a surface plate. A tooling surface or surface plate is normally considered accurate enough to be used as a *datum simulator*.

CYLINDRICAL FEATURES (AXES)

Cylindrical features, such as shafts and holes, are commonly used as datum features. See Figure 6-8. A datum feature symbol placed in line with the diameter dimension for a cylinder defines that cylinder as a datum feature.

When a cylinder is used as a datum feature, the cylindrical feature locates a datum axis. This theoretical datum axis

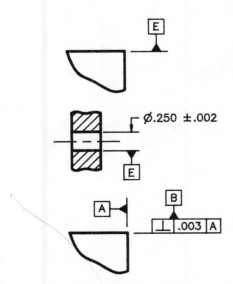

Figure 6-5. A datum identifier can be placed on an extension line in line with a dimension, combined with a feature control frame, or attached to a feature.

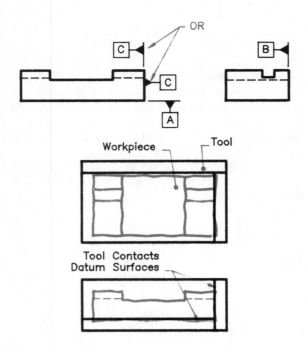

Figure 6-7. Application of a datum feature symbol on either an extension line or a leader has the same meaning.

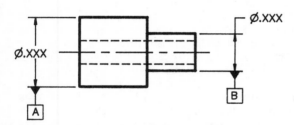

Figure 6-8. Datum feature symbols are applied to physical features and not to centerlines.

is not to be identified by placing a datum feature symbol on the centerline on the drawing. The cylinder must be identified as the datum feature since this physical feature will be used in production and inspection. The centerline can't be used since it isn't a physical feature. As an example, a lathe chuck clamps on a physical feature such as a cylinder; it does not clamp on the centerline of a drawing.

Holding a cylinder in a chuck or collet locates the axis of the cylinder, and thereby locates the datum axis. The centerline is not identified as the datum feature since it cannot be physically contacted.

TARGETS

It is not always practical to identify an entire feature as a datum feature. A very large feature, such as the outside diameter of a missile section, is not practical for use as a datum feature. Features subject to distortion, such as castings and weldments, are also poor features to use in their entirety.

Targets are used when the whole feature is not to be used as a datum feature. See Figure 6-9. The types of targets that may be used are points, lines, and areas.

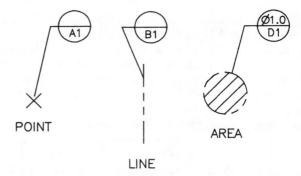

Figure 6-9. There are three datum target types—point, line, and area.

Targets are identified with a *datum target symbol*. This symbol is a circle bisected by a horizontal line. A target label is placed in the bottom half of the circle. Target labels include a letter and a number. The letter identifies the plane or axis. Each target used to establish a single plane or axis is given a unique number. If datum plane D is established by three points, they are labelled D1, D2, and D3.

A leader is extended from the target identification symbol to the target. See Figure 6-10. The leader does not have an arrowhead. A solid line is used for the leader when the target is on the near side of the feature. A dashed line is used for the leader when the target is on the far side.

In the given figure, the datum feature symbol is shown on an extension line from the surface even though targets are also shown. This is a preferred practice. The datum feature symbol is applied to supplement the targets. It helps clarify the drawing by showing the location of the features on which the datum targets are located.

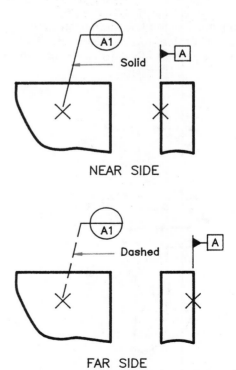

Figure 6-10. A dashed leader from a datum target symbol indicates the target is on the far side of the feature.

Points

Target points on a surface indicate the locations where tooling must make point contact with the workpiece surface. See Figure 6-11. It is possible to create a complete datum reference frame by using targets on three surfaces.

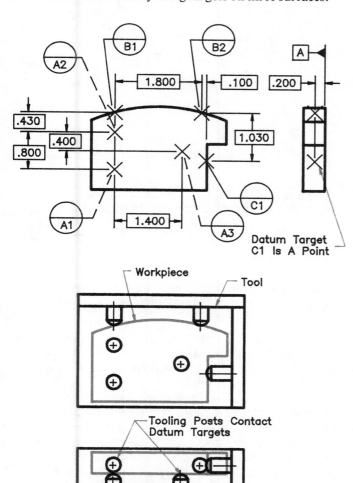

Figure 6-11. Target points can be used to establish all the datums for a part. Target points are commonly picked up by spherical-ended tool posts.

Each target point must be located and labelled. Points forming a single datum reference frame should be located relative to one another whenever possible. In the given example, the dimensions that are provided show the relationship between the targets.

Dimensions used to locate targets may be basic or toleranced dimensions. If target locating dimensions are basic, the national standard indicates that standard toolmaker tolerances apply to the target locations. Some people consider this guideline to be vague since toolmaking standards vary from one company to another. Although not defined by the current standard, a positional tolerance is sometimes placed on the target point positions. If and when this is done, a note should be placed on the drawing to explain what the requirement means. A note is needed since the standard does not define the meaning of a tolerance ap-

plied to a datum target.

A production tool can be used to locate the workpiece by contacting the datum target points. Contacting the datum target points locates the datum planes. When tooling is used to locate the workpiece, the tool must have a feature to pick up each of the identified targets.

Since the targets on the given drawing are points, point contact must be made. This requires the use of a special-shaped tool post. Usually, spherical-ended tool posts are used since they result in point contact, and are not likely to damage the workpiece.

A conical point could also be used to establish point contact, but conical points are typically avoided since a sharp point might *Brinell* (indent) the workpiece. This is especially true if the workpiece is clamped into the tool with a significant amount of force.

Standard spherical-ended tool posts and tooling balls are available and should be used. The usage of standard tooling components is less expensive than requiring fabrication of special tooling parts.

The example in Figure 6-11 has six targets on the workpiece. Six targets are adequate to completely locate and stabilize the workpiece. Targets A1, A2, and A3 establish a plane upon which the workpiece rests. Targets B1 and B2 stop rotation of the workpiece on the datum A targets. Target C1 stops the workpiece from sliding along the datum B targets.

The three point contact on the primary plane, two point contact on the secondary plane, and one point contact on the tertiary plane is standard when using flat surfaces to establish a datum reference frame.

Lines

Target lines on a surface indicate the locations where line contact must be made with the workpiece surface. See Figure 6-12. The given example is identical to the previous figure except that datum target C1 is now a target line. The datum target line appears as a phantom line in the right side view. The end view of the line, as shown in the front view, looks like a target point. Since the end view of a target line appears the same as a target point, it is essential to look at all views of a drawing before deciding if targets are points or lines.

Target lines, like target points, must be located and labelled on the drawing. Target lines used to form part of a datum reference frame should be located relative to other datum targets when possible. In the given example, the dimensions that are provided do show the relationship between the targets.

The tool used for this workpiece is similar to the one in the previous figure, except for the feature that locates datum target C1. Datum target C1 requires a tooling feature that makes line contact. Line contact is achieved in the shown tool by a dowel pin. Contact between the side of the round dowel pin and the flat side of the workpiece results in a line contact.

Line contact along the full length of the target line is only required when the workpiece is perfect. Since it is not likely for the workpiece to be perfect, the dowel pin will not touch all points along the line. Contact along the full length of the

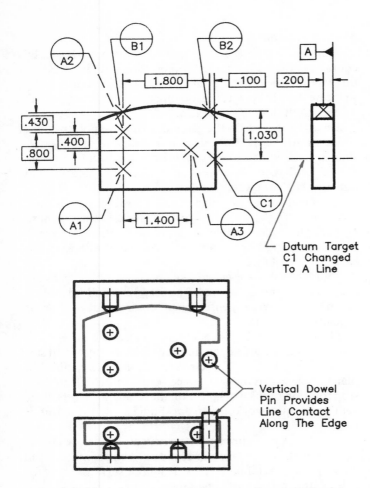

Figure 6-12. A datum target line appears like a target point when seen as an end view. Target lines are picked up by tooling features that make line contact.

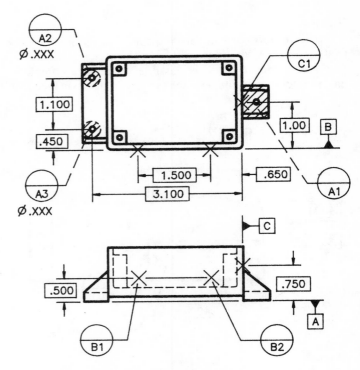

Figure 6-13. Datum target areas are outlined with a phantom line and crosshatched.

line is not required. It is only required that the dowel be located so that contact can be made along the full length if the workpiece is theoretically perfect.

Comparison of the tools in Figures 6-11 and 6-12 shows the one in Figure 6-12 to be simpler in design. This design is also less expensive to produce. When the function of a design allows optional target usage, then target selection should be made in a manner that will lower product cost.

Areas

Target areas on a workpiece indicate the locations where tooling surfaces must contact the workpiece surface. See Figure 6-13. The given example shows three surface areas. They are on the bottom of the mounting feet. The three datum target areas define datum A.

Phantom lines are normally drawn around the perimeter of target areas as shown for targets A2 and A3 on the given part. If the target areas are small, the perimeters may be defined by a continuous line. Perimeter lines must be drawn when the target is smaller than the surface on which it is located.

A phantom line is not needed for target area A1 on the given part since its perimeter is defined by the geometry of the foot. Datum target A1 is all of the bottom surface on the mounting foot. Its perimeter is therefore shown with object lines.

Target areas are normally located by dimensions, just as is required for target points and lines. An exception is when the areas are defined by the limits of a feature, such as target area A1 in the given example.

In addition to giving dimensions to locate target areas, it is necessary to define the size of the area. Round target areas are dimensioned by placing the diameter inside the top half of the datum target symbol. If the diameter dimension does not fit in the available space, it may be directly applied to the area or placed adjacent to the datum target symbol.

Noncircular target area shapes are sized by direct application of the dimensions. The size and location must be clearly defined regardless of the target area shape.

Target areas A1, A2, and A3 in the given example are on the far side of the object, so the leader lines are dashed. As explained earlier, the leader must be dashed to indicate the far side of the object.

Target areas may be any shape or size needed to properly control the design. However, standard tooling components must be considered when selecting the shape and size of the target area. Specifying target areas compatible with standard tooling components will minimize the tool design and tool fabrication costs. As an example, specifying a .500″ diameter target area permits the use of a standard tool post. Specifying a .523″ diameter target area forces the fabrication of a special tool post.

SURFACES AS DATUM FEATURES

Flat surfaces are commonly used as datum features. The means for identifying a flat surface as a datum feature is shown in Figure 6-7. Although continuous flat surfaces do sometimes occur on parts, it is very common for surfaces to have holes and slots in them. These interrupted, noncontinuous flat surfaces are often the features that must serve as datum features.

An interrupted surface can be identified as a datum feature. See Figure 6-14. The datum feature symbol is placed on an extension line from the surface. If interruptions of the surface are very large, an extension line should be placed across the interruption to indicate the continuation of the datum feature.

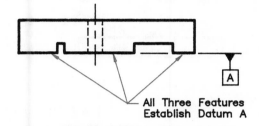

Figure 6-14. Surface interruptions may exist in a datum feature.

A note indicating the number of surfaces may also be placed adjacent to the datum feature symbol when the datum feature is interrupted. The note applicable to Figure 6-14 would state 3 SURFACES.

Using a surface to establish a datum depends on two things. First, the surface must be identified as a datum feature. Second, it must be referenced in a feature control frame or note. When a datum feature surface is identified, that identification does not indicate how the feature is used. It is the reference to the datum, in a feature control frame or note, that determines how the datum is used.

ORDER OF PRECEDENCE

The order in which datum references are made establishes an order of precedence for the datums. Refer back to Figure 6-2. The datum references are always read from left to right. The first datum is primary, the second is secondary, and the third is tertiary.

Datums referenced in a feature control frame establish a datum reference frame. The datum reference frame is located and oriented according to the order of precedence specified for the datums. It is important to remember that the datum reference frame is going to be perfect, regardless of the condition of the datum features.

Two datum reference frames are established if separate feature control frames refer to different datums. Even if one or two common references are shown, one changed datum reference creates a new datum reference frame. As an example, two different datum reference frames are created if one feature control frame references datums A, B, and C, and another references datums B, D, and E.

In addition, two datum reference frames are created when the same datums are referenced in a different order of precedence. As an example, datums A, B, and C establish a different datum reference frame than datums B, A, and C.

Primary Datum

The first datum reference in a feature control frame is to

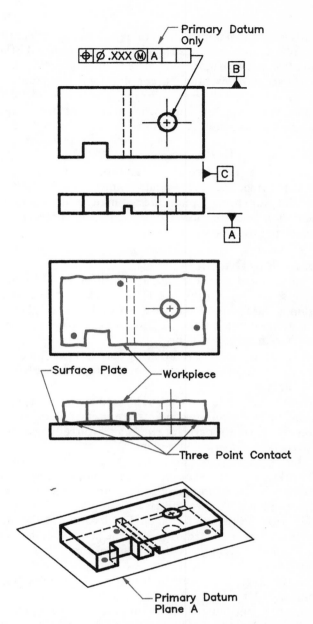

Figure 6-15. A primary datum plane is located by the primary datum feature on a part.

the *primary datum.* See Figure 6-15. The primary datum establishes the location of the first plane of the datum reference frame. The location of the primary plane is important since the other two planes of the datum reference frame are perpendicular to it.

A position tolerance referencing datum A as the primary datum is shown in the given figure. Datum feature A is the bottom surface. A small slot extends across the bottom surface, but the datum feature is assumed to be continuous.

The primary datum plane is located by three high points on the surface. *Three points define a plane.* The location of the three highest points is unknown, so the workpiece is positioned on a flat plate. The workpiece comes to rest on the three high points. The location of the three high points can't be predicted, therefore the flat plate must be at least as large as the datum feature. A smaller plate may not contact the three highest points.

It is possible that more than three points coincide in the same plane, and more than three points may contact the flat plate. Contact with more than three points is acceptable.

Contact with fewer than three points is not acceptable. A convex workpiece may be within tolerance, but would not be stable on the flat plate. If the workpiece is not stable when placed on a flat plate, it is permissible to use shims or other means to stabilize the workpiece. The workpiece must be stabilized on the primary plane.

The workpiece is free to slide and rotate on the primary plane, provided that three point contact is maintained. A secondary and tertiary plane are required to fully establish orientation and location requirements that prevent any movement on the primary plane.

Secondary Datum

The second datum reference in a feature control frame is to the *secondary datum*. See Figure 6-16. The secondary datum feature establishes the location of the second plane of the datum reference frame. The secondary plane is perpendicular to the primary plane.

A position tolerance referencing datum B as the secondary datum is shown on the given part. Datum feature B is the back surface.

The secondary datum plane is located by two high points on the surface. Only two points are needed to orient the plane relative to the part since the plane is already known to be perpendicular to primary plane A.

The locations of the two highest points are unknown, so the workpiece is pushed against an angle block or perpendicular plate while maintaining three point contact with the primary plane. The workpiece will stop when contact is made with the two high points. The location of the high points can't be predicted, therefore the perpendicular plate must be at least as large as the datum feature.

It is possible that more than two points will contact the plate. Contact with more than two points is acceptable. Contact with fewer than two points is not acceptable. Contacting only one point would permit the workpiece to rock, therefore orientation of the workpiece would not be established. If surface curvature results in only one point making contact, or should the two points be close together, then steps must be taken to stabilize the part. It is permissible to use shims or other means for stabilization. Stabilizing the workpiece simulates having two point contact.

Workpiece orientation is defined when the primary and secondary planes are established. Although workpiece orientation is defined, it is still free to move while in that orientation. The tertiary datum will establish location of the workpiece.

Tertiary Datum

The third datum reference in a feature control frame is to the *tertiary datum*. See Figure 6-17. The tertiary datum feature establishes the location of the third plane of the datum reference frame. The locations of the primary and secondary planes are already established, and the tertiary plane is perpendicular to both of them.

A position tolerance referencing datum C as the tertiary

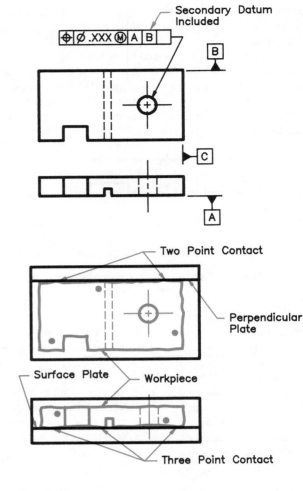

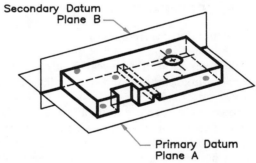

Figure 6-16. The secondary datum plane is at a 90° angle to the primary datum plane, and is located by the secondary datum feature on the part.

datum is shown on the given part. Datum feature C is the right end.

The tertiary datum plane is located by one high point on the surface. Only one point is needed since tertiary plane C only needs to be located. It is already known to be perpendicular to the primary and secondary planes.

The location of the highest point is unknown, so the workpiece is pushed against an angle block or perpendicular plate while maintaining three point contact with the primary plane and two point contact with the secondary plane. The workpiece will stop when contact is made with the highest point. The location of the high point can't be predicted,

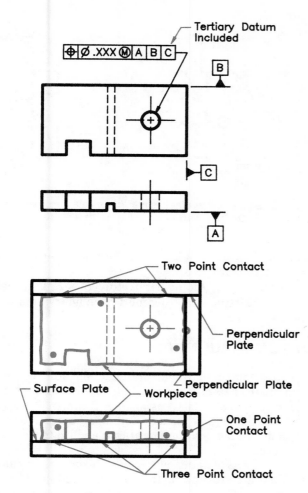

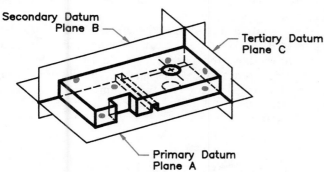

Figure 6-17. The tertiary datum plane is perpendicular to the primary and secondary planes, and is located by the tertiary datum feature on the part.

therefore the perpendicular plate must be at least as large as the datum feature.

It is possible that more than one point will contact the plate. Contact with more than one point is acceptable.

Complete location and orientation of the workpiece is obtained when the primary, secondary, and tertiary datums are established.

DATUM SIMULATION

Datum simulation methods must be used to locate theoretical datums from datum features. Simulation is required since fabrication and inspection operations must be made in relationship to the specified datums. The simulation can be accomplished with a tool-quality piece of hardware, or it may be done through computer calculations based on data obtained with coordinate measuring machines (CMM) or numerical-controlled (NC) machines.

Datum simulation is a means of approximating the theoretical location of the datums. Datum simulation is meant to locate the datum on the basis of the workpiece datum feature. See Figure 6-18. The datum feature has variations caused by inaccuracies in fabrication methods. Because of the inaccuracies, the feature itself is not an exact equivalent of the datum.

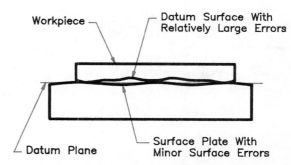

Figure 6-18. Accurate tooling components placed against the datum features on a workpiece simulate the theoretical datums.

Datum features on a workpiece are used to locate datum planes. A datum plane is located by the high points on a datum surface (feature). Location of the plane is approximated by resting the datum surface on a datum simulator. The datum simulator is a highly accurate device such as a surface plate or angle block. For most situations, the high degree of accuracy in these devices permits measurements from them as if the simulators are the datum planes.

DATUM FEATURES OF SIZE

Features of size are often identified as datum features. It is common to use cylindrical features, such as holes and shafts as datum features of size. Rectangular slots, rails, and tabs are also used as datum features of size.

Placement of the datum feature symbol in line with a size dimension indicates that the feature of size is the datum feature. When a datum feature of size is identified, the theoretical datum is an axis, centerline, or center plane.

CYLINDRICAL FEATURES

Shafts, holes, counterbores, and other cylindrical items can be identified as datum features of size. See Figure 6-19. The datum feature symbol is normally placed in line with the diameter dimension of the cylindrical feature, but it may also be placed on a leader that extends to the circular view of the feature. The datum feature symbol should not be placed on an extension line from the surface.

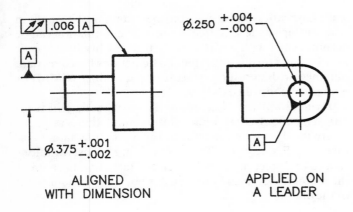

ALIGNED
WITH DIMENSION

APPLIED ON
A LEADER

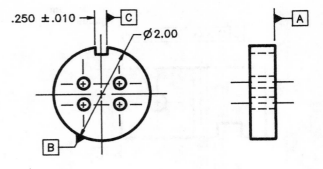

Figure 6-20. A center plane is identified as a datum by placing the datum feature symbol in line with the feature width dimension.

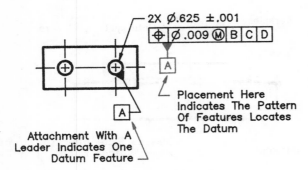

LEADER POINTING TO ONE
FEATURE OF SIZE

Figure 6-19. A datum feature of size is identified by associating the datum feature symbol with the size of the feature.

The given figure shows three examples of cylindrical datum features. In the first example, the small diameter on a stepped shaft is identified as datum feature A. Datum feature A establishes the datum axis.

The second example shows a hole that is identified as a datum feature of size. The datum identification triangle is attached to the hole. This indicates that the hole establishes a datum axis.

The third example shows a part with two holes. A datum feature symbol is connected to one hole with a leader. This identifies only one hole as a datum feature. Placement of the datum feature symbol on the leader of the callout for the two holes or on the feature control frame would establish the hole pattern as the datum feature instead of a single hole.

PARALLEL SURFACES

Slots, rails, and tabs are often used as datum features of size. See Figure 6-20. The given figure shows the width of a keyseat as a datum feature of size. The datum is the center plane of the keyseat.

The datum identifier is placed in line with the dimension. It is incorrect to show the datum identifier on the centerline.

Placement of the datum identification symbol on an extension line from one side of the keyseat would not have the same meaning as its placement on the dimension. Place-

ment on the extension line would indicate a single surface as a datum feature.

To identify a rectangular feature of size as a datum feature, the datum identifier must be in line with or on the dimension line.

DATUM SIMULATION

The method used to simulate a datum feature of size depends on the datum reference. Two important factors must be considered. One factor is whether the reference is a primary, secondary, or tertiary datum. The other factor is which material condition modifier is applied to the datum reference.

The impacts of datum precedence and material condition modifiers are explained in the following sections. Figure 6-21 provides a brief look at how a shaft referenced as a datum feature might be simulated.

If the shaft in Figure 6-21 is referenced as a primary datum at RFS, then the shaft can be clamped in a machine chuck to establish the axis location. Another option is to clamp the shaft in a collet. In either case, the axis of rotation for the clamping device acts as the datum axis.

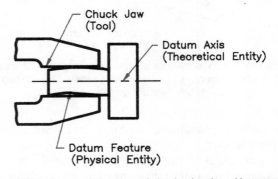

Figure 6-21. A datum feature of size is simulated by a tooling device that picks up the feature of size.

Tolerance requirements on the workpiece must be evaluated to determine the needed level of accuracy for datum simulation equipment. Although a lathe chuck or collet is not perfect, their accuracy is sufficient for many datum simulation needs.

DATUM REFERENCES IN A FEATURE CONTROL FRAME

Identification of a datum feature has no impact on fabrication requirements unless a reference is made to the datum. Generally, datum feature symbols should not be placed on a drawing unless references to the datums are made. Datum references may appear in feature control frames or in notes.

The number of datum references shown in a feature control frame depends on the type of tolerance specification and the level of control being specified. Some tolerance types, such as parallelism, usually need only one datum reference. Position tolerances generally require more than one datum reference. Regardless of the tolerance type, the number of datum references depends on the desired control. Datum reference guidelines provide flexibility for identifying the needed number and type of datums.

Selection of the appropriate number of datums is one step in completing the datum reference portion of a feature control frame. The order in which to reference them must also be determined. The order of precedence impacts how the part is located in the datum reference frame.

ORDER OF PRECEDENCE

References to multiple datums establishes a datum reference frame. If a feature control frame includes references to three datums–A, B, and C–then a single unique datum reference frame is created. Any tolerance specification that references datums A, B, and C in the same order of precedence will be measured relative to the same datum reference frame.

Any tolerance specification that changes the order of precedence for the datums results in a different datum reference frame. See Figure 6-22. By varying the order of three datum references, it is possible to create six datum reference frames. Only one of the six possible datum reference frames is optimum for a specific design application.

Care must be taken to select the correct datum features and to reference them in the most appropriate order of precedence. Datum selection and the order of precedence are determined on the basis of design function and manufacturing considerations.

The given figure shows datum features A, B, and C on a relatively simple part. The design function requires that datum surface A sit flat on a mating part when two bolts are passed through the holes. Making datum A primary ensures that at least three points will be in contact with the mating part. This surface being primary also makes sense for fabrication and inspection of the two holes.

Edges of the part are identified as secondary and tertiary datum features. Datum feature B is selected to be secondary since it is the longest surface and will ensure alignment of parts in the assembly. This surface is also better than datum feature C for stabilizing the part in fabrication.

Selection of datum feature A as primary, and B as secondary, results in datum feature C being the tertiary reference.

The best order of datum precedence for this part is A, B, and C. There are five alternate orders of precedence that could be used, but none of them meet the design or manufacturing needs.

Based on the selection of Datum A as primary, Datum B as secondary, and Datum C as tertiary, the workpiece can be located in a tool as shown in Figure 6-23. Three mutually perpendicular datum simulators are set up for locating the workpiece. Primary datum A must make contact on at least three points. Secondary datum B must make contact on at least two points. Tertiary datum C must make contact on at least one point.

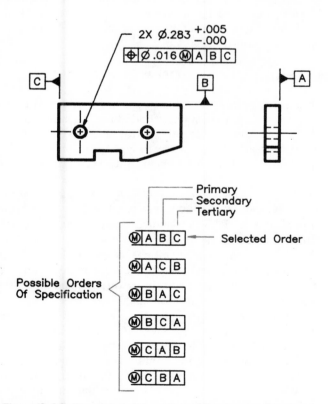

Figure 6-22. The order of precedence for datum references affects the requirements shown in the feature control frame.

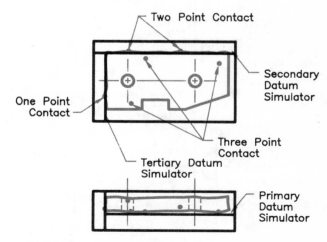

Figure 6-23. The datum order of precedence sets the requirements for how the workpiece must be located in the datum reference frame.

Number Of Datum References

If the datum features on a part are flat features, then it is necessary to reference three datums to establish the three planes in a datum reference frame. See Figure 6-24. Proper specification of a position tolerance for the holes in the given part requires that three datum references be shown. Three surfaces on the part are selected to serve as datum features. The selection of surfaces is made through the process already described. References to the datums are inserted in the position tolerance specification.

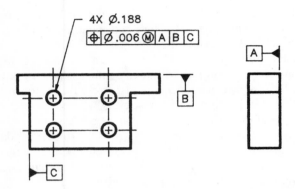

Figure 6-24. Three datum surfaces can be used to establish a datum reference frame.

If one of the datum features is a feature of size, then it isn't always necessary to reference three datums to establish the three planes in the datum reference frame. See Figure 6-25. The given part has a pilot shaft that is identified as datum feature B. Referencing datum axis B as the primary datum establishes two intersecting planes at the axis. Referencing a second datum feature, which is perpendicular to the axis, establishes the third plane in the datum reference frame. In the given figure, only two datum references are used to establish all three planes in the datum reference frame.

Figure 6-26 shows a part on which one surface and two features of size are identified and referenced as datum fea-

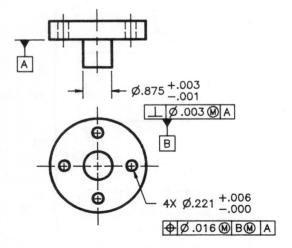

Figure 6-25. One datum feature of size and one datum surface can be used to establish a datum reference frame.

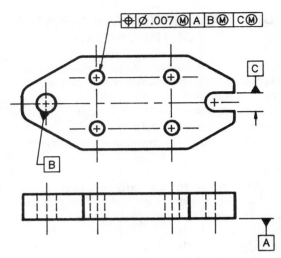

Figure 6-26. One datum surface and two datum features of size can be used to establish a datum reference frame.

tures. In this example, the primary datum feature is a surface. It locates the primary datum plane of the datum reference frame.

The secondary datum feature is a hole. The second and third planes of the datum reference frame are located on the datum axis created by the hole. These planes are perpendicular to the primary plane.

The tertiary datum feature is only needed to establish the rotation position of the planes that pass through the secondary datum axis.

The number of references shown in a feature control frame is affected by the type of datum features and the order of precedence. The required number of references is affected by the type of tolerance being specified, and the desired level of control.

MATERIAL CONDITION MODIFIERS

Material condition modifiers are not used when references are made to datum features that are flat surfaces. See Figure 6-27. Since a flat surface does not have a material condition, it would be incorrect to specify an applicable material condition when referencing the surface as a datum feature.

Features of size do have a material condition. Material condition modifiers are appropriate on references to datum features of size. RFS is assumed on references to datum features of size unless shown otherwise.

Figure 6-27 shows a positional tolerance that includes three datum references. All three datum features are flat surfaces. No material condition modifier is shown on the datum references since they are surfaces.

Regardless of the size of any feature on the part, the surfaces will be placed in contact with flat datum simulators to locate datum planes. The given part will be placed on the surfaces so that three point contact is made on datum feature A, two point contact on datum feature B, and one point contact on datum feature C. The three datum planes are mutually perpendicular, even if one of the datum features is made with some error.

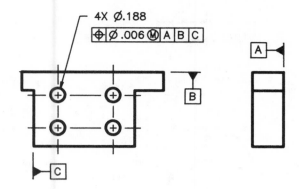

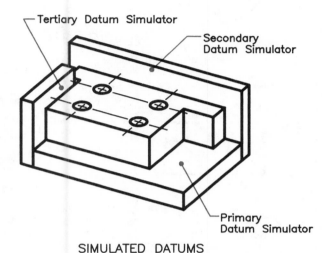

SIMULATED DATUMS

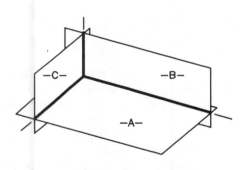

DATUM REFERENCE FRAME

Figure 6-27. Three flat tooling surfaces can be used to simulate the datum reference frame that is created by three datum surfaces.

The three datum planes for the given part are located by the high points on the given surface. Datum plane C is located such that it extends through a portion of the part. This is acceptable since there isn't a requirement to use the outermost surfaces as datum features.

Rule #2

Rule #2 of the national standard requires that the RFS material condition modifier be assumed on all tolerances. This means that a modifier must be shown when references to datum features of size are to apply at MMC or LMC.

Careful attention to the appropriate material condition modifier is needed for each reference to a datum feature of size. Generally, the application of MMC in tolerance specifications can reduce product cost. When design function allows the use of the MMC modifier, it should be used.

RFS On A Primary Datum Reference

Placing a datum feature symbol on a diameter dimension establishes a feature of size as a datum feature. See Figure 6-28. The shaft on the given part is identified as datum feature B. Datum feature B provides a physical feature from which the location of theoretical datum axis B can be determined.

The given figure shows one surface of the flange as datum feature A. The shaft has a perpendicularity tolerance of .003″ diameter relative to datum A, with the tolerance applicable at MMC. No material condition modifier is shown on the datum reference, and none is applicable since datum feature A is a surface.

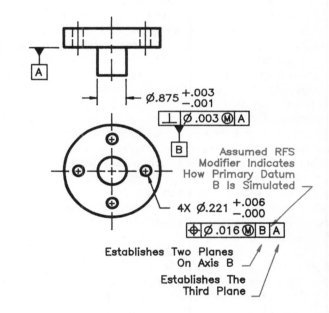

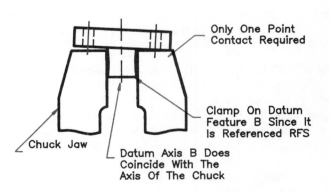

Figure 6-28. A primary datum referenced at RFS requires the datum to be simulated by contacting the surface of the datum feature.

The symbol for datum feature B is attached to the feature control frame that shows the perpendicularity requirement. Showing the datum feature symbol in this manner indicates the same thing as if the datum feature symbol is shown separately.

Four holes have a position tolerance of .016″ diameter at MMC relative to primary datum B at RFS and secondary datum A. RFS is assumed since datum feature B is a feature of size. Showing datum B as primary at RFS has a specific effect. Changing either the order of datum precedence or the applicable material condition modifier does have a different effect.

Datum axis B is located by datum feature B. Since datum B is primary, the axis locates two of the planes in the datum reference frame. Since datum B is referenced at RFS, the datum axis must be located by a device that actually clamps on the surface of the shaft.

A machine chuck (or collet) can be used to simulate the datum axis. The jaws of the chuck are adjusted to make contact with the shaft regardless of its size (RFS). The axis of the shaft will coincide with the axis of the chuck when the shaft is clamped in place.

Hole locations on the shown part can be inspected with the shaft clamped in a chuck, provided the secondary datum is also located. To locate the secondary datum, the shaft would be moved into the chuck far enough for secondary datum feature A to make contact. Point contact with datum feature A might be all that is possible since the shaft and flange may not be perfectly perpendicular.

It is important to realize that secondary datum plane A is perpendicular to primary datum axis B. This is true even if datum feature A is not perpendicular to the shaft. The datum planes must be perpendicular since a datum reference frame is always made of perpendicular planes that define a coordinate system.

Datum axis B on the given part determines the orientation of all three planes in the datum reference frame, and establishes the location for two of them. Datum feature A only locates the third plane.

The given figure shows only one of the several possible position tolerance specifications that could be provided for the four holes. Selecting the proper specification requires consideration of the application for the part. The primary reference to datum B at RFS is appropriate for a situation where the shaft presses into a hole. RFS is required for a press fit between parts, because regardless of the feature sizes, the axes of a hole and shaft will coincide when the parts are pressed together.

MMC On A Primary Datum Reference

Figure 6-29 shows the same part as the previous figure, except the reference to datum B has an MMC modifier. The MMC modifier applied to a primary datum reference has a significant impact on how the datum axis is established. The result is that the datum simulator for datum feature B has a fixed diameter equal to the MMC size.

Three conditions must be met for a datum simulator to be equal to the MMC size of a feature. One condition is that the datum must be referenced as a primary datum. The second condition is that the reference must include an MMC modifier. The third condition is that no axis straightness

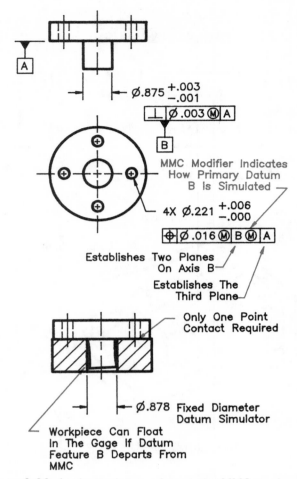

Figure 6-29. A primary datum referenced at MMC requires the datum be simulated by a tool that has a fixed size equal to the MMC of the datum feature.

tolerance be applied to create a virtual condition for the feature. In the given figure, datum B is referenced as a primary datum and the reference includes an MMC modifier. No axis straightness tolerance is shown.

Datum axis B is located by the MMC size of the shaft. The MMC size is .878″ diameter. To locate the axis, the shaft is inserted in a .878″ diameter hole. A shaft made .878″ diameter will fill the hole, and the axis of the hole and shaft will coincide. Any shaft made at a smaller diameter is free to move inside the hole. The amount of movement depends on the amount of departure from the MMC size.

The MMC modifier allows simulation of the datum axis with a fixed diameter gage. This can be easier and less expensive than the axis simulation methods required by a datum reference made at RFS.

The datum references in the given position tolerance in Figure 6-29 define all three planes in the datum reference frame. Datum axis B establishes the location of two planes. Datum A locates the third plane, which must be perpendicular to datum axis B.

Depending on the produced diameter of the shaft, datum feature B may or may not establish orientation of the datum reference frame before datum feature A comes into contact with the datum simulator. Consider a shaft made at .878″ diameter. If the .878″ diameter shaft is made with a perpendicularity error of .003″, datum surface A cannot sit flat on its simulator. This is an acceptable condition based on the tolerance specification showing datum B as the primary datum.

A reference to a primary datum feature of size at MMC is appropriate for some of the possible applications of the given part. An MMC reference is appropriate when a clearance fit exists between mating parts, and assembly of the parts is the main concern.

RFS On A Secondary Or Tertiary Datum Reference

A reference to a datum feature of size includes a material condition modifier regardless of whether the referenced datum is primary, secondary, or tertiary. RFS is assumed for all tolerances.

The given part has a position tolerance with a reference to primary datum A. It does not include a material condi-

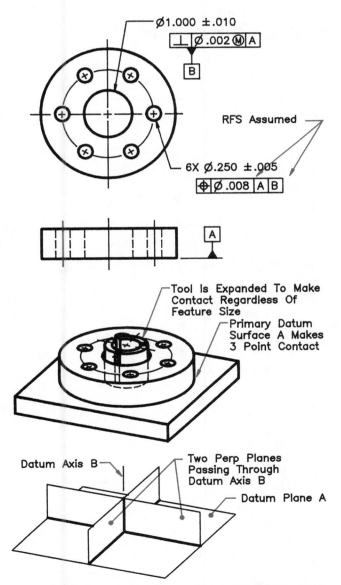

Figure 6-30. A secondary datum referenced at RFS requires the datum be simulated by contacting the surface of the datum feature.

tion modifier since datum feature A is a surface. The reference to the secondary datum B is assumed to be at RFS. See Figure 6-30.

The first plane in the datum reference frame for the shown position tolerance is established by datum surface A. The orientation of the other two planes in the datum reference frame are required to be perpendicular to datum plane A.

Datum feature B is a hole passing through the part. Datum axis B is perpendicular to datum A and located by the hole. Datum axis B locates the two remaining planes in the datum reference frame. Datum axis B must be perpendicular to datum plane A regardless of any errors in the perpendicularity of the hole.

The datum reference frame for the given part can be established through the following datum simulation methods. Primary datum feature A is placed against a flat surface to locate datum plane A. An expanding mandrel extending perpendicular to the flat plate is expanded until it makes contact with the hole. It must expand only until the part cannot slide on the flat plate. This establishes the location of datum axis B. Datum axis B may not coincide with the centerline of the hole.

Care must be taken to expand the mandrel just far enough to stop the part from moving. Expanding it too far will raise the primary datum feature off the datum simulator. If this happens, the datum reference frame isn't properly established.

Secondary datum references at RFS are simulated differently than primary datum references at RFS. A primary datum reference at RFS both locates and orients the datum. A secondary or tertiary datum reference at RFS may only serve to establish datum location. Any datum reference made at RFS requires that the feature surface be contacted to establish the datum reference frame.

MMC On A Secondary Or Tertiary Datum Reference

A reference to a datum feature of size includes a material condition modifier regardless of whether the referenced datum is primary, secondary, or tertiary. If the datum is to apply at MMC, the modifier must be shown.

The given part has a position tolerance with a reference to primary datum A. It does not include a material condition modifier since datum A is a surface. The reference to secondary datum B does include an MMC modifier. See Figure 6-31.

The first plane in the datum reference frame for the shown position tolerance is established by datum surface A. The second and third planes are established by datum axis B. Datum axis B is perpendicular to datum A.

Datum plane A is simulated by a flat plate. Datum axis B is simulated by placing a fixed-diameter pin through the hole. The pin must be perpendicular to the plate, and it must have a diameter equal to the virtual condition of the hole. *A reference to a secondary or tertiary datum at MMC requires that the feature be picked up at its virtual condition, not its MMC size.* The virtual condition is the combined effect of the MMC size and any applied tolerance.

Virtual condition for datum feature B on the given part is

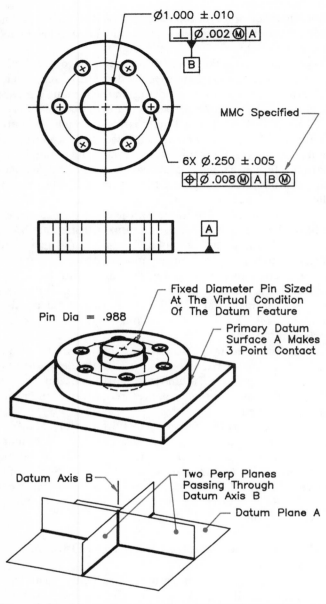

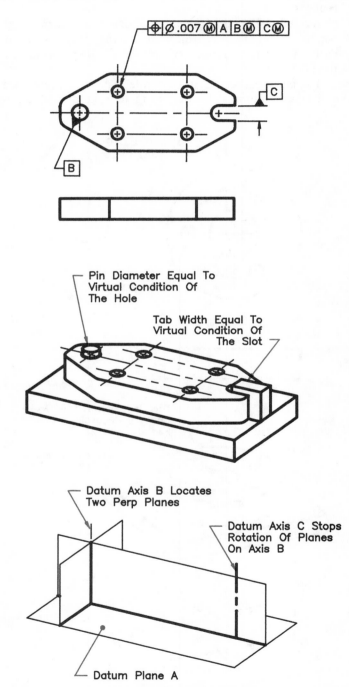

ence frame. See Figure 6-32. Four holes are shown with a position tolerance that is related to datums A, B, and C. Primary datum feature A is a flat surface. Secondary datum feature B is a hole. Datum feature C is the width of the slot. The references to datums B and C include MMC modifiers.

The three datums must be simulated to establish the datum reference frame for this part. Datum A is simulated by a flat surface. Datum axis B is simulated by a fixed-diameter pin. The diameter of the pin must be equal in size to the virtual condition of the hole, since the hole is referenced as a

Figure 6-31. A secondary or tertiary datum reference showing an MMC modifier requires the datum to be simulated by a tool that has a size equal to the virtual condition of the datum feature.

Figure 6-32. A primary datum surface and secondary datum feature of size can be used to establish the datum reference frame with a tertiary datum feature of size used to set the orientation of the datum reference frame.

.988″ diameter. This is determined by subtracting the .002″ perpendicularity tolerance from the .990″ MMC of the hole. The tolerance is subtracted since an out-of-perpendicular hole would appear to be smaller when extending a pin through it.

A condition similar to the previous description for the virtual condition of a hole also exists for external features such as shafts. The virtual condition for a shaft is caused by the effect of the MMC size and any applicable tolerance that tends to make the shaft appear to have an increased diameter. To determine the virtual condition for a shaft, any applicable tolerance is applied to the MMC size of the shaft. Virtual condition is described in detail in following chapters.

It is sometimes necessary to reference one plane and two features of size in order to establish the needed datum refer-

secondary datum at MMC. The pin must also be perpendicular to datum simulator A.

Tertiary datum C is simulated by a flat key. The key must have a width equal to the virtual condition of the slot. It must also be perpendicular to the datum A simulator.

All three of the shown datum references are required to completely stabilize the given part. Datum A only locates the part in one direction. Datum axis B locates two planes that intersect at the axis. These planes are perpendicular to datum A, but are free to rotate on datum axis B. Datum feature C serves to set a required orientation for the two planes.

MULTIPLE FEATURE CONTROL FRAMES RELATED TO THE SAME DATUM REFERENCE FRAME

Multiple groups of features, controlled by single-segment position tolerance specifications, act as a single pattern when the following conditions are met:

1. The feature control frames reference the same datums.
2. The order of precedence for the datums are identical.
3. Each datum reference has the same material condition modifier.

Features may be any size or shape and the single pattern requirement applies if the above conditions are met.

Figure 6-33 shows two drawings of the same part. Both drawings have exactly the same interpretation although the tolerances are specified differently. In the first drawing, the slot is controlled relative to datum A, and datum B at MMC. The slot width is identified as datum feature C. The four hole pattern is controlled by a position tolerance referenced to datum A, datum B at MMC, and datum C at MMC. The

first two datum references for the four hole pattern are identical to those in the feature control frame attached to the slot.

The position tolerance on the four holes includes a reference to datum C. The intent is to establish a rotational position of the holes relative to the slot. Establishing a rotational position is sometimes referred to as *clocking*.

The reference to datum C is not wrong, but it is not necessary. A rotational relationship between the holes and the slot can be achieved without referencing the slot as datum C.

The second drawing shows the same part. Datum C has been omitted from this drawing. Although datum C has been omitted, the relationship between the holes and slot is controlled.

Since the tolerances applied to the slot and to the four holes have the same datum references, in the same order of precedence, and with the same material condition modifiers, the slot and the holes act as a single pattern. Since they act as a single pattern, the rotational position of the holes relative to the slot is controlled.

The principles of groups acting as a single pattern are covered in more detail in the chapter on position tolerances.

DATUM TARGETS BASED ON ORDER OF PRECEDENCE

The datum precedence expressed in a feature control frame must be considered when defining datum targets on a drawing. See Figure 6-34. A two hole pattern on the given part has a position tolerance that references primary datum A, secondary datum B, and tertiary datum C.

Datum targets are used to identify datums on the given part since the cast surfaces are considered too uneven for use

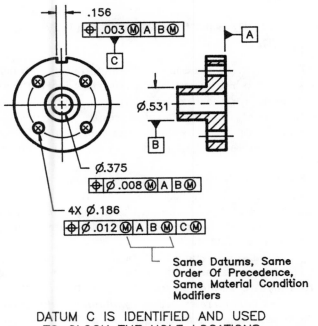

DATUM C IS IDENTIFIED AND USED
TO CLOCK THE HOLE LOCATIONS

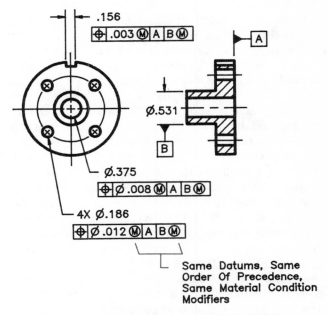

DATUM C IS OMITTED AND THE SINGLE
PATTERN REQUIREMENT ESTABLISHES
CLOCKING LOCATIONS

Figure 6-33. Groups of features toleranced relative to the same datum reference frame act as a single pattern.

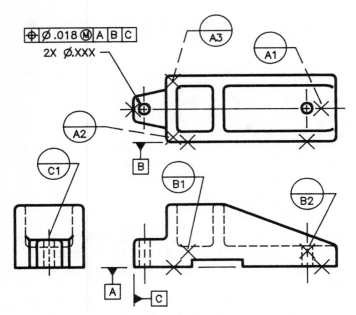

Figure 6-34. The number of targets used for each datum feature depends on the order of precedence for the datum.

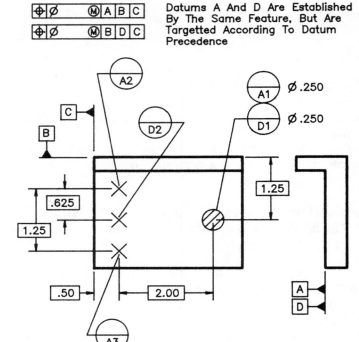

Datums A And D Are Established By The Same Feature, But Are Targetted According To Datum Precedence

Figure 6-35. The same surface may be identified with more than one datum letter when using datum targets.

as datum surfaces. The feature control frames on the given drawings include a reference to primary datum A. Since target points are being used, a minimum of three targets are required for primary datum A.

The given feature control frame includes references to secondary datum B and tertiary datum C. Secondary datum B is identified by two datum target points. Tertiary datum C is identified by one datum target line. The target line could be replaced by a target point if desired.

Regardless of whether target points, lines, or areas are used, the datum precedence must be considered prior to deciding the number of targets to show on a feature.

It is sometimes necessary to use a feature as a primary datum in one tolerance specification, and a secondary datum in another. If datum targets are being specified, the dual use of one datum features requires some special attention. See Figure 6-35. One way of using a single feature for two levels of datum is to identify the surface with two datum letters. One datum letter is to identify the primary reference and the second datum letter is for the secondary reference. **This should only be done when using datum targets.**

Figure 6-35 shows two feature control frames. In the first, datum A is shown as a primary datum. Datum A must therefore be identified by at least three datum target points. The second feature control frame requires the same surface be used as a secondary datum. To accomplish this, the surface is given a duplicate identification–datum D. Datum D is defined using two target points, one of which shares a common location with target point A1.

Although the same datum feature can be referenced at two levels of datum precedence, the same datum letter isn't normally used to indicate both the primary and secondary references when using datum targets. The reason is simple. If two targets are used to identify datum D, and it is mistakenly referenced as a primary datum; there wouldn't be enough points for establishing the primary datum plane.

SPECIAL APPLICATIONS

Datum surfaces, datum features of size, and datum references (as explained in the previous sections of this chapter) meet the needs of many design requirements. However, there are times when more complex design needs must be dimensioned and toleranced. Some special applications of datums and datum references are described in the remaining portion of this chapter.

COMPOUND (SIMULTANEOUS) DATUM FEATURES

Rotating shafts are commonly supported by two bearings. The two bearing surfaces on the shaft locate the axis of rotation for the shaft. Specifying compound datum features permits the design tolerances to reflect the functional needs of a rotating shaft. *Compound datum features* are two features used at the same time to establish one datum.

Figure 6-36 shows how compound datums can be used. Centerdrilled holes on each end of the shaft are identified as datum features. Each centerdrill has a datum feature symbol attached. One is identified as datum feature A, and the other as datum feature B.

Three runout tolerance specifications reference datums A and B in a manner that establishes a single datum axis from the two datum features. This is done by showing a compound datum reference. A *compound datum reference* shows two datum letters separated by a dash. The two datum letters and the dash are contained in a single cell within the feature control frame. Neither datum feature has precedence over the other.

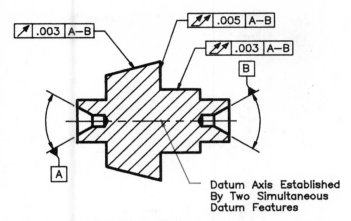

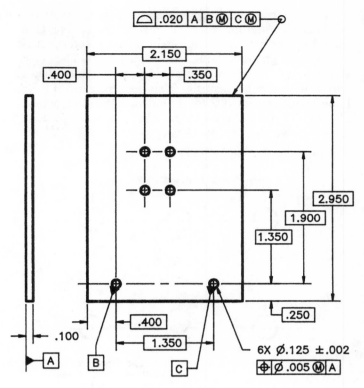

Figure 6-36. Compound datum features establish a single datum.

The datum axis for this part can be simulated by placing a machine center in each of the centerdrilled holes. The axis between the two machine centers acts as the datum axis.

Compound datums can also be used to establish a datum plane. See Figure 6-37. The given part has two slots, one cut in each end of the part. Each slot is identified as a datum feature. The profile tolerance on the curved surfaces reference simultaneous datum B-C as a secondary datum. The two slots are picked up simultaneously to establish the center plane through them.

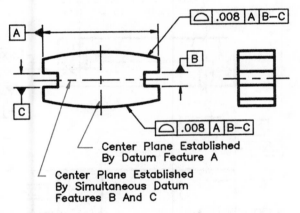

Figure 6-37. Rectangular features of size can be used as compound datum features to establish a datum plane.

A SURFACE AND TWO HOLES

Printed wiring boards and flat panels often include two tooling holes that locate the part during fabrication. See Figure 6-38. These holes can be identified as datum features, and tolerances referenced to them.

The given part has six holes. They are all toleranced relative to datum A. The holes act as a single pattern. Two of the holes are identified as datum features B and C. The perimeter of the part has a profile tolerance that is referenced to datums A, B, and C. This tolerance completely controls the location, orientation, and size of the perimeter with respect to the specified datums. In effect, the outline of the part is controlled to the positions of the holes.

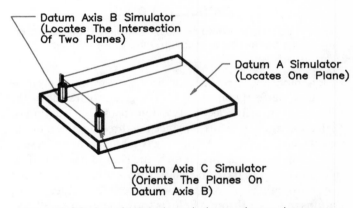

Figure 6-38. A surface and two holes can be used to create and orient a datum reference frame.

The primary and secondary datum in the given example locates all planes in the datum reference frame, but the rotation of the second and third planes are not defined until the tertiary datum is defined. This was previously explained with respect to Figure 6-32.

Figure 6-38 shows that a datum reference frame can be completely defined by one surface and two holes. The surface and holes are physical features on the part. Notice that the datum reference frame is created by referencing features on the part as datum features. Datum feature symbols are not placed on the centerlines.

One significant reason for not placing datum feature symbols on centerlines is that the standard specifically says not to do it. There is no defined meaning for a datum feature symbol placed on a centerline. The absence of a defined meaning should be adequate for not placing datum identifiers on centerlines.

A reason for prohibiting the placement of datum feature symbols on centerlines is the confusion it can cause. Consider the placement of a datum identifier on a centerline that passes through three holes. See Figure 6-39. The figure shows a centerline through three holes, and it is identified as datum A. Since the identifier is on the centerline, it is not known which holes to use for the location of datum plane A. Should all three holes be used, or should only two be selected? If two are used, which two?

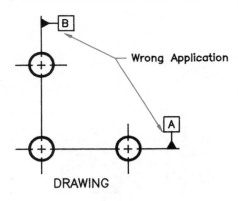

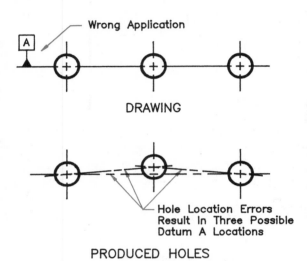

Figure 6-39. It is incorrect to show a datum feature symbol on a centerline that passes through a group of holes.

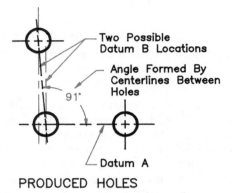

Figure 6-40. Drawing requirements are not clear when datum feature symbols are placed on two perpendicular centerlines.

It is important to realize that three holes are not likely to be produced perfectly on one line. The given figure shows the effect of arbitrarily selecting two holes to establish datum A. (Selection must be arbitrary since the drawing doesn't indicate which holes to use.) Selecting two holes at a time results in three possible datums. Which one should be used? This question does not have an answer since the datum identifier is not properly applied. The correct method for identifying datums that pass through holes is shown in Figure 6-38.

Another common and confusing mistake is the application of datum feature symbols on two perpendicular centerlines. See Figure 6-40. In the given example, two perpendicular centerlines pass through three holes. The centerlines on the drawing intersect at one of the holes, and each centerline passes through one of the other two holes.

Identification of two perpendicular lines that are supposed to run through three holes causes some problems. Holes are seldom, if ever, produced in the exact dimensioned locations relative to one another. It is likely that some error will exist in the location of the produced holes.

The given figure shows three produced holes. They are within the allowable position tolerance relative to one another, but lines drawn between the holes do not form a perfect 90° angle. Since a coordinate system has 90° between axes, all three holes can't be used to establish two planes of the coordinate system.

Two holes are all that are needed to establish and orient the coordinate system. This was shown in Figure 6-38.

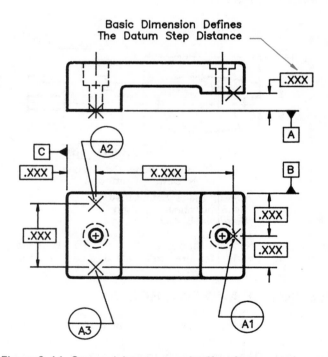

Figure 6-41. Stepped datums permit offset features to be used to locate a single datum plane.

Since two holes establish the origin and orientation of a coordinate system, attempting to use a third hole only raises questions. What is done if the third hole does not rest on the established coordinate system? Should the coordinate system be distorted to some angle other than 90° just to use the specified hole?

It is best to avoid using three holes to define a coordinate system. Using three holes can raise questions for which there are no defined answers in the dimensioning and tolerancing standard.

The problems and ambiguity of using three holes to establish a coordinate system can be avoided by correctly using the methods illustrated in Figure 6-38.

STEPPED DATUM SURFACES

Parts are often designed that must mount on stepped surfaces. Since datum features are normally selected on the basis of how a part is assembled, it is desirable to have a means for using offset or *stepped datum* surfaces. See the example in Figure 6-41.

When stepped surfaces are used to define a datum plane, the step distance is given with a basic dimension. The basic dimension defines the step distance for the tooling that establishes the datum plane. In the given example, datum target points are used to define the stepped datum requirements.

The distance between the surfaces on which the stepped datum targets are located may be indicated with the same basic dimension that defines the distance between the stepped datum targets. The distance between the two surfaces may also be dimensioned with a toleranced dimension.

If the basic dimension defines both the target step distance and the distance between surfaces, a profile tolerance is typically required to define the surface location tolerance. The target location tolerance is not defined on the drawing. Standard toolmaking tolerances are understood to apply for the location of targets dimensioned with basic numbers.

If a toleranced dimension defines the distance between the surfaces, the allowable location of the surfaces is defined by the dimension tolerance. The step distance for the target locations is still shown as a basic dimension, and standard toolmaker tolerances apply for the location of the datum target simulators.

EQUALIZING TARGETS TO LOCATE DATUM PLANES

Castings and forgings are often designed such that datum planes need to be located through the use of equalizing datum targets. *Equalizing datum targets* are those that cause the workpiece to be centered when the targets are contacted. See Figure 6-42. Locating a datum plane with equalizing datum targets may take more than three targets.

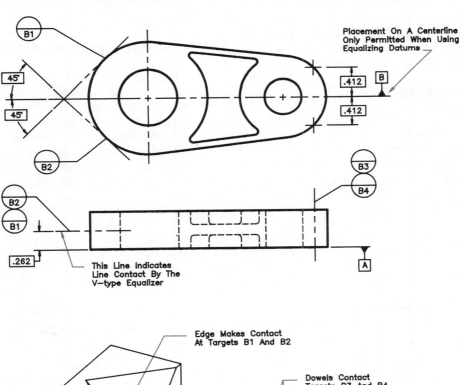

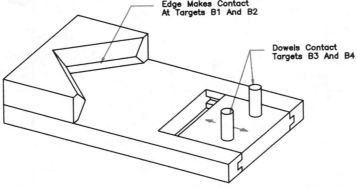

Figure 6-42. Equalizing datum targets are used to locate a center plane or axis.

Four datum targets are used to locate datum B on the given part. Targets B1 and B2 are horizontal target lines at a 90° angle to each other. Datum plane B bisects the 90° angle formed by the target lines. Targets B3 and B4 are vertical target lines. These targets are located equal distances from datum B, and a line running through the two targets is perpendicular to the datum.

Targets B1 and B2 are known to be target lines because of the horizontal lines shown in the front view. Removing the horizontal lines from the front view would change the target requirements to planes at a 90° angle.

The tool used to locate the given part, assuming datum A is primary and datum B secondary, is shown in the given figure. The flat plate simulates datum A. The two vertical dowel pins serve to pick up datum targets B3 and B4. The 90° V-block is free to slide and pick up the two horizontal target lines. The V-block surfaces have horizontal knife edges since line contact is required.

If the horizontal target lines were removed from the front view of the given drawing, then flat surfaces on the V-block would be required.

Drawings containing equalizing datums may include a datum feature symbol on the center plane. This is an exception to the general requirements for application of the datum feature symbol. The datum feature symbol is not to be placed on a centerline or center plane unless equalizing datum targets are used. The targets define the features that establish the datum plane. The datum feature symbol is shown on the centerline only for reference information to clarify the location of the equalized datum.

TARGETS ON DIAMETERS

It is not always possible to use the entire surface of a cylindrical part as a datum feature of size. Irregular parts such as castings or weldments make the utilization of the entire surface difficult. Also, it isn't practical to attempt to use all of a very large surface as a datum feature.

Datum targets can be identified on cylindrical features. See Figure 6-43. The given part has six datum target lines, and they are used to define datum axis A. There are three target lines equally spaced on each end of the part. The targets are identified as A1 through A6. They are shown in both views to provide a clear definition of the target definition.

Targets placed on a cylinder must take into consideration how the targets can be contacted. The part in this figure can be clamped on each end with a three-jaw chuck. Each jaw will make contact with a line along the targeted surface.

When the datum axis is defined by two features having different diameters, then the targets on each feature are given separate datum feature letters. See Figure 6-44. It is possible to use datum target points, lines, or areas to establish a datum axis. They may also be used in combination.

The given part has datum targets B1, B2, and B3 on one feature. On the other end of the part, datum target area A1 is shown. The datum axis for this part can be referenced by placing the simultaneous datum feature reference A-B in a tolerance specification.

An alternate datum targeting method using a target line is shown in Figure 6-44. The target line is shown across the cylinder, but the target actually is a circular element on the surface of the part.

The targets shown in Figures 6-43 and 6-44 are on external features. The same techniques can be applied to internal features.

COMBINED TARGET AREAS AND POINTS

Datums are defined by the means that best reflects the functional requirements of the part and facilitates manufacturing. This often means using datum targets, and sometimes results in combined target types.

Target areas may be combined with points as shown in Figure 6-45. Large target areas may result in multiple point contact at random locations across the area, similar to when an entire surface is identified as a datum feature. Datum

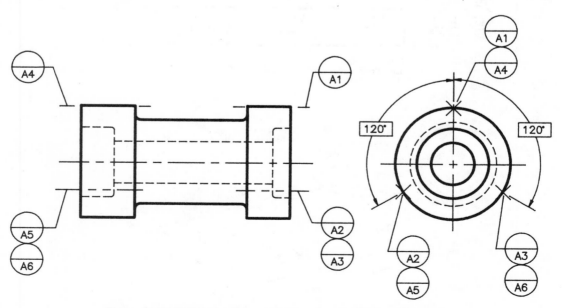

Figure 6-43. Datum targets can be used to establish a datum axis.

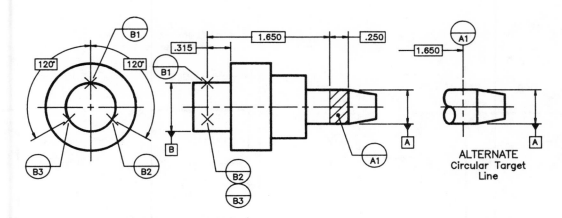

Figure 6-44. Datum target areas can be used to control the area of a cylinder that must be contacted to locate a datum axis.

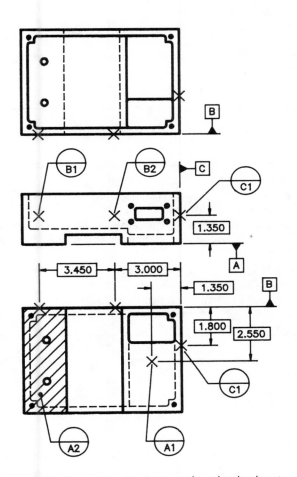

Figure 6-45. Datum target types can be mixed to locate a single datum plane.

target area A2 on the given part is relatively large. It spans the full depth of the part. Combining target area A2 with target point A1 ensures at least three points of contact, should datum A be referenced as primary.

Target lines can be combined with areas, and points can be combined with lines. Whatever combination best represents the functional requirements is appropriate. It is necessary for any selected combination to provide the proper datum location based on the order of precedence for the datum.

DIMENSIONS RELATED TO A DATUM REFERENCE FRAME

As explained in Chapter 4, Rule #1 of ASME Y14.5 requires perfect form of a feature when it is at maximum material condition. The requirement for an individual feature to have perfect form does not require the feature to be in any specific relationship to other features. Relationships between features are normally controlled by orientation, position, runout, or general angularity tolerances.

It is also possible to control feature size and orientation relative to a datum reference frame. See Figure 6-46. This is done by including a note on the drawing. The note can be general and apply to all dimensions, or it can be a local note that is applied to specific dimensions. If a general note is used, make sure that dimensions extend from the datum features. The note must specify the order of precedence for datums to which dimensions are referenced.

A local note is applied to the given drawing. It states that "noted dimensions are related to datum A primary, datum B secondary, and datum C tertiary". The note is placed in a notes list, and the note number is placed adjacent to the applicable dimensions.

The effect of dimensions related to a datum reference frame is the creation of an *envelope*. Envelope measurements are made relative to the datum reference frame. Envelope size is based on the maximum material condition for the dimensioned feature. The envelope has perfect orientation with respect to the datum reference frame.

The requirement for perfect orientation to the datum reference frame can increase product cost significantly above the requirements of Rule #1, which does not by itself force perfect orientation relative to the datum reference frame. Relating dimensions to a datum reference frame should only be used when design requirements dictate that it is necessary.

The impact on the least material condition requirements when dimensions are referenced to a datum reference frame is somewhat controversial. A generally accepted approach is to verify least material limits by direct two point measurement between surfaces. This approach ensures that the part features do not violate the minimum size limit allowed by the dimension tolerance.

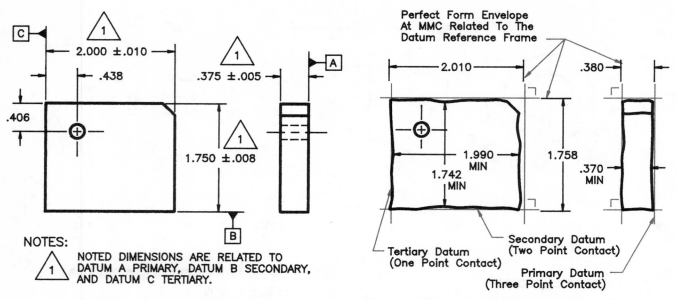

Figure 6-46. Dimensions related by note to the datum reference frame establish an envelope at the MMC size of the dimensioned feature.

CHAPTER SUMMARY

Datum references make it possible to define toleranced relationships to true geometric planes and axes.

A datum feature is a physical feature on a part.

Datum features are identified on drawings.

Physical datum features are used to locate theoretical datum planes and axes.

References shown in a feature control frame are made to the theoretical datum planes and axes.

Datum simulators (tools) are used to locate the theoretical datums from the physical locations of the datum features on a part or object.

Datum references to datum features of size include material condition modifiers. Rule #2 of the current standard defines requirements for showing material condition modifiers on drawings.

Datum feature symbols are not applied to centerlines except when equalizing targets are used.

The number of datum targets used is dependent on the order of precedence for the datum.

Multiple feature control frames that refer to the same datum reference frame establish a single pattern of toleranced features.

A secondary or tertiary datum reference including an MMC modifier results in a requirement that the datum be simulated at its virtual condition.

REVIEW QUESTIONS

Answer the following questions on a separate sheet of paper. Do not write in this book. Accurately complete any required sketches.

MULTIPLE CHOICE

1. _____ mutually perpendicular planes form a datum reference frame.
   A. Two
   B. Three
   C. Four
   D. Both A and B.

2. A datum reference frame can be thought of as the theoretical planes and axes that set up a _____.
   A. coordinate system
   B. inspection system
   C. tooling system
   D. feature control frame

3. A datum surface is used to establish a datum _____.
   A. point
   B. line
   C. area
   D. plane

4. A datum axis of a feature is established from a _____.
   A. hole
   B. shaft
   C. centerline
   D. Both A and B.

5. The leader on a datum target symbol is made _____ if the target is on the far side of the feature.
   A. dashed
   B. solid
   C. phantom
   D. None of the above.

6. A _____ datum is always the first one shown in a feature control frame.
   A. letter A
   B. primary
   C. secondary
   D. tertiary
7. When a datum plane and datum axis are referenced in the same feature control frame, _____.
   A. the plane must be the primary datum
   B. the axis must be the primary datum
   C. one of the two is selected as primary based on design function
   D. None of the above.
8. Identifying a cylindrical feature of size as a datum feature establishes a datum _____.
   A. axis
   B. plane
   C. target
   D. symbol
9. Rule #2 of the dimensioning and tolerancing standard requires that material condition modifiers be shown on _____ tolerances.
   A. form
   B. orientation
   C. all
   D. runout
10. A primary datum reference including an MMC modifier requires the datum feature be simulated at its _____.
   A. virtual condition
   B. maximum material condition
   C. least material condition
   D. produced size
11. When using datum target points to establish a primary datum, at least _____ points must be used.
   A. one
   B. two
   C. three
   D. six
12. Two datum letters separated by a dash within a single cell of a feature control frame indicates _____.
   A. an error
   B. primary and secondary datums
   C. compound datums
   D. all the letters have been used and double letters are now in use
13. A _____ dimension must be shown between offset surfaces that are used to establish a single datum plane.
   A. basic
   B. close toleranced
   C. reference
   D. None of the above.

TRUE/FALSE

14. Regardless of surface errors on a part, the planes forming the datum reference frame are always mutually perpendicular. (A)True or (B)False?
15. A datum axis must not be identified by placing the datum feature symbol on a centerline. (A)True or (B)False?
16. A datum target area is outlined with a dashed line. (A)True or (B)False?
17. The order of precedence for datums in a datum reference frame depends on the order shown in the feature control frame. (A)True or (B)False?
18. The datum references in a feature control frame must be shown in alphabetical order. (A)True or (B)False?
19. A hole can be used as a datum feature of size, but this practice has no practical application. (A)True or (B)False?
20. Rule #2 of the dimensioning and tolerancing standard requires that material condition modifiers be assumed as RFS on all tolerances. (A)True or (B)False?
21. A reference to compound datums means the two features are used to establish two datums. (A)True or (B)False?
22. A datum feature symbol may be applied to a centerline only if the datum is identified by equalizing targets. (A)True or (B)False?
23. The three datum target types are not to be combined to establish a single datum. (A)True or (B)False?

FILL IN THE BLANK

24. The intersections of the planes in the datum reference frame establish __axis__.
25. A __datum__ symbol is used to identify a feature from which a datum is established.
26. Datum targets may be __point__, __line__, or __area__.
27. Datum _____ is determined by the order in which datums are referenced in the feature control frame.
28. To identify a cylinder as a datum feature, a datum feature symbol is placed on the _____ dimension.
29. A datum feature that is a surface can be identified by placing the datum feature symbol on a(n) __extension__ line from the surface.
30. A datum feature symbol applied to the width dimension on a rectangular slot establishes a datum __plane__ at the center of the slot.
31. A secondary datum reference including an MMC modifier requires the datum feature be simulated at its _____ condition.
32. A minimum of __2__ points are required to establish a secondary datum plane when using datum target points.
33. Dimensions may be related to a datum reference frame using a __note__.

SHORT ANSWER

34. Explain the difference between a datum feature and a datum.
35. When basic dimensions are shown between datum targets, what is the tolerance on the location of those targets?
36. How can a datum plane be simulated when the datum feature is a flat external surface?

37. When is it necessary to show material condition modifiers on datum references? *Features of Size*

38. Explain how to establish a datum axis when a hole has been referenced at MMC.

39. The size of a datum target area should not be arbitrary. Explain how at least one consideration impacts the size of a target area.

40. Explain why two targets labelled E1 and E2 shouldn't be referenced as a primary datum.

APPLICATION PROBLEMS

Some of the following problems require that a sketch be made. All sketches should be neat and accurate. Each problem description requires the addition of some dimensions for completion of the problem. Apply all required dimensions in compliance with dimensioning and tolerancing requirements.

41. Identify the top surface on the given part as datum feature A.

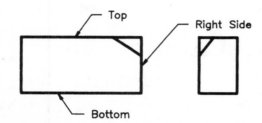

42. Identify the bottom surface on the given part as datum feature B.

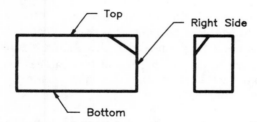

43. Identify the right side surface on the given part as datum feature C.

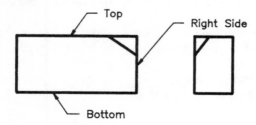

44. Identify the small diameter as datum feature A to establish a datum axis.

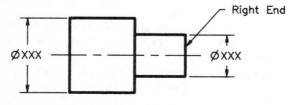

45. Identify the large diameter as datum feature B to establish a datum axis.

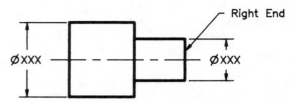

46. Identify the right end surface on the given shaft as datum feature C.

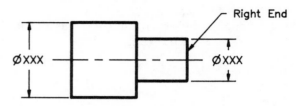

47. Label the primary, secondary, and tertiary datum references in the given feature control frame.

48. Sketch a tool to simulate the three required datum planes and show the part in the tool. Indicate the number of points that must be in contact with each datum simulator.

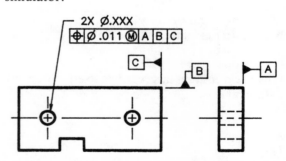

49. Revise the given drawing to include datum target points for all three required datums. Locate the targets in positions that you feel best stabilize the part. Show dimensions and target symbols.

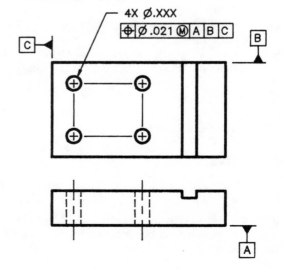

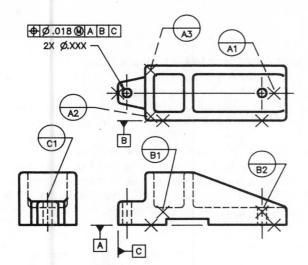

Chapter 7

Orientation Tolerances

INTRODUCTION

Orientation tolerances are used to control the parallelism, perpendicularity, or angularity of a feature relative to one or more datums. They are specified using a feature control frame, and may be applied to surfaces and to features of size. As with other tolerances, the method of application determines whether the tolerance controls the surface or feature of size.

Orientations between features are sometimes controlled by general angularity tolerances that are shown in the title block of a drawing. When this general angularity tolerance does not result in the desired control of a feature, a tolerance can be applied directly to the angle dimension. It is also possible to show the orientation tolerances in a feature control frame. A feature control frame is commonly used to indicate an orientation tolerance that is more restrictive than the level of control already defined by general tolerances that are shown in the title block. It is also possible to use a feature control frame to increase the amount of permitted variation or to simply make the orientation requirement more clear.

Orientation tolerances are always referenced to datums, and therefore have a well-defined origin for completion of measurements.

ORIENTATION TOLERANCE CATEGORIES

Parallelism, perpendicularity, and *angularity* are orientation tolerances. There is a symbol that is designated for each of these tolerance categories. When an orientation tolerance is applied to a flat surface, it produces a tolerance zone that is bounded by parallel planes. The planes forming the tolerance zone are in perfect orientation relative to the referenced datums.

SYMBOLS

Parallelism is indicated by two parallel lines drawn at a 60° angle. See Figure 7-1. Parallelism is only used to indicate a tolerance between parallel features.

Perpendicularity is defined by two perpendicular lines. Perpendicularity is only used on features at a 90° angle.

The angularity symbol is made of two lines drawn at a 30° angle. Angularity is applied to features at any orientation other than parallel or perpendicular.

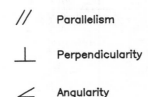

Figure 7-1. There are three orientation tolerances.

FEATURE CONTROL FRAMES

Feature control frames for orientation tolerances are generally drawn in the same format as other tolerances. See Figure 7-2. The tolerance symbol (characteristic) is shown on the left, followed by the tolerance value, and then the datum references. There must always be at least one datum reference in an orientation tolerance specification.

Orientation tolerances can be combined with other tolerances in a multiple-line feature control frame. Figure 7-2 shows a two line feature control frame in which a perpendicularity tolerance is used to refine a position tolerance. The given example has a .012″ perpendicularity tolerance and a .032″ position tolerance.

Although position tolerances have not yet been discussed, a position tolerance controls both location and orientation. If orientation is more critical than location, then an orientation tolerance can be used to refine the control provided by the position tolerance. This permits the position tolerance to be maximized and only controls orientation to the smaller value.

Multiple-line feature control frames are limited only by the needed control for the feature. The previously described feature control frame included a position tolerance that was refined by a smaller orientation tolerance. The orientation tolerance can be further refined with a form tolerance. This is done by adding a third line that shows the form tolerance. Figure 7-2 shows a three line feature control frame. The first two lines are identical to the previously explained example. The third line adds a requirement for the feature to be straight within .004″.

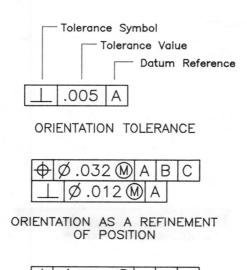

ORIENTATION TOLERANCE

ORIENTATION AS A REFINEMENT
OF POSITION

FORM AS A REFINEMENT
OF ORIENTATION

Figure 7-2. Orientation tolerances can be used to refine positional tolerance requirements. Form tolerances can refine the orientation requirements.

Multiple-line feature control frames may look overly restrictive when first seen. In reality, they maximize each of the tolerance categories (if properly used). Using a single feature control frame to achieve all the controls of the shown three line specification would require a position tolerance of .004″. A position tolerance of .004″ is very small and significantly increases part cost. The given three line feature control frame permits the maximum amount of tolerance for each of the three controls. It permits .032″ of location tolerance provided the perpendicularity is within .012″ and the feature is straight within .004″. It does not unnecessarily control the position to .004″ as a means of controlling straightness.

Single Or Multiple Datum References

Orientation tolerances must reference at least one datum. See Figure 7-3. There are applications when referencing two datums is necessary to obtain the proper level of control.

A reference to a single datum establishes orientation only to that one datum. There is no implied relationship to any other feature. When an orientation tolerance references two datums, the primary datum is the one from which the required angle is measured, and the secondary datum establishes a fixed rotation on the primary datum. This is illustrated in a later segment of this chapter.

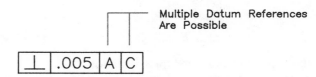

Figure 7-3. Orientation tolerances may include more than one datum reference.

Material Condition Modifier

The tolerance value includes a material condition modifier, either implied or explicitly shown, if an orientation tolerance is applied to a feature of size. See Figure 7-4. *The tolerance is assumed to apply Regardless of Feature Size (RFS) unless a modifier is shown.* Modifiers that may be shown are Maximum Material Condition (MMC) and Least Material Condition (LMC). It is not necessary to show an RFS modifier on orientation tolerances since it is automatically implied when the MMC and LMC modifiers are not shown.

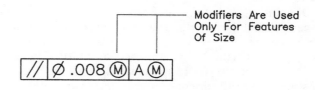

Figure 7-4. A material condition modifier is applicable when the controlled feature or the datum reference is to a feature of size.

Datum references to features of size include a material condition modifier. The datum reference is assumed to apply RFS if the MMC or LMC modifier are not shown.

Material condition modifiers are not shown on tolerances applied to surfaces, nor are they shown on datum references that apply to surfaces. Since a surface does not have a material condition, the application of a modifier would be incorrect.

VIRTUAL CONDITION

When an orientation tolerance includes an MMC modifier and is applied to a feature of size, a virtual condition of the part is established. *The virtual condition is the combined effect of the tolerance and the MMC size of the feature.* The virtual conditions of mating parts must be equal or provide a clearance if the two parts are to freely fit together.

An orientation tolerance applied to an internal feature creates a virtual condition that is smaller than the MMC size. See Figure 7-5. As the orientation error on an internal feature increases, the apparent size (virtual condition) of the opening decreases. The given example shows how a slot appears to get smaller as it is rotated out of the perpendicular orientation.

The virtual condition of an internal feature is calculated by subtracting the orientation tolerance from the MMC size. The virtual condition is, in effect, an envelope inside of which no segment of the internal feature can extend. No part of the sides on the shown slot extend inside the virtual condition boundary.

An orientation tolerance applied to an external feature creates a virtual condition that is larger than the maximum material condition size. This is opposite of the effect caused on internal features. See Figure 7-5. The given example shows how a rail appears to get larger as it is rotated out of the perpendicular orientation.

The virtual condition of an external feature is calculated by adding the orientation tolerance to the MMC size. The virtual condition is, in effect, an envelope outside of which no segment of the external feature can extend.

PARALLELISM

Two features drawn parallel are assumed to have an implied angle of 0° between them unless dimensions are given to specify otherwise. Almost any drawing will disclose several features that are parallel. The tolerances on these parallel features may be controlled by a general angular tolerance, limits of size, or parallelism tolerances. *A parallelism tolerance controls the orientation between two features and does not affect location or size of the features.*

Control of parallel features through a general angular tolerance is acceptable when there is only one set of parallel features. It can become ambiguous when several features are parallel to one another. As an example, several slots on one surface are parallel to each other. Does the angular tolerance accumulate between each slot, or is each tolerance measured relative to some implied reference feature?

The same ambiguity can exist if parallel features are dimensioned with plus and minus tolerances for location and size control. If the limits on the dimensions are used to control orientation, an accumulation of tolerances can occur. Careful dimensioning can reduce the tolerance accumulation and improve orientation control, but this may force a smaller size or location tolerance than is necessary for the function of the part.

Parallelism tolerances permit control of the orientation between specific features without accumulation of tolerances. They specify the required level of control for parallelism without restricting size or location tolerances. Parallelism requirements can be completely separated from size and location tolerances. This is a significant advantage since increased cost due to restrictive size and location requirements is avoided.

The 0° angle between parallel features is assumed to be a basic dimension when a parallelism tolerance is applied.

PARALLELISM APPLIED TO A FLAT SURFACE

Orientation between a flat surface and a datum plane can be controlled with a parallelism tolerance. See Figure 7-6. A feature control frame must be used to specify the parallelism tolerance for each surface that is to be controlled relative to a datum. Application of a parallelism tolerance requires

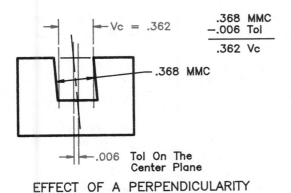

ORIENTATION APPLIED TO AN
INTERNAL FEATURE OF SIZE

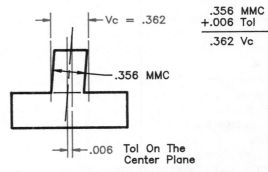

ORIENTATION APPLIED TO AN
EXTERNAL FEATURE OF SIZE

Figure 7-5. A virtual condition is created when an orientation tolerance is applied to a feature of size.

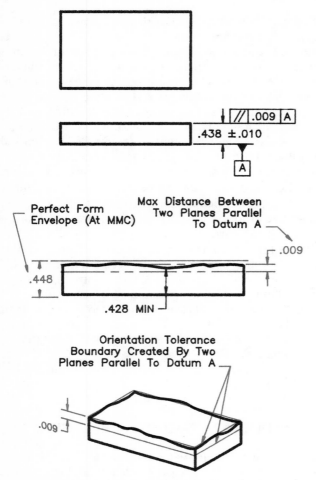

Figure 7-6. Two planes form the boundary for a parallelism tolerance.

that a datum feature be identified, since all parallelism tolerances must reference a datum.

In the given example, the bottom surface is identified as datum feature A. The top surface has a parallelism tolerance of .009″ with respect to datum plane A.

The parallelism tolerance of .009″ is less than the size tolerance. It would be wrong to apply a parallelism tolerance greater than the size tolerance.

A size tolerance between parallel flat surfaces also controls the orientation between those surfaces. This is true since Rule #1 requires perfect form of an individual feature at MMC.

The size tolerance between the top and bottom surface of the part in Figure 7-6 is ±.010″. The total size tolerance is .020″. Since Rule #1 requires perfect form at MMC, the two surfaces must be perfectly flat and parallel when at MMC. If the bottom surface is produced perfectly flat, then the size tolerance permits a maximum surface error of .020″ on the top surface. This means the form and orientation errors are limited to .020″.

If .020″ form and orientation errors are acceptable, then no control other than the size tolerance is needed. Although it is not wrong, it would be redundant to apply a parallelism tolerance that is equal to the size tolerance.

The given figure shows a part on which the parallelism

tolerance is less than the size tolerance (.009″ compared to .020″). *The parallelism tolerance only controls the accuracy of the surface orientation and form. It does not affect the size tolerance.*

The given figure shows that a perfect form boundary is established at the .448″ maximum material size. No portion of the surface may protrude outside this boundary. Datum plane A is located by the bottom surface of the part. The three high points on the surface will locate the datum plane. The parallelism tolerance zone is bounded by two planes separated by .009″. These planes must be parallel to datum A.

The part is acceptable if the surface remains within the perfect form boundary, and also remains within the parallelism tolerance zone. The tolerance zone is permitted to float to any location, provided the minimum size limit is also maintained.

The minimum size limit is verified by direct measurements across the feature. Minimum size measurements are not made from the datum plane, but rather from the part surface.

VERIFICATION OF PARALLELISM ON A FLAT SURFACE

One method for verifying parallelism of a flat surface is to use a dial indicator on a height stand and surface plate. See Figure 7-7. A surface plate is used to simulate datum A. The workpiece is set on the surface plate to establish the datum plane location.

A height stand with a dial indicator is moved across the controlled surface. The dial indicator is kept at a fixed height as it is moved across the entire surface area. If the full indicator movement is less than or equal to .009″, the parallelism requirement is met.

This method for verifying parallelism only checks the orientation requirement. It does not check the size tolerance. Before the part can be accepted, size tolerances must also be verified.

PARALLELISM APPLIED TO A CYLINDRICAL FEATURE OF SIZE

Parallelism can be used to control the orientation of an axis relative to either a datum axis or a datum plane. Figure

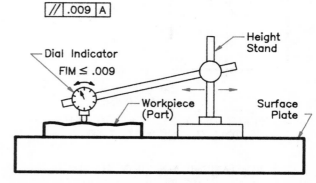

Figure 7-7. Parallelism is only a surface control. It controls orientation, but does not control location.

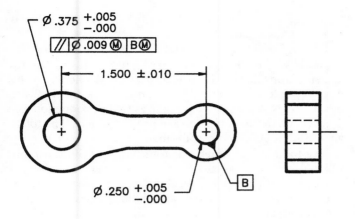

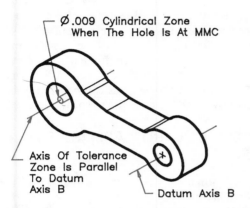

Figure 7-8. Parallelism can be used to control the orientation of one axis to another.

7-8 shows a parallelism tolerance used to control the axis of one hole relative to the datum axis established by another hole.

The given linkage has two holes. One hole is identified as datum feature B. A parallelism tolerance having a diameter of .009″ at MMC relative to datum B at MMC is applied to the second hole. There is a .020″ total location tolerance between the two holes.

The orientation tolerance is separate from the location requirement. Both must be met, but one does not directly affect the other. In addition to the orientation and location requirements, both holes also have limits of size.

The datum axis is established by the hole identified as datum feature B. The tolerance zone is a .009″ diameter cylinder that extends the length of the hole. The cylinder is parallel to the datum axis. The diameter of the tolerance zone is permitted to increase if either the controlled hole or the datum feature depart from MMC.

VERIFICATION OF PARALLELISM APPLIED TO A FEATURE OF SIZE

Functional gages can often be used to check orientation tolerances applied to features of size. Careful attention must be given to the proper design of the gage so as not to force unintended controls on the workpiece. As an example, a gage for the previously shown linkage must check parallel-

ism within .009″ while permitting hole location to vary by .020″.

One possible functional gage for this linkage is shown in Figure 7-9. The gage has two parallel pins; one pin is mounted on an adjustable slide. The slide permits the distance between the pins to be varied, but the pin parallelism is not affected.

The gage pins have fixed diameters since the parallelism tolerance and datum reference include MMC modifiers. The following two paragraphs explain how the pin diameters are determined.

Datum B is simulated by a pin equal in diameter to the MMC of the datum feature (the .250″ hole). On the given linkage, the MMC size of the datum hole is .250″ diameter. (The .250″ diameter is the smallest permitted hole size, and the smallest hole is the MMC size.) A datum feature of size is simulated at the MMC size if the following conditions are met: the datum reference is a primary datum, the MMC modifier is applied to the datum reference, and a form tolerance is not applied to the datum feature.

The second pin on the gage has a diameter equal to the virtual condition of the controlled hole. Any controlled feature of size can be checked by a gage that is sized to the feature's virtual condition if the tolerance value includes an MMC modifier. For a hole, the virtual condition is determined by subtracting the parallelism tolerance from the MMC size. On the given part, the virtual condition of the hole is .366″ diameter. This is equal to .375″ diameter (MMC) minus .009″ (the parallelism tolerance).

The pins on the gage can be inserted into the holes if the linkage is made in compliance with the size limits and the parallelism tolerance. When both holes on the linkage are at MMC, the pins will fit into the workpiece if the parallelism error is no more than .009″.

The effects of the MMC modifiers can be seen through the examination of what happens when the gage in Figure 7-9 is used on several different linkages. Consider a linkage

$$Vc = MMC - Tolerance$$
$$= .375 - .009$$
$$Vc = .366$$

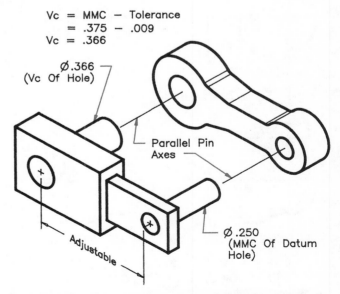

Figure 7-9. Gages used to verify parallelism must not force unrequired location requirements.

that is produced with both holes at MMC. The datum B hole is at .250″ diameter (MMC), and the other hole has a .375″ diameter, which is its MMC size. When the gage is inserted into this part, the datum B simulator fits snugly into the .250″ diameter hole. This snug fit establishes the orientation of the gage. The .366″ diameter pin is .009″ smaller than the .375″ diameter hole, and fits into it, provided there is no more than .009″ parallelism error.

The second example linkage is produced with the datum B hole at MMC, and the other hole has departed from MMC. The datum B hole has a .250″ diameter. The second hole has a .380″ diameter. The .380″ diameter is a .005″ departure from the MMC size of .375″. When the gage is inserted into this part, the datum B simulator fits snugly. The .366″ diameter pin is .014″ smaller than the produced hole, and fits into it, provided there is no more than a .014″ parallelism error. *The departure of a feature from its MMC size results in increased tolerance when an MMC modifier is applied to the tolerance value.*

The third linkage to consider has both holes at a size larger than MMC. The datum B hole has a .255″ diameter, and the other hole has a .380″ diameter. When the gage is inserted into the part, the datum B simulator (pin) is free to move within the datum B hole. The movement of the datum B simulator within the datum B hole increases the amount of allowable parallelism tolerance. The increased diameter of both holes impacts the allowable variation since the tolerance and datum reference includes MMC modifiers.

PERPENDICULARITY

Perpendicularity tolerances are used to specify the orientation requirements of a feature relative to a datum plane or axis that is at a 90° angle to the feature. They are applied to surfaces and to features of size.

Drawing conventions do not require application of 90° angle dimensions on drawings. See Figure 7-10. Any feature drawn to appear at 90° is assumed to be perpendicular on the drawing unless an angle dimension is applied to indicate otherwise. No assumption is made as to the production tolerances for the 90° angle on the produced part. The tolerance must be specified in some manner.

The following are methods for specifying the allowable variation on perpendicular surfaces. A general angular tolerance may be applied in the title block or in a drawing note. A note can also be used to relate all dimensions to a datum reference frame, thereby establishing perfect form and interrelationships at MMC, but this practice is not recommended. Perpendicularity tolerances can also be shown through the application of feature control frames.

The general angular tolerance in the title block is assumed to apply to 90° angles if no feature control frame is applied to specify the orientation requirement. If a feature control frame is applied to control the perpendicularity, then the 90° angle is implied to be a basic dimension. It is not necessary to draw the 90° dimension to show that it is basic.

PERPENDICULARITY APPLIED TO FLAT SURFACES

A perpendicularity tolerance is applied to a surface by placing the feature control frame on an extension line or on a leader from the surface. See Figure 7-11. The tolerance must reference at least one datum. No material condition modifier is applied to the tolerance value when the controlled feature is a surface.

The given figure shows a perpendicularity tolerance applied to the right end of the given part. The tolerance only affects the end of the part to which it is attached. Since the shown tolerance is applied to the right end of the part, it has no direct effect on the left end, top, bottom, front, or back.

The shown perpendicularity tolerance references datum A. The bottom surface is identified as datum feature A. The tolerance specification and identification of the datum feature establishes an orientation control between the indicated surface and the referenced datum plane.

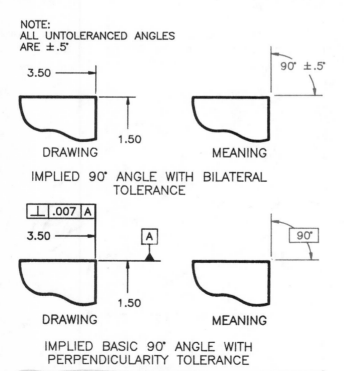

Figure 7-10. Implied 90° angles are assumed to be basic when a perpendicularity tolerance is applied.

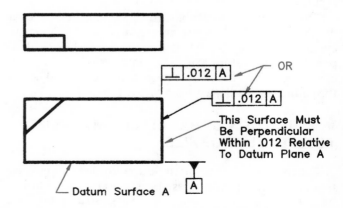

Figure 7-11. Perpendicularity can be specified on a surface by placing the feature control frame on an extension line or leader.

Referenced To One Datum

A perpendicularity tolerance must include one or more datum references. A single datum reference is adequate for many applications. However, a reference to one datum only controls orientation to the indicated datum. There are not any assumed relationships to other datums or features.

Figure 7-12 illustrates a perpendicularity tolerance that references a single datum. The feature control frame is applied to the right side of the part in the given figure. It references datum A, and the bottom surface of the part is identified as datum feature A.

The length of the part is controlled by a size dimension of 2.00″ ± .03″. The perpendicularity tolerance is .005″. The size tolerance and perpendicularity tolerance are separate requirements.

There is a perfect form envelope for the two ends at the MMC dimension of 2.03″. This perfect form envelope only controls parallelism of the two ends, and it does not control perpendicularity of the end surface relative to datum A. Based on only the size dimension, the two ends can have any shape or orientation, provided the ends do not violate the perfect form envelope or the minimum size dimension.

The right end of the part is controlled by more than the 2.00″ size dimension. The shown perpendicularity tolerance establishes orientation control of the right end. This tolerance requires that the surface fall between two parallel planes separated by .005″. The two planes must be perpendicular to datum plane A. Datum plane A is located by the high points on datum surface A.

Perpendicularity of a datum surface can be verified using the method shown in Figure 7-13. An angle block is set on a surface plate. The workpiece is clamped with datum surface A against the angle block.

A height stand is positioned to hold a dial indicator on the controlled surface. The height stand is moved across the surface plate to move the dial indicator across the workpiece surface. Full indicator movement of less than or equal to .005″ shows that the perpendicularity tolerance has been met. The workpiece must be clamped in a fixed location while the surface measurement is being made.

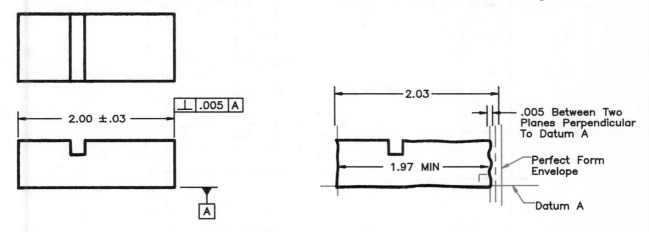

Figure 7-12. The tolerance zone for a perpendicularity tolerance on a flat surface is bounded by two planes.

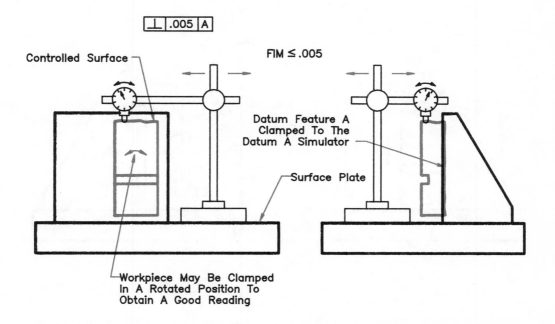

149

If the workpiece fails the first attempt to check the perpendicularity tolerance, it doesn't mean the part is bad. Since only one datum is referenced in the feature control frame, the workpiece can be rotated on the angle block and clamped in place for another measurement.

The workpiece can be rotated on the angle block since perpendicularity is only controlled relative to datum A. The proper relationship relative to datum A is maintained if the workpiece is clamped to the angle block. Perpendicularity to datum A is not affected by rotation of the workpiece on datum A.

If the full indicator movement can't be minimized to .005″ or less, the part is bad. If the readings are brought to within .005″, the perpendicularity requirement has been met.

The amount of rotation on datum A may be controlled by a general angularity tolerance. The general angularity tolerance will control the angle between the sides of the part. If control relative to the sides of the part is to be equal to the perpendicularity tolerance, a second datum must be specified in the feature control frame.

Any surface that lies within an orientation tolerance zone also has a form that is equal to or better than the orientation tolerance. If the surface lies between two planes separated by .005″, flatness has also been controlled to within .005″. It is not necessary to apply a form tolerance to this surface unless a form tolerance less than .005″ is needed.

Referenced To Two Datums

A perpendicularity tolerance can be referenced to two datums. See Figure 7-14. Referencing two datums in a perpendicularity tolerance establishes complete control of the surface orientation. Referencing two datums is a more strin-

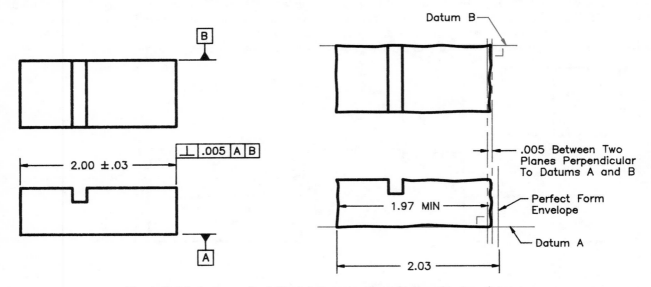

Figure 7-14. A perpendicularity tolerance can be referenced to two datums.

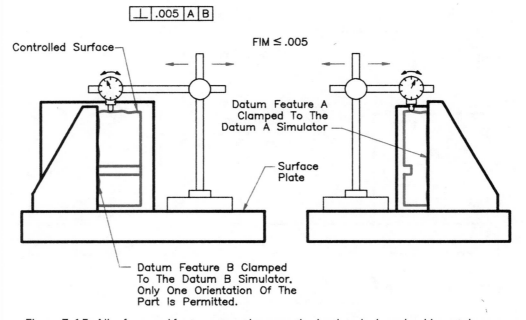

Figure 7-15. All referenced features must be properly simulated when checking a tolerance specification.

gent control than referencing one datum.

It is necessary to identify two datum features on a part if two datums are referenced in a feature control frame. The given figure has datum features A and B identified. The feature control frame for the perpendicularity tolerance references datum A as primary and datum B as secondary. For a perpendicularity tolerance, the primary datum feature must be at a 90° angle to the controlled feature. The secondary datum may be established by any feature that prevents rotation of the part while it is located on the primary datum.

The perpendicularity tolerance zone for the toleranced surface on the given part is bounded by two planes separated by .005″. The planes are perpendicular to both datum planes A and B.

The tolerance zone can be verified by clamping datum surface A of the workpiece against an angle block (datum A simulator) that is set on a surface plate. See Figure 7-15. The workpiece must be clamped against the datum A simulator with datum feature B making two point contact with a second angle block (datum B simulator). Clamping the workpiece against the two angle blocks establishes the required orientation of the part. The workpiece must remain in this position while the perpendicularity measurements are made.

A dial indicator mounted in a height stand can be moved across the controlled surface. A full indicator movement of .005″ or less indicates the perpendicularity tolerance has been met. It is not acceptable to rotate the part in an attempt to improve the indicator reading since primary and secondary datums are referenced. A reading of more than .005″ indicates a bad part.

Face Surface Elements On A Shaft

Perpendicularity of the face surface on a shaft can be controlled relative to the shaft axis. The feature control frame is attached to an extension line or leader from the surface. See Figure 7-16. The shown feature control frame references datum A.

The shaft diameter is identified as datum feature A. Datum feature A is a feature of size. The datum reference in the feature control frame does not show a material condition modifier, therefore it is assumed to apply regardless of feature size.

Two planes establish the tolerance zone for the given part. They are separated by .007″ and are perpendicular to the datum axis. The surface may lie anywhere between the two planes.

PERPENDICULARITY APPLIED TO FEATURES OF SIZE

Perpendicularity applied to a feature of size establishes control of the center plane or axis of the feature. The feature control frame must be associated with the size dimension. It can be placed adjacent to the dimension, or it can be attached to the dimension line. See Figure 7-17.

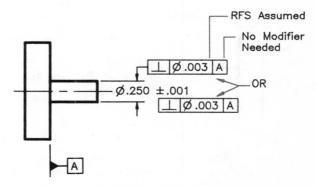

Figure 7-17. An axis of a feature can be specified to be perpendicular to a plane surface.

Material Condition Modifiers And Virtual Condition

A material condition modifier applies to each orientation tolerance that is applied to a feature of size. See Figure 7-18. If no material condition modifier is shown, then RFS is assumed. Either the MMC or LMC modifier must be shown if they are to apply.

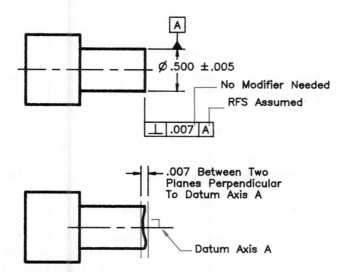

Figure 7-16. Perpendicularity of a flat surface relative to a datum axis can be specified.

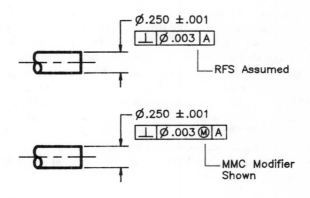

Figure 7-18. Orientation tolerances applied to features of size are assumed to apply RFS unless specified otherwise.

Controlling Axis Orientation Of A Pin

The perpendicularity of a pin can be controlled relative to another feature. See Figure 7-19. This figure shows a pin controlled relative to a flat surface. The feature control frame is placed adjacent to the diameter dimension.

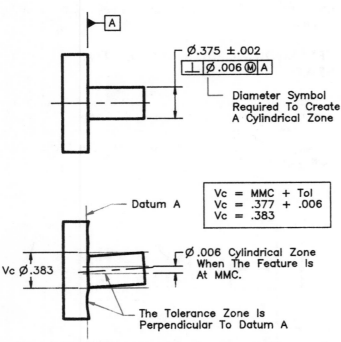

Figure 7-19. The virtual condition for an external feature of size is larger than the MMC of the feature.

A diameter symbol is shown in front of the tolerance value. The diameter symbol is generally required when the tolerance applies to a cylindrical feature of size.

In the given figure, the perpendicularity of the pin is controlled only by the value in the feature control frame. If the feature control frame was not shown, then a general angularity tolerance would be required on the drawing to establish some level of control. The perpendicularity tolerance or general angularity tolerance is required since the size tolerance on the pin only controls size and form; it doesn't control orientation.

A cylindrical tolerance zone with a .006″ diameter extending perpendicular to datum A defines the boundary inside of which the axis of the pin must fall. The axis may be straight or irregular provided the requirement for perfect form at MMC is met. Any shape axis is permitted if the axis is completely contained within the cylindrical tolerance zone.

Since the tolerance includes an MMC modifier, the cylindrical tolerance zone has a diameter of .006″ when the pin is at MMC (.377″). As the pin departs in size from the .377″ MMC, the tolerance zone is permitted to increase. For every .001″ departure from MMC, there is an equal amount of increase in the tolerance zone.

Virtual Condition Of A Pin

Tolerances with the MMC modifier result in a virtual condition. When a perpendicularity tolerance is applied to a pin, the virtual condition is equal to the apparent size of the pin when it is at MMC and the maximum permitted orientation error exists.

Virtual condition can be thought of as the required size for a mating part, assuming the mating part is perfect. In the case of a pin that includes an orientation tolerance, the virtual condition is the size of a perfect hole that fits over the pin when the pin is at MMC and is in the worst permitted orientation.

Figure 7-20 shows the virtual condition for the pin in the previous figure. The virtual condition is determined by adding the MMC size and the orientation error. A pin produced at the .377″ MMC size, with a perpendicularity error of .006″, has a virtual condition of .383″ diameter.

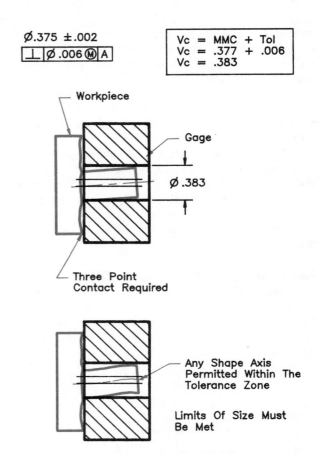

Figure 7-20. A functional gage for an external feature is sized on the basis of the virtual condition of the feature.

A gage can be produced at the virtual condition size to check the orientation of the pin. Any acceptable orientation will fit into the gage. Acceptable orientations are those that fall within the cylindrical zone. The axis may be straight but inclined, or it may be irregular. There is no requirement other than for the axis to fall within the tolerance zone.

It was previously explained that a tolerance with the MMC modifier is permitted to increase as the controlled feature departs from MMC. On the given part, the orientation tolerance of .006″ at MMC, increases if the pin departs from MMC. When the pin is at a diameter of .376″, there is a .001″ departure from MMC. This permits the orientation tolerance to increase by .001″ to a total permitted error of .007″. A produced pin diameter of .376″ and orientation tolerance of .007″ combine for a total apparent diameter of .383″ diameter. This is equal to the virtual condition. Since the allowable apparent size of the pin is always equal to the virtual condition, one gage sized to the virtual condition can be used to inspect the orientation tolerance on any pin that is made within the specified size tolerance. NOTE: The fixed diameter gage may only be used if the tolerance includes the MMC modifier.

A mating hole must have a minimum diameter equal to the virtual condition of the pin. This ensures that the parts will assemble when a worst-case condition exists.

Controlling Axis Orientation Of A Hole

The perpendicularity of a hole can be controlled relative to another feature. See Figure 7-21. This figure shows a hole controlled relative to a flat surface. The feature control frame is placed adjacent to the diameter dimension.

A diameter symbol is shown in front of the tolerance value. The diameter symbol is generally required when the tolerance applies to a cylindrical feature of size.

In the given figure, the perpendicularity of the hole is controlled only by the value in the feature control frame. If the feature control frame was not shown, a general angularity tolerance or another control would be required on the drawing. The perpendicularity tolerance or other control is required since the size tolerance on the hole controls only the size and form: it does not control orientation.

A cylindrical tolerance zone measuring .008″ diameter and extending perpendicular to datum A defines the boundary inside of which the axis of the hole must fall. The axis may be straight or irregular. Any shape axis is permitted if the axis is completely contained within the cylindrical tolerance zone, but the requirement for perfect form at MMC is required.

Since the tolerance includes an MMC modifier, the cylindrical tolerance zone has a diameter of .008″ when the hole is at MMC (.391″). As the hole departs in size from the .391″ MMC, the tolerance zone is permitted to increase. For every .001″ departure from MMC, there is an equal amount of increase in the tolerance zone.

Virtual Condition Of A Hole

Tolerances that include an MMC modifier result in a virtual condition. When a perpendicularity tolerance is applied to a hole, the virtual condition is equal to the apparent size of the hole when it is at MMC and the maximum permitted orientation error exists.

The virtual condition of a hole is the size of a perfect pin that would fit into the leaning hole. A mating pin must have a virtual diameter equal to the virtual condition of the hole. This ensures that the parts will assemble when the worst-case condition exists.

Figure 7-22 shows the virtual condition for the part in the previous figure. The virtual condition is determined by subtracting the orientation error from the MMC size. A hole produced at the .391″ MMC size with a perpendicularity error of .008″ has a virtual condition of .383″ diameter.

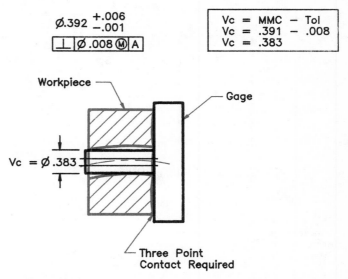

Figure 7-22. A functional gage for an internal feature is sized on the basis of the virtual condition of the hole.

A gage can be produced at the virtual condition size to check the orientation of the hole. Any acceptable orientation will fit onto the gage. Acceptable axis orientations are those that fall within the cylindrical tolerance zone. The axis may be straight but inclined, or it may be irregular if the hole is not at MMC.

It was previously explained that a tolerance with the MMC modifier is permitted to increase as the controlled feature departs from MMC. On the given part, the

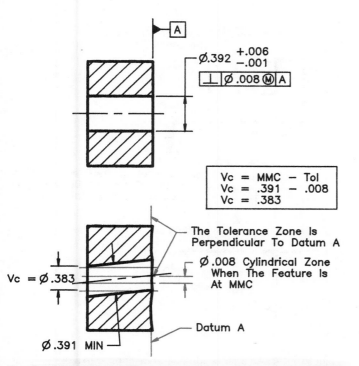

Figure 7-21. The virtual condition for an internal feature of size is smaller than the MMC of the feature.

orientation tolerance of .008" at MMC increases if the hole departs from MMC. When the hole is at a diameter of .392", there is a .001" departure from MMC. This permits the orientation tolerance to increase by .001" to a total permitted error of .009". A produced hole diameter of .392" and orientation tolerance of .009" result in an apparent hole diameter of .383" diameter. This is equal to the virtual condition. Since the allowable apparent size of the hole is always equal to the virtual condition, one gage at the virtual condition can be used to inspect the orientation tolerance on any hole that is made within the specified size tolerance. NOTE: The fixed diameter gage can only be used if the tolerance includes the MMC modifier.

Controlling Two Perpendicular Cylinders

There are exceptions to the requirement for a diameter symbol on perpendicularity tolerances applied to cylinders. See Figure 7-23. No diameter symbol is shown on a perpendicularity tolerance when applied to a cylindrical feature and referenced to a datum axis.

The shaft diameter in the given figure is identified as datum feature A. The hole has a perpendicularity tolerance of .007" referenced to datum A. There isn't a diameter symbol on the tolerance value. The tolerance value and the datum reference both have MMC modifiers.

The tolerance zone is bounded by two planes that are perpendicular to datum axis A. The planes are separated by .007" when both features, the hole and datum feature A, are at MMC. The axis of the hole can be anywhere within the two planes and it will meet the perpendicularity requirement relative to the datum axis.

The orientation tolerance does not control the location of the hole relative to the datum axis. It only controls the orientation. If location of the hole needs to be controlled, then a position tolerance must be applied.

Controlling Perpendicularity Of A Center Plane

It is sometimes necessary to control the perpendicularity of the center plane of a rectangular feature of size. See Figure 7-24. In the given figure, the center plane of a slot is controlled. The same method of controlling a center plane can be applied to external features such as rails and tabs.

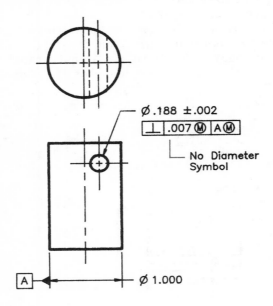

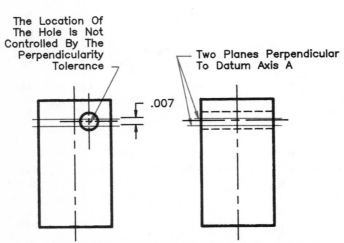

Figure 7-23. When a perpendicularity tolerance is applied to control the orientation between two cylindrical features, the tolerance zone is bounded by two parallel planes.

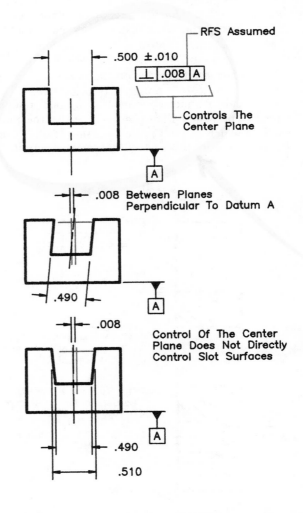

Figure 7-24. An orientation tolerance applied to a rectangular feature of size controls the orientation of the center plane of the feature.

The shown feature control frame is placed adjacent to the size dimension for the slot. This indicates that the feature of size is controlled, which in effect controls the center plane. The specified tolerance is .008″ RFS relative to datum A. Datum feature A is the bottom surface of the part.

Only the center plane of the slot is controlled by the given perpendicularity tolerance. Neither the sides nor the bottom of the slot are controlled by this tolerance.

The sides of the slot may be in any condition provided that Rule #1 and the .490″ and .510″ limits of size are met. This means that the sides of the slot must be parallel to one another when produced at a size of .490″. The size limits do not require the sides of the slot to be perpendicular to datum A. On the given drawing, only the feature control frame invokes a perpendicularity requirement.

Regardless of the produced size of the slot, the center plane must be within a boundary established by two planes separated .008″. Any orientation within the boundary is acceptable. The two planes forming the boundary are perpendicular to datum A. The tolerance zone extends the full height of the slot.

Two possible, and acceptable conditions for the slot are shown in the given figure. The first example shows a slot produced at the MMC size of .490″. The two sides of the slot are parallel, but oriented such that the center plane just does stay inside the tolerance boundary. Since the slot is at MMC, the perpendicularity tolerance indirectly affects the sides of the slot.

The second example shows the slot produced with a .490″ width at the bottom and a .510″ width at the top. In this example, the perpendicularity of the center plane is perfect, but the sides of the slot are at considerable angles to datum A. This part meets both the size and center plane perpendicularity requirements.

Control of the orientation of a center plane does not directly affect the surfaces of the controlled feature. Orientation tolerances should only be applied to rectangular features of size when the center plane is what needs to be controlled.

Perpendicularity of the sides of a slot can be controlled through the application of a feature control frame to the slot surfaces. See Figure 7-25. The feature control frame is attached to an extension line from the surface to be controlled.

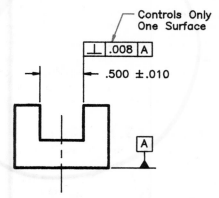

Figure 7-25. A feature control frame placed on an extension line only controls the surface from which the line extends.

The shown feature control frame is applied to only one surface. It only controls the surface to which it is applied. The other side of the slot is controlled by the limits of size on the slot. If direct control of both surfaces is desired, a feature control frame can be applied to each of them.

ANGULARITY

Control of any flat surface in an orientation other than parallel or perpendicular can be established with an angularity tolerance. Angularity tolerances are always referenced to one or more datums. See Figure 7-26. The feature control frame is attached to a leader or extension line from the surface to be controlled.

The angle dimension that defines the surface orientation must be basic. No angles are assumed other than perpendicularity and parallelism as already explained.

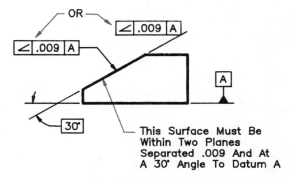

Figure 7-26. A feature control frame for an angularity tolerance may be placed on an extension line or leader extending from the surface to be controlled.

ANGULARITY APPLIED TO FLAT SURFACES

Application of an angularity tolerance is shown in Figure 7-27. The vertex of the angle is located with a toleranced dimension. The angle of 30° is shown as basic. Datum feature A is identified. An angularity tolerance of .009″ relative to datum A is applied on the extension line from the inclined surface.

The tolerance zone for the given figure is bounded by two parallel planes separated .009″. These planes are at an angle of 30° relative to datum A.

Surface conditions are only controlled to the extent that they must fall within the .009″ boundary. This boundary is free to float under the condition that it remain at a 30° angle to datum A, and the corner on the part fall somewhere within the limits defined. The corner of the given part may fall anywhere within the limits of .240″ and .260″.

Some controversy exists regarding the allowable location of the boundary planes. The following explanation is based on the premise that orientation tolerances only control orientation when applied to a surface: they do not control location. Should a desire exist to simultaneously control location and orientation, a profile tolerance may be specified.

It is possible for the corner of the part shown in Figure 7-27 to be located at .260″, and have the bottom plane of the angularity boundary pass through the corner. It is also

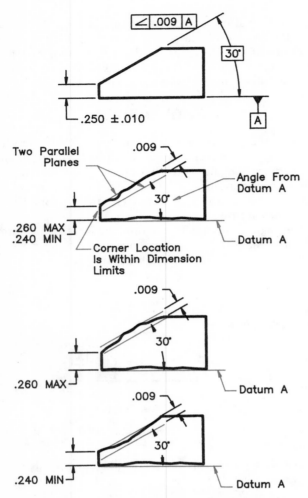

Figure 7-27. The angle dimension defining the surface orientation must be basic when an angularity tolerance is applied.

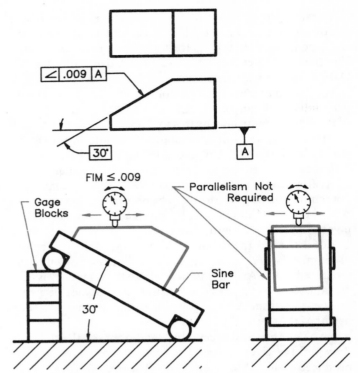

Figure 7-28. As for perpendicularity, angularity must be verified relative to the datums that are referenced. No datums are to be assumed.

possible to have the corner at .240″, and the top plane of the boundary pass through the corner. Anything between these two extremes is also permitted.

Any surface that lies within the orientation tolerance boundary also meets a form tolerance that is equal to the orientation tolerance. The given figure shows this effect. The boundary is defined by two parallel planes. If the surface is between the planes, then form is controlled.

Referenced To One Datum

An angularity tolerance referenced to one datum only controls orientation relative to that datum. See Figure 7-28. The given angularity tolerance specifies an orientation requirement of .009″ relative to datum A. The angularity tolerance does not control the surface relative to any feature other than datum A, nor does it control the location of the surface. Location requirements are defined by a toleranced dimension or a profile tolerance. Profile tolerances are defined in a later chapter.

The given angularity tolerance specification permits rotation of the controlled surface around any axis that is perpendicular to datum A. Rotation around such an axis does not affect the 30° angle relative to datum A.

The shown method of verification for the given angular-

ity tolerance illustrates the permitted rotation of the surface. A sine bar is set on gage blocks to orient it at a 30° angle. The workpiece is placed on the sine bar in a position that results in the specified surface being horizontal.

A dial indicator is passed across the workpiece to determine if the orientation requirement is met. A full indicator movement of .009″ or less indicates the part is good.

The workpiece in the given figure is not aligned with the edge of the sine bar. This is an acceptable position since only one datum is referenced. Control of angularity is specified relative to only datum A, and resting the bottom surface of the part on the sine bar establishes orientation relative to datum A.

Although the feature control frame doesn't control orientation other than to datum A, there is some limit to the permitted rotation of the surface. A complete drawing of the given part would include size dimensions for all features and general angularity tolerances. These controls would provide a limit to the amount of rotation of the subject surface.

The degree of orientation control established by size dimensions on angled surfaces is difficult to calculate. If the orientation control needs to be well defined, an angularity tolerance should be specified. It may need to be referenced to two datums to establish the needed level of control.

Referenced To Two Datums

Figure 7-29 shows the same part as the previous figure except the feature control frame references two datums and a second datum feature is identified. The feature control frame shows datum A as primary and datum B as secondary. Datum feature A is the bottom of the part and is the surface from which the angle is dimensioned. Datum feature

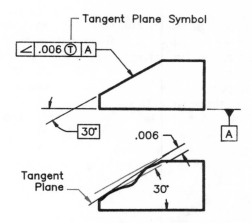

Figure 7-30. Angularity may be specified to control the plane that is tangent to surface high points.

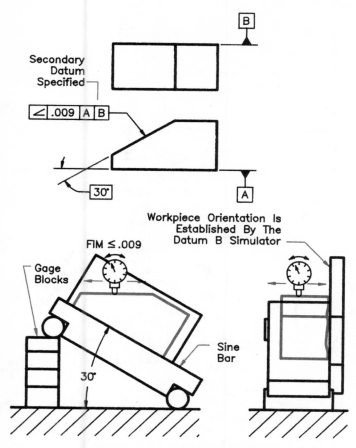

Figure 7-29. All referenced datums must be simulated when checking angularity tolerances.

B is one side of the part.

The workpiece is set on the same sine bar that was shown in the previous figure. The difference in verification methods is that datum feature B must be properly oriented on the sine bar. For this part, datum feature B is aligned with the edge of the sine bar. With the workpiece properly oriented on the sine bar, a dial indicator is moved across the surface. A full indicator movement of .009″ or less indicates a good part.

If a reading greater than .009″ is obtained, the part should be rejected or reworked. The orientation of the part cannot be changed to reduce the indicator movement as it can when only one datum is referenced.

TANGENT PLANE

Each orientation tolerance, when applied to a surface, establishes a tolerance boundary inside of which the controlled surface must lie. Any surface condition within the boundary is permitted. One possible surface condition permitted by a .010″ orientation tolerance is a .010″ step at one location on the surface. Another permitted condition is for the surface to be flat and at an angle that extends across the tolerance zone.

Significantly different effects are caused when a flat plate is located against the two surface conditions described in the previous paragraph. The angle of the plate is affected by the amount and location of the surface errors.

It is possible to control the effects of surface conditions

when applying an angularity tolerance. One method of control is to further refine the surface control with a form tolerance. Another method is to specify the orientation tolerance to apply to a tangent plane. See Figure 7-30. This is done by placing the tangent plane symbol in the tolerance specification. This may be done for parallelism, perpendicularity, and angularity tolerances. Prior to 1994, a tangent plane notation was required.

Specifying tolerances for a tangent plane should only be done when it is important to control the tangent plane. An orientation tolerance on a tangent plane is more difficult to verify than an orientation tolerance on a surface. If control of the part surface is adequate, the tangent plane requirement should not be used.

The angularity tolerance zone for a tangent plane is identical to any other angularity tolerance zone, except the surface of the controlled feature is not required to fall within the tolerance zone boundary. Only a plane tangent to the high points on the surface must be within the tolerance zone.

The given figure shows a tolerance zone that contains a tangent plane. The surface itself extends outside the tolerance zone.

Orientation tolerances controlling a tangent plane do not control the surface form. This means that a separate form tolerance must be applied to the surface if form control is desired.

COMBINED FORM AND ORIENTATION TOLERANCES

Multiple feature control frames can be applied to a single feature. This is often done to control orientation to one value, and form to a smaller value.

Combined orientation and form tolerances are only required on surfaces when the maximum allowable form errors are less than the allowable orientation tolerance. This is due to the fact that orientation tolerances control the form of a surface to a value equal to the orientation tolerance.

ORIENTATION ALSO CONTROLS FORM

The current standard specifically states that orientation tolerances applied to surfaces also control form. This means that any orientation tolerance on a flat surface also controls the flatness of the surface to a value equal to the orientation tolerance. As an example, application of a .006″ perpendicularity tolerance also controls flatness within .006″. Exception to this requirement is taken when the tolerance is applied to a tangent plane.

The standard does not specifically state that orientation tolerances applied to features of size also control form. It does state, however, that any straightness of an axis requirement must be less than any orientation tolerance on the axis.

ORIENTATION AND FLATNESS

Multiple feature control frames can be applied to an individual feature. See Figure 7-31. This figure shows a perpendicularity tolerance and flatness tolerance applied to the same surface. A perpendicularity tolerance of .012″ relative to datum A is specified. In addition, a flatness tolerance of .004″ is specified.

Both feature control frames are placed on one extension line in the given figure. This is not required, but is a common practice. Placement of all tolerances for a surface on one extension line has the advantage of locating all specifications for a given surface in one place. This helps to ensure that all requirements are easily seen.

The perpendicularity tolerance in the given figure is shown above the form tolerance. Feature control frames are commonly organized to place the larger tolerance zone first. The subsequent feature control frames indicate tolerances that refine the previous tolerance. This arrangement makes the tolerance requirements easier to read and keeps them in a logical order.

The given figure shows one possible condition for the toleranced surface. There is a .012″ wide perpendicularity tolerance zone. The surface can lie anywhere within this zone, provided it is also flat within .004″ The figure shows a surface that is flat within .004″, and the surface is oriented to extend all the way across the .012″ perpendicularity tolerance zone. It is permissible for the tolerance zone boundaries to overlap provided the surface of the part does not violate either zone.

ORIENTATION AND AXIS STRAIGHTNESS

Figure 7-32 shows a perpendicularity and straightness tolerance applied to a pin. The perpendicularity tolerance requires the axis to lie within a .015″ diameter cylinder that is perpendicular to datum A. The straightness tolerance requires the axis to be within a .005″ diameter cylinder. There is no orientation requirement on the straightness tolerance, other than the requirement that the pin axis meet the .015″ perpendicularity tolerance.

One possible condition for the pin is shown in the given figure. The tolerance zones overlap in this example, but the axis of the pin remains within both tolerance zones for the full length of the pin. This is an acceptable condition.

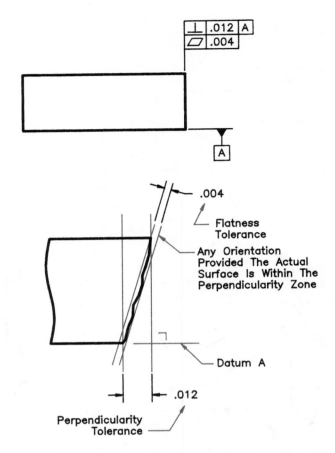

Figure 7-31. A form tolerance can be applied to a surface to refine the orientation requirement.

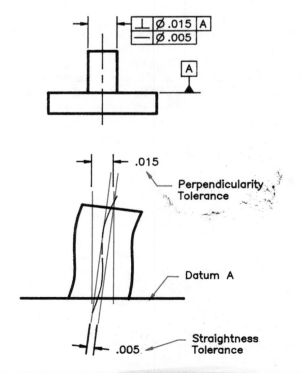

Figure 7-32. A form tolerance can be applied to a feature of size to refine the orientation requirement.

Bonus Tolerances Applies To Both.

Orientation tolerances are parallelism, perpendicularity, and angularity.

Orientation tolerances can be used to refine the orientation requirements imposed by tolerances such as position.

Surface condition requirements can be refined by adding a form tolerance to the orientation tolerance.

A virtual condition exists when an orientation tolerance is applied to a feature of size and an MMC modifier is included in the tolerance specification.

Parallelism does not impact size requirements, but it does refine orientation within the size limits.

Functional gaging is made easier when orientation tolerances are applied at MMC on features of size.

An orientation tolerance applied to a cylindrical feature of size normally requires a diameter symbol.

All orientation tolerance specifications must include at least one datum reference.

Two datum references can be used for an orientation tolerance specification.

The center plane is controlled when an orientation tolerance is applied to a rectangular feature of size.

Orientation tolerances control surface conditions when applied to a surface. The tolerance can be specified to control the tangent plane when special circumstances require this type of control.

REVIEW QUESTIONS

Answer the following questions on a separate sheet of paper. Do not write in this book. Accurately complete any required sketches.

MULTIPLE CHOICE

1. Orientation tolerances control _____ as well as orientation.
   A. form
   B. position
   C. concentricity
   D. runout

2. A feature control frame must include a minimum of _____ datum reference(s) for an orientation tolerance.
   A. one
   C. two
   C. three
   D. None of the above.

3. A material condition modifier applies to a datum reference when the _____.
   A. datum feature is a surface
   B. datum feature is a feature of size
   C. datum reference is primary
   D. tolerance value includes a modifier

4. A parallelism tolerance applied to a flat surface also controls _____.
   A. size
   B. position
   C. flatness
   D. Both A and C.

5. A tolerance applied to control parallelism of two holes _____ the location tolerance between the holes.
   A. does affect
   B. does not affect
   C. eliminates
   D. Both A and C.

6. When a perpendicularity tolerance is applied to a surface, the controlled surface must _____.
   A. lie within the perpendicularity tolerance zone
   B. lie within the size tolerance
   C. Neither A nor B.
   D. Both A and B.

7. A(n) _____ symbol is used to indicate that the orientation tolerance zone is cylindrical.
   A. MMC
   B. LMC
   C. circularity
   D. diameter

8. An orientation tolerance is applied to a _____ by placing the feature control frame on the dimension.
   A. surface
   B. centerline
   C. feature of size
   D. None of the above.

9. The virtual condition for a .376″ MMC pin with a perpendicularity tolerance of .008″ diameter at MMC is _____″.
   A. .008
   B. .368
   C. .376
   D. .384

10. The virtual condition for a .223″ MMC hole with a perpendicularity tolerance of .004″ diameter at MMC is _____″.
    A. .004
    B. .219
    C. .223
    D. .227

11. An angularity tolerance is used for features at _____.
    A. any angle
    B. any angle other than 90°
    C. any angle other than parallel
    D. Both B and C.

12. An orientation tolerance can be specified to control a(n) _____ instead of surface conditions.
    A. tangent plane
    B. circumscribing cylinder
    C. enclosing envelope
    D. None of the above.

13. A _____ tolerance can be applied to refine the surface control required by an orientation tolerance.

A. position
B. concentricity
C. form
D. Both A and C.

TRUE/FALSE

14. Size dimensions on a drawing control the tolerance on 90° angles.(A)True or (B)False?
15. Two sides of a part must be perfectly flat and parallel when the part is at its least material condition. (A)True or (B)False?
16. Multiple datum reference frames may be connected to one feature. (A)True or (B)False?
17. Regardless of the feature type, virtual condition is always larger than the MMC size. (A)True or (B)False?
18. Parallelism applied to a flat surface controls flatness to the same value as the parallelism tolerance. (A)True or (B)False?
19. Angles measuring 90° must be dimensioned. (A)True or (B)False?
20. A perpendicularity tolerance referenced to one datum only controls the orientation of the surface relative to the one datum. (A)True or (B)False?
21. Controlling the center plane of a slot with a perpendicularity tolerance does not impose a direct control of the slot sides. (A)True or (B)False?
22. An angularity tolerance controls feature orientation and location. (A)True or (B)False?

FILL IN THE BLANK

23. Two sides of a part must be perfectly parallel when the feature is at its __MMC__.
24. If a form tolerance is applied to the same surface as an orientation tolerance, the form tolerance must be _____ than the orientation tolerance.
25. Virtual condition is the combined effect of the _____ size and any applicable tolerance.
26. An implied 90° angle is understood to be _____ when a perpendicularity tolerance is applied to perpendicular features.

SHORT ANSWER

27. Explain why it is sometimes necessary to place an orientation and form tolerance on the same surface.
28. Sketch the symbol for each of the following:
    A. Parallelism.
    B. Perpendicularity.
    C. Angularity.
29. Describe a method for checking parallelism between two flat surfaces.
30. Describe two methods for showing that a perpendicularity tolerance applies to a surface.
31. What is achieved by referencing a secondary datum in an orientation tolerance specification?
32. How is the size of a functional gage calculated for checking perpendicularity of a hole?
33. Define a tangent plane as the term relates to an orientation tolerance.

APPLICATION PROBLEMS

Some of the following problems require that a sketch be made. All sketches should be neat and accurate. Each problem description requires the addition of some dimensions for completion of the problem. Apply all required dimensions in compliance with dimensioning and tolerancing requirements. Show any required calculations.

34. Label each segment of the given feature control frame and define the requirements of each segment. Assume the feature control frame is attached to a surface and that the datum feature is a surface.

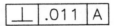

35. Complete a two line feature control frame that requires parallelism within .015″ relative to datum D and a flatness of .008″.

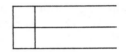

36. Apply a parallelism requirement of .009″.

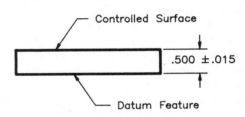

37. Apply a perpendicularity requirement of .007″.

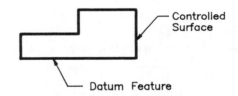

38. Apply a perpendicularity requirement of .010″ relative to two datums.

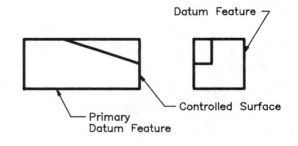

39. Apply an angularity tolerance of .012″ relative to two datums.

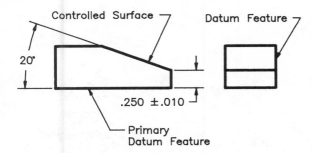

40. Sketch an inspection setup for the part in Problem 39, and describe the requirements for acceptance of the part.

41. Apply a perpendicularity tolerance to control the pin within a .005″ cylindrical tolerance zone at MMC relative to the datum surface.

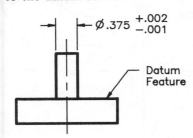

42. Sketch and dimension a functional gage to check the part in Problem 41. Show your calculations.

43. Apply a perpendicularity tolerance to control the hole within a .007″ cylindrical tolerance zone relative to the datum surface.

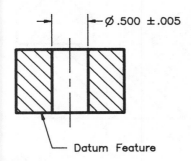

44. Sketch and dimension a functional gage to check the part in Problem 43. Show your calculations.

45. Sketch one acceptable configuration for a part fabricated to the shown dimensions. Some departure from the nominal dimensions is to be shown.

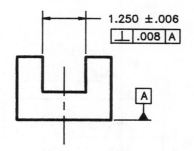

46. Make the necessary changes to apply the shown tolerance to a tangent plane.

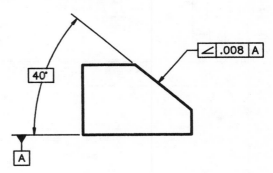

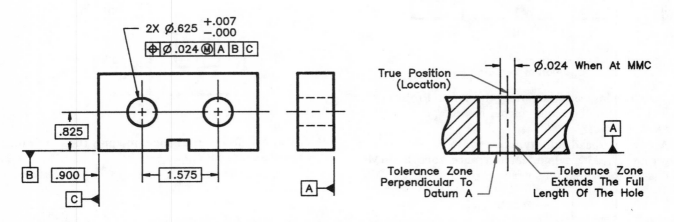

POSITION TOLERANCE ZONES

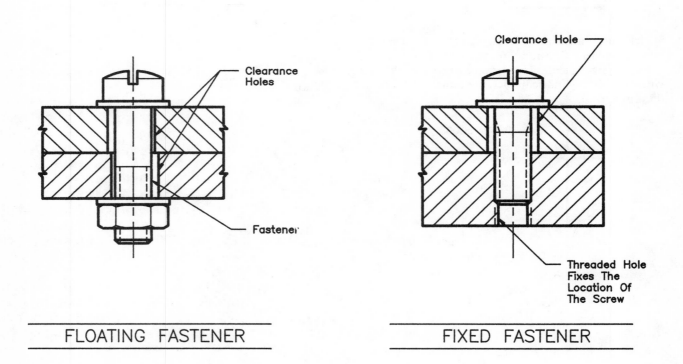

FLOATING FASTENER

FIXED FASTENER

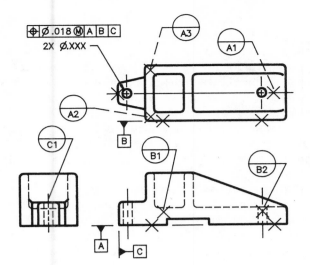

Chapter 8

Position Tolerancing-Fundamentals

CHAPTER TOPICS

☐ Construction of a feature control frame for position tolerances.

☐ The location and shape of a position tolerance zone.

☐ The requirements regarding material condition modifiers and the effect of those modifiers on position tolerances.

☐ Advantage of using the maximum material condition modifier.

☐ Proof of the validity in the maximum material condition concept.

☐ Calculating position tolerances for fixed and floating fastener conditions.

☐ Calculating the amount of bonus tolerance due to departure of a part feature from the maximum material condition.

☐ Paper gaging of position tolerances.

☐ Advantages of position tolerances.

☐ Special applications for position tolerances, including counterbored holes, threads, projected tolerance zones, bidirectional tolerances, and noncylindrical features.

INTRODUCTION

Drawings of production parts include tolerances that define the acceptable amount of variation on dimensions. There must be an allowable variation on all aspects of the part geometry, including the locations for features of size. As an example, the required location for a hole must include a tolerance. One method for defining location tolerances on holes and other features of size is to apply a position tolerance.

Drawings must define the acceptable amount of variation on sizes and locations since it is not possible to produce perfect parts. This chapter describes how to use position tolerances for specification of allowable location errors.

POSITION

Position tolerances are used to specify the required location accuracy for features of size. Tolerance specification methods make it possible to control the location of as many features as necessary. One feature control frame can be used to show the position tolerance on a single hole, or it can be used to control an entire pattern of holes.

Any drawing that includes a position tolerance on holes must include several characteristics. See Figure 8-1. Datum features are identified. Basic dimensions are used to define hole locations. There is at least one feature control frame that specifies a position tolerance.

Each of these characteristics has a purpose. The datum features serve to locate the datum reference frame from which the hole locations are measured. The basic dimensions define the true positions for the holes. The feature control frame defines the size of the tolerance zone that is located on each of the true positions.

Figure 8-1 shows two feature control frames that are used to define the position tolerances for five holes. The feature control frame in the top view is placed adjacent to a noted size for two holes. Therefore, it gives the position tolerance for both of these holes. The other feature control frame is placed adjacent to the noted size for the three hole pattern. This feature control frame specifies the required position tolerance for each of the three holes.

Both feature control frames in the given figure have a diameter symbol in front of the tolerance value. A diameter symbol in the tolerance specification indicates a round tolerance zone. Omitting the diameter symbol would be incorrect for the given part.

The shape specified for a tolerance zone is determined from the functional requirements for the part. Holes are typically given round tolerance zones. Rectangular features are given rectangular tolerance zones.

Feature control frames are used to specify position tolerances. The composition of the single-line position tolerance feature control frame defines the tolerance zone shape and size. It also defines the datum reference frame to which the tolerance zone must be located and oriented. Multiple-line

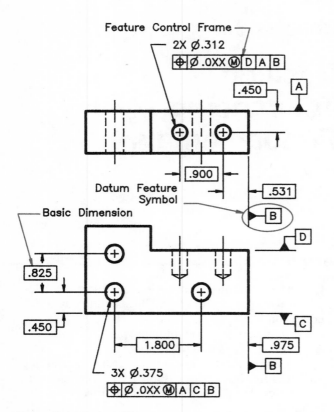

Figure 8-1. Basic dimensions, datum features, and position tolerance specifications are required to completely define position tolerance requirements.

and composite position tolerance feature control frames are more complex. They are explained in the next chapter.

Position tolerances include an implied RFS material condition modifier if no symbol is shown. When a material condition modifier other than RFS is to apply, either the MMC or LMC modifier must be shown. Material condition modifiers are appropriate, because position tolerances are only applied to features of size. Maximum material condition (MMC) and least material condition (LMC) modifiers permit the tolerance zones to vary according to feature size. This can significantly increase the producibility of the parts. The use of modifiers is one advantage of the position tolerancing system.

FEATURE CONTROL FRAME

Information shown in a position tolerance feature control frame must be shown in the correct sequence. See Figure 8-2. The tolerance symbol (characteristic) must be shown

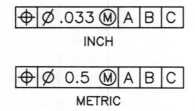

INCH

METRIC

Figure 8-2. A position tolerance feature control frame must include datum references.

first. It is followed by the tolerance value, including a diameter symbol if the zone is circular. The tolerance value will be followed by a material condition modifier if MMC or LMC is applicable. No modifier is shown if RFS is applicable.

If tolerances are determined using statistical calculations, the statistical symbol may be inserted in the feature control frame. It is placed after any shown material condition modifier.

Datum references follow the tolerance value. Implied datum references are not permitted on position tolerances. As with all other feature control frames, the datum references are read from left to right. The primary datum is shown first, followed by the secondary and tertiary datums.

Drawings completed to standards prior to ANSI Y14.5M-1982 generally showed datum references on all position tolerances, although it wasn't required. Early dimensioning and tolerancing standards permitted implied datums. The utilization of implied datums is no longer permitted since implied datums can cause confusion. Implied datums can result in confusion since a datum reference frame must be assumed. Being required to assume a datum reference frame can introduce inconsistency since two people might assume different datum reference frames. It must be realized that two different datum reference frames will not create the same results when fabricating parts.

Position tolerances are only applied to features of size. The feature control frame can be connected to the appropriate feature with a leader, it can be placed adjacent to the feature dimension, or it can be placed on the dimension line for the size dimension. Feature control frames for position tolerances are never attached to centerlines.

Many position tolerancing applications can be met using the single-line feature control frame already described. More complex applications sometimes require the use of a multiple-line or composite feature control frame. The formats of these feature control frames is shown in Figure 8-3. These are explained in the next chapter.

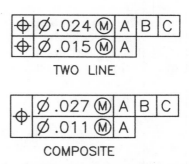

TWO LINE

COMPOSITE

Figure 8-3. Single-line, two-line, or composite position tolerance specifications can be used to achieve the required level of control.

Material Condition Modifier Application

Rule #2 of the current standard requires that all tolerances be assumed to include the RFS material condition modifier. See Figure 8-4. Maximum material condition (MMC) and

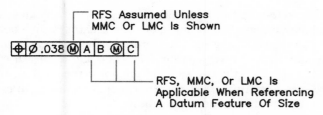

Figure 8-4. Material condition modifiers are not assumed on position tolerances. The applicable modifiers must always be shown.

least material condition (LMC) must be shown if they are to apply. Selection of the appropriate modifier depends on the design application.

Each modifier has a significantly different effect on the requirements of the specified tolerance. The effect of each modifier is explained in this chapter.

Datum references to features of size also include a material condition modifier. The modifiers used on datum references can be different than the modifier used on the tolerance value. Datum references to surfaces should not include a material condition modifier since a surface does not have a material condition.

POSITION TOLERANCE ZONE

A position tolerance zone will usually have a defined shape, size, location, and orientation. See Figure 8-5. The shape is generally determined by whether or not a diameter symbol is placed in the feature control frame. The three-dimensional size of the tolerance zone is typically defined by the tolerance value shown in the feature control frame and by the length of the hole (or whatever feature of size is being controlled). Basic dimensions and references to datums define the location and orientation requirements for the tolerance zone. Orientation dimensions may be any specified angle, but are generally defined by drawing convention to be a basic 90° angle. Any angle other than 90° must be dimensioned.

A typical position tolerance specification and the resulting tolerance zone are shown in Figure 8-5. The shape, size, location, and orientation of the tolerance zone is defined. The tolerance specification has a diameter symbol applied to the .024″ tolerance value. This indicates a requirement for the tolerance zone to be a .024″ diameter cylinder. The cylindrical zone extends completely through the hole. It terminates flush with the two surfaces. The tolerance zone cylinder is centered on the true position that is defined by the basic dimensions. The tolerance zone must be perpendicular to primary datum A.

Basic Dimensions and True Positions

Basic dimensions must be used to define the locations of features to which position tolerances are applied. See Figure 8-6. The given figure shows two holes located by three basic dimensions.

The basic dimensions define the *true positions* for the two holes. *True position is a theoretically exact location defined by basic dimensions shown on a drawing.* The true position is an exact location and should not be interpreted as the required location for a hole. The true position is only the location on which to center the specified position tolerance. The axis of the hole can be located anywhere within the tolerance zone that is centered on the true position.

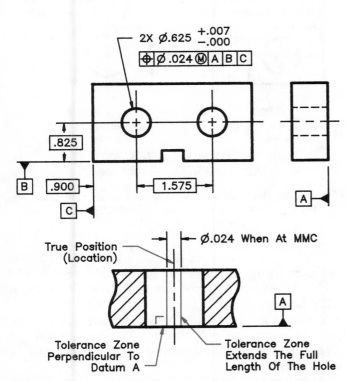

Figure 8-5. The position tolerance on a hole is cylindrical and extends all the way through it.

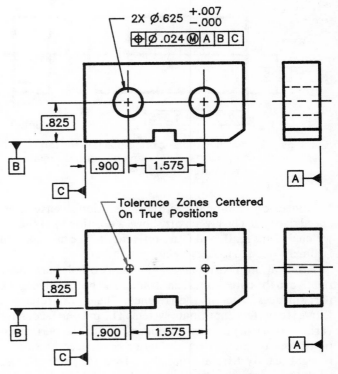

Figure 8-6. Tolerance zones for holes are centered on the theoretically true positions for the holes.

Figure 8-6 shows two horizontal dimensions chained together to locate the true positions of the two holes. There would be no difference in the true positions if both holes were dimensioned with baseline dimensions extending from the datum feature. There wouldn't be a difference since there is no tolerance accumulation on basic dimensions. There is no tolerance applied to the basic dimension itself; the position tolerance is applied to the feature.

Relationship to the Datum Reference Frame

Position tolerances must include datum references. The datum references establish the datum reference frame from which measurements must be made. See Figure 8-7. A datum reference frame provides a known origin and orientation for measurements. Review Chapter 6 for additional information about datum reference frames.

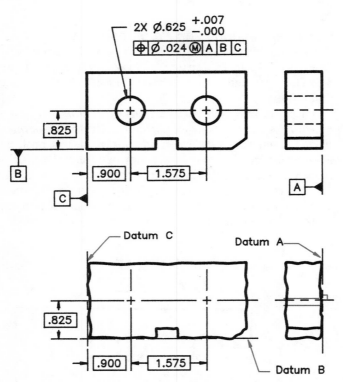

Figure 8-7. Datum references in a position tolerance specification require that measurements be made relative to a datum reference frame.

Dimensions are applied with consideration given to referenced datums. The given part shows a position tolerance in which datums A, B, and C are referenced. The hole pattern is dimensioned to locate it relative to these datums.

All location measurements for the two holes are made relative to the datum reference frame, and not relative to the part surfaces. There is a difference. The datum reference frame is a perfect coordinate system. The part surfaces have variations. Depending on the magnitude of the part variations, measurements from the datum reference frame could be significantly different than those from the part surfaces.

Variations on the shown part indicate the importance of the datum reference frame. When working relative to the

datum reference frame, all measurements are in a single coordinate system. Remaining in one coordinate system is not likely when working off the feature surfaces. Because of angular errors between datum features B and C, measurements made off the two surfaces are not perpendicular. If the measurements aren't perpendicular, then an accurate check of the dimensions cannot be made.

EFFECT OF MATERIAL CONDITION MODIFIERS

Each of the three material condition modifiers have distinctly different effects when applied to a position tolerance. One of the modifiers is optimum for any given design application. Selection of the correct modifier is relatively easy when their effects are clearly understood.

Regardless of Feature Size

No Regardless of Feature Size (RFS) modifier symbol is needed because RFS is implied. See Figure 8-8. The assumed RFS modifier indicates that the specified tolerance value is to remain constant, regardless of the produced size of the toleranced part.

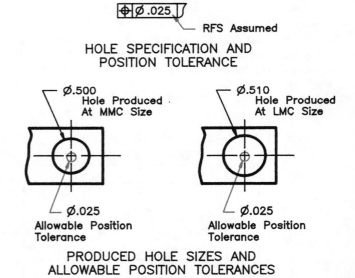

Figure 8-8. The position tolerance zone is not affected by the produced feature size when the RFS modifier is applicable to the tolerance specification.

A hole specification and position tolerance are shown in the given figure. Limits of size for the hole are .500″ and .510″ inch diameter. The position tolerance is specified as .025″ diameter RFS.

Any hole produced within the limits of size is permitted a position tolerance of .025″ diameter. The size of the hole has no effect on the allowable position tolerance since the RFS modifier is shown.

Two examples of produced holes are shown. One hole is produced at .500″ diameter, the maximum material condition for the hole. The position tolerance for this hole is .025″ diameter. The other hole is produced at .510″ diameter, the least material condition for the hole. The position tolerance for this hole is also .025″ diameter.

The RFS modifier is typically used for applications where holes have zero clearance or press fits. The allowable diameter variation on a press fit hole does not provide any additional position tolerance, therefore RFS is required.

Maximum Material Condition

The Maximum Material Condition (MMC) modifier is a circled "M". See Figure 8-9. The MMC modifier indicates that the specified tolerance value applies only when the controlled feature is actually produced at MMC. If the controlled feature is produced at any allowable size other than MMC, the position tolerance value increases an amount equal to the size departure from MMC.

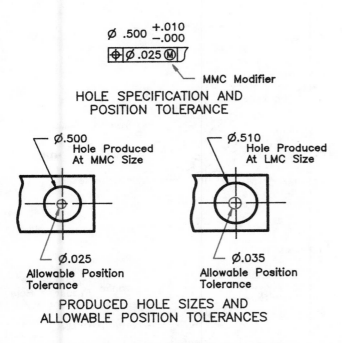

Figure 8-9. The position tolerance zone increases in size when the produced feature size departs from MMC and the MMC modifier is applied to the tolerance specification.

A hole specification and position tolerance are shown in the given figure. Limits of size for the hole are .500″ and .510″ inch diameter. The position tolerance is specified as .025″ diameter MMC.

Any hole produced at the exact MMC size of .500″ is permitted a position tolerance of .025″ diameter. Any hole produced at a larger diameter is permitted a larger amount of position tolerance. For every .001″ diameter increase in hole size, there is an allowable .001″ increase in the diameter of the position tolerance zone.

Two examples of produced holes are shown. One hole is produced at .500″ diameter, the maximum material condition for the hole. The position tolerance for this hole is .025″ diameter. The other hole is produced at .510″ diameter, the

least material condition for the hole. The position tolerance for this hole is increased to .035″ diameter.

The MMC modifier does not permit violation of the limits of size. The example hole must not be produced at a size smaller than .500″ or greater than .510″.

The MMC modifier is typically used for clearance hole applications. An example is a clearance hole through which a bolt is passed. It makes sense to allow the position tolerance to increase as hole size increases on such applications. If a clearance hole is made larger, it can be mislocated by a greater amount and still let the bolt pass through.

Least Material Condition

The Least Material Condition (LMC) modifier is a circled "L". See Figure 8-10. The LMC modifier indicates that the specified tolerance value applies only when the controlled feature is actually produced at LMC. If the controlled feature is produced at any allowable size other than LMC, the position tolerance value increases an amount equal to the size departure from LMC. This is similar to the MMC concept, except it uses the LMC size as the size to which the original tolerance value is applied.

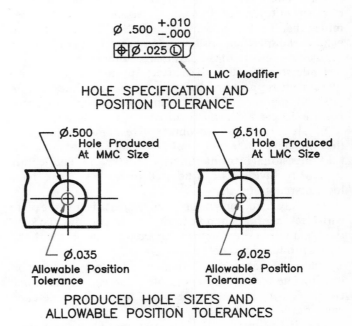

Figure 8-10. The position tolerance zone increases in size when the produced feature size departs from LMC and the LMC modifier is applied to the tolerance specification.

A hole specification and position tolerance are shown in the given figure. Limits of size for the hole are .500″ and .510″ inch diameter. The position tolerance is specified as .025″ diameter LMC.

Any hole produced at the exact LMC size of .510″ is permitted a position tolerance of .025″ diameter. Any hole produced at a smaller diameter is permitted a larger amount of position tolerance. For every .001″ diameter decrease in hole size, there is an allowable .001″ increase in the diameter of the position tolerance zone.

Two examples of produced holes are shown. One hole is produced at .510″ diameter, the least material condition for the hole. The position tolerance for this hole is .025″ diameter. The other hole is produced at .500″ diameter, the maximum material condition for the hole. The position tolerance for this hole is increased to .035″ diameter.

The LMC modifier is typically used for applications where edge distance is of concern. Examples are a clearance hole centered in a boss or a structural hole near the edge of a plate.

Proof of the MMC Concept

The MMC modifier should be used in a majority of clearance hole applications where the main concern is free assembly of the parts. Application of MMC permits greater freedom in how the part is produced.

Application of an MMC modifier on a position tolerance permits an increase in location error as the hole size is increased. This provides additional flexibility during the production of the part.

When the MMC modifier is on the tolerance, the machinist can determine how to best utilize the total permitted tolerance. It may be determined that it is best to drill a hole near its least material condition. This allows a maximum amount of position tolerance. In some situations, it may be determined that it is best to work near the MMC size, and try to achieve the specified position tolerance. This permits more latitude for rework of the part if an error is made.

Understanding how size affects position tolerance requirements supports the utilization of the MMC modifier on position tolerances. The following explanation shows how feature size can affect the required position tolerance.

Figure 8-11 shows two simple plates. One has two pins, the other has two clearance holes. It is not necessary to know the dimensional information for the two parts. The geometry and relative sizes can be used to see how size affects location requirements.

The only requirement for the given parts is that the plate with clearance holes be able to fit over the two pins. To keep the explanation simple, it is assumed that the plate with the pins is produced as a perfect part.

One of the clearance holes is assumed to have a snug fit on its pin. The second clearance hole is larger, and has some clearance between it and the pin. There are two detailed views of the second hole.

The first detail view shows an extreme possible position for the clearance hole when the hole is at its smallest limit of size (the MMC size). Only a small amount of clearance exists between the hole and the pin. The acceptable location error for the hole is directly related to the amount of clearance between the pin and the hole.

The second detail view shows an extreme possible position for the clearance hole when the hole is at the maximum limit of size (departed from MMC). The amount of clearance between the hole and the pin has increased due to the increase in the hole size. The increased clearance permits a greater amount of location error.

Increased hole size in the previous explanation resulted in an increase in the amount of permitted location error. The characteristics described show that departure from MMC

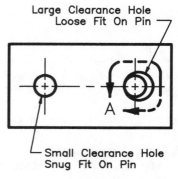

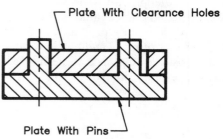

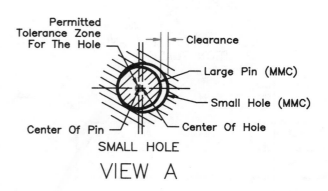

SMALL HOLE

VIEW A

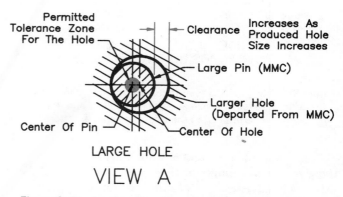

LARGE HOLE

VIEW A

Figure 8-11. A hole with a large amount of clearance can tolerate a larger location error than can a close-fitting hole.

does permit increased location errors when clearance fits are used in an assembly. This provides proof that the MMC concept is valid.

Application of an MMC modifier on a tolerance value has a well-defined impact. *When the MMC modifier is applied to a tolerance value, the permitted position tolerance increases by exactly the same amount as the feature departs from MMC.*

TOLERANCE CALCULATION

Tolerance values should always be calculated. Calculation is the only means of making sure the tolerances will always result in produced parts that will assemble properly. Two simple formulas are all that must be remembered to complete the calculations for many common tolerancing applications. These two formulas and their proper use are defined in the following paragraphs.

These formulas are used to calculate tolerances for the assembly of parts in either of two fastener conditions. These conditions are the floating fastener condition and the fixed fastener condition.

Floating Fastener Condition

A *floating fastener condition* exists when a fastener passes through multiple holes and none of those holes fix the location of the fastener. A floating fastener condition exists when all the holes have a diameter larger than the fastener. See Figure 8-12. Two plates are shown; each has a clearance hole that is larger than the bolt that passes through them. Since the holes are larger than the bolt, it is free to float within them.

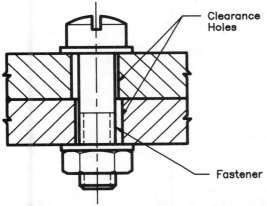

Figure 8-12. A floating fastener condition exists when a bolt or shaft passes through two clearance holes.

When designing parts that assemble in a floating fastener condition, it is common to use the same size clearance hole in each of the parts. When using the same size holes, the following simple formula can be used to calculate a position tolerance:

$$T = H - F$$

T = **TOLERANCE** applied to each hole

H = **HOLE** (MMC size)

F = **FASTENER** (MMC size)

Care must be taken to only use this formula for the holes in a floating fastener application. The calculated tolerance value is the number that must be applied to the holes on each of the parts.

The hole and fastener sizes used in the calculation must be the MMC sizes of each feature, since the two MMC sizes when combined result in the worst condition fit between the two features. Since the position tolerance value is determined using the MMC sizes for the parts having a clearance fit, the tolerance can be specified with the MMC modifier.

Figure 8-13 shows information that validates the floating fastener formula. The figure shows two plates. The edges of the plates are aligned and held in a fixed location relative to one another.

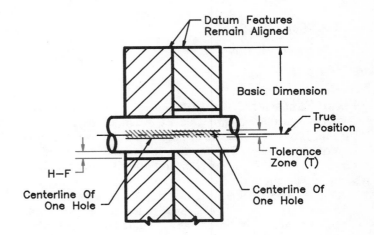

$$T = H - F$$
T Is The Tolerance
At MMC For Both Parts

Figure 8-13. A simple figure is shown to prove the validity of the floating fastener formula.

A hole is shown in each of the plates. The two holes are equal in size. A shaft is passed through the two holes. The two holes and the shaft are all assumed to be at MMC.

The shaft is shown centered on the true positions for the holes. The holes are offset to the maximum extent possible without moving or interfering with the shaft. Offsetting the holes in opposite directions locates the centerlines of the holes on opposite sides of the true position. The distance between the two offset centerlines is equal to **T**, which is equal to **H - F**.

The condition shown is the worst case. Any other position of either hole would be better than the condition shown. Since the tolerance **T = H - F** works for the worst case condition, it will work for all others.

Example Calculation Problem:

Required: Calculate the position tolerance for a floating fastener application.

Given information: Specified Hole Diameter .328″ ± .004″

Fastener Diameter: .312″ MMC

Solution:

T = H - F

T = .324″ - .312″

T = .012″

The solution of a floating fastener problem requires that at least two parameters be known. If there is only one known parameter, such as the size of the fastener, then either a tolerance or a hole size must be selected. Usually a standard hole size can be selected that results in a position tolerance that is large enough to be practical for fabrication.

The floating fastener formula can easily be used to solve for a hole size when the fastener and a preselected tolerance are known. Simple mathematics is used to solve for the hole size.

$$T = H - F$$

$$T + F = H - F + F$$

$$H = T + F$$

Fixed Fastener Condition

A *fixed fastener condition* exists when a fastener passes through multiple holes and one of those holes fixes the location of the fastener. The fastener location can be fixed by threads, a press fit, a taper, or any other feature that prevents movement of the fastener.

Only one part is permitted to fix the fastener location. The other parts include clearance holes. See Figure 8-14. Two plates are shown. One has a threaded hole. The other has a clearance hole. The fastener has a location that is fixed by the threaded hole. If the threaded hole is produced somewhere off true position, then the fastener will be pulled off true position with the hole.

When designing parts that assemble in a fixed fastener condition, the following simple formula can be used to calculate a position tolerance:

T = (H - F)/2

T = **TOLERANCE** applied to each hole

H = **HOLE** (MMC size)

F = **FASTENER** (MMC size)

CAUTION: This formula assumes the tolerance on the fixed fastener feature will be applied using a projected tolerance zone.

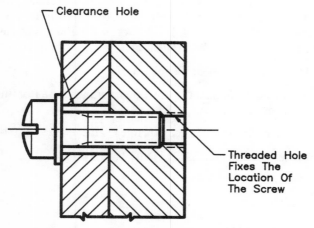

Figure 8-14. A fixed fastener condition exists whenever the location of the fastener is determined by one of the parts.

Care must be taken to only use this formula for the holes in a fixed fastener application. It should only be used if the tolerance applied to the fixed fastener feature is specified as a projected tolerance zone. The calculated tolerance value is the number that is applied to the holes on each of the parts.

Figure 8-15 shows information that validates the fixed fastener formula. The figure is simplified for illustration purposes, and shows a tolerance zone within the threaded hole. In reality, the tolerance zone for the threaded hole would be specified to project into the clearance hole.

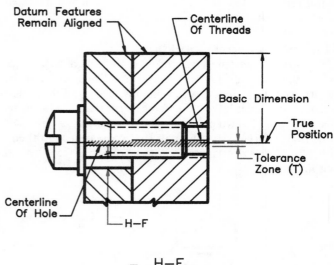

$$T = \frac{H - F}{2}$$

T Is The Tolerance
Applied To Both Parts

Figure 8-15. A simple figure is shown to prove the validity of the fixed fastener formula.

The figure shows two parts. The edges of the parts are aligned and held in a fixed location relative to one another.

A threaded hole is shown in one part; a clearance hole is shown in the other. A fastener is passed through the clearance hole and threaded into the second part. The fastener and the threaded hole have the same centerline since screw threads are self-centering. The clearance hole and the fastener are shown to be at their MMC size.

Neither the fastener nor the clearance hole are shown centered on the true positions for the holes. The threaded hole and the clearance hole are offset in opposite directions. Since moving the threaded hole to one side also moved the fastener, the clearance hole could not be moved very far before one side of the clearance hole made contact with the fastener. In effect, the permitted movement of the threaded and clearance holes depends on the clearance created by the clearance hole diameter.

The diametral difference between the clearance hole and the fastener is the total amount of position tolerance that can be split between the threaded hole and the clearance hole. When evenly divided, the two features can move equal distances on opposite sides of the true position centerline just as the figure shows.

Example Calculation Problem:

Required: Calculate the position tolerance for a fixed fastener application.

Given information: Specified Hole Diameter .296″ ± .003″

Fastener Diameter: .250″ MMC

Solution:

$T = (H - F)/2$

$T = (.293″ - .250″)/2$

$T = .043″/2$

$T = .0215″$

Just as for a floating fastener condition, the solution of a fixed fastener problem requires that at least two parameters be known. The fixed fastener formula can easily be used to solve for a hole size when the fastener and a preselected tolerance are known. Simple mathematics is used to solve for the hole size.

$T = (H - F)/2$

$2T = H - F$

$2T + F = H - F + F$

$H = 2T + F$

BONUS TOLERANCES

Formulas used for calculating tolerances result in a numerical value. The numerical value is the amount of tolerance placed in the feature control frame.

Depending on the design function, the feature control frame will include one of the three material condition modifiers: MMC, LMC, or RFS. When the MMC modifier is used, the value in the feature control frame only applies when the controlled feature is produced at the MMC size.

At any size other than MMC, the allowable tolerance zone is increased. The allowable increase in the tolerance is directly related to the departure of the controlled feature from its MMC size. *The allowable increase in the tolerance is known as the bonus tolerance.*

The amount of bonus tolerance depends on the actual produced size. It is possible to calculate the bonus tolerance for each of the possible produced sizes, but the exact allowable bonus for a particular feature is not known until the feature is produced.

Figure 8-16 shows the affect of the MMC modifier on the position tolerance for a hole. Limits of size for the hole are .500″ and .506″ diameter. The specified position tolerance of .035″ diameter applies when the hole is at its MMC size of .500″ diameter. The chart shows that for every .001″ increase in the hole diameter, there is a .001″ increase in the allowable position tolerance.

The allowable position tolerance can easily be calculated. The allowable tolerance is equal to the sum of the specified tolerance and the bonus tolerance. The bonus tolerance is determined by finding the difference between the produced hole size and the MMC hole size.

The figure shows how the allowable tolerance is calculated for a hole produced at a .503″ diameter. The difference between the produced size (.503″) and the specified MMC size (.500″) is found to determine the bonus tolerance (.003″). The bonus tolerance (.003″) and the specified position tolerance (.035″) are added together to determine the allowable position tolerance (.038″).

Position tolerances applied to external features are affected by modifiers in the same manner as the tolerances on

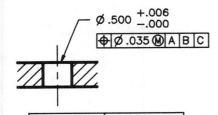

Figure 8-16. The MMC modifier, when used on the position tolerance for a hole, can significantly increase the allowable tolerance on the produced holes.

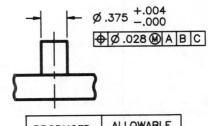

Figure 8-17. The MMC modifier when used on the position tolerance for a pin or shaft can significantly increase the allowable tolerance on the produced parts.

internal features. See Figure 8-17. The pin in the given figure has limits of size from .375" to .379" diameter. A position tolerance of .028" diameter at MMC is applied to the pin.

When the pin is produced at the MMC size of .379" diameter, the allowable position tolerance is .028" diameter. There is an allowable increase in the position tolerance if the pin departs from the MMC size. The given chart shows that for every .001" decrease in the pin diameter, there is a .001" increase in the allowable position tolerance.

The allowable position tolerance can just as easily be calculated for a pin as has already been demonstrated for a hole. The figure shows how the allowable tolerance is calculated for a pin produced at a .376" diameter. The difference between the produced size (.376") and the specified MMC size (.379") is found to determine the bonus tolerance (.003"). The bonus tolerance (.003") and the specified position tolerance (.028") are added together to determine the allowable position tolerance (.031").

PAPER GAGING

Regardless of how parts are dimensioned and toleranced, it is necessary to inspect completed parts to see if they met the drawing requirements. It is relatively simple to verify position tolerances. Several methods can be used.

Some companies have computer-driven coordinate measuring machines that automatically check hole locations and verify position tolerances. It is not necessary to have one of these expensive machines to complete an accurate check of hole locations.

Position tolerances can be verified using relatively simple equipment for measuring coordinates and a sheet of paper for recording data. It is only necessary for the equipment to be good enough to make accurate coordinate measurements relative to the datum reference frame. It is also necessary to have a means for accurately measuring the diameters of produced holes.

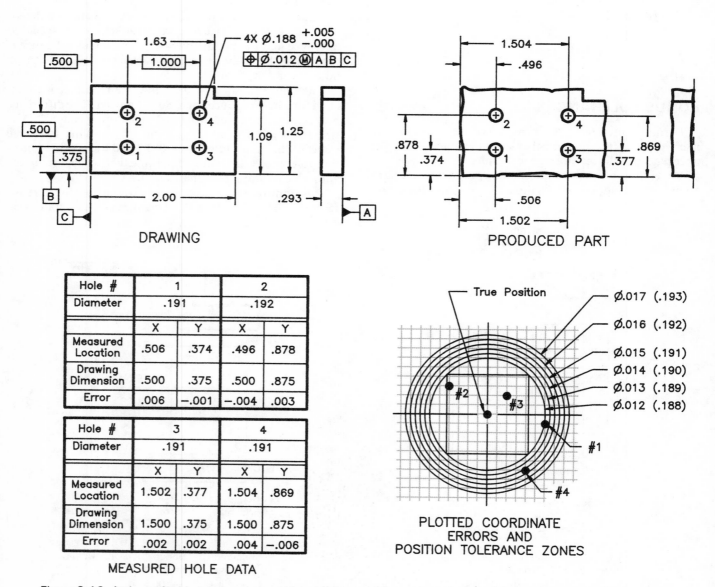

Hole #	1		2	
Diameter	.191		.192	
	X	Y	X	Y
Measured Location	.506	.374	.496	.878
Drawing Dimension	.500	.375	.500	.875
Error	.006	−.001	−.004	.003

Hole #	3		4	
Diameter	.191		.191	
	X	Y	X	Y
Measured Location	1.502	.377	1.504	.869
Drawing Dimension	1.500	.375	1.500	.875
Error	.002	.002	.004	−.006

MEASURED HOLE DATA

Figure 8-18. A piece of grid paper can be used for plotting coordinate errors from a produced hole pattern. A series of concentric circles can be overlaid on the grid to determine the position errors.

Figure 8-18 illustrates how a hole pattern can be verified to meet the required position tolerances. The four holes in the given drawing include a position tolerance of .012″ at MMC, relative to datums A, B, and C.

Before starting to inspect the part, a document (paper) needs to be set up to record the data. A piece of graph paper is adequate if preprinted forms are not available. A table should be created for showing both the requirements and the measured data. This permits an easy comparison between the specified values and the actual produced part.

In addition to checking hole positions, the hole diameters are measured. This can be done with pin gages or an inside micrometer. The produced diameters are recorded on the sheet of paper. If any hole is oversize, the part must be scrapped or accepted out-of-tolerance. If any hole is undersize, the part can be reworked to increase the size of the hole.

The first step in verifying the hole positions is to set the part up on the three referenced datums. The equipment used for this can be relatively simple. For this part, a surface plate and two angle blocks is adequate. Care must be taken to locate the workpiece to maintain the proper surface as the primary, secondary, and tertiary datums.

Coordinate measurements to each of the holes is made relative to the datum reference frame. The coordinate data is entered in the table. The coordinate data is compared to the required coordinates to determine the amount of X and Y error in the location of each hole.

Hole #1 has been measured at a location of .506″ in the X direction, .374″ in the Y direction. The dimensions had indicated a location of .500″ and .375″. The amount of coordinate error is therefore .006″ and -.001″, respectively.

The coordinate error can be used in either of two ways to determine the position tolerance error. Both methods are relatively simple. The method shown in the figure is to plot the coordinate errors on a graph.

The shown graph includes a series of concentric circles. The smallest circle is drawn to represent a diameter of .012″. Another circle is drawn at every .001″ increase in diameter up to a diameter of .017″. The circles in this figure are drawn at a scale of 100:1 to minimize any errors that might be made. In actual practice, a scale of 250:1 is often used.

The concentric circles are superimposed on a grid. Coordinate errors for the four holes are plotted on this grid, using the center of the circles as the origin. The plotted location for hole number one is at .006, -.001. The point falls slightly outside the .012″ diameter circle. Since hole #1 was measured at a diameter of .191″, its position tolerance is .015″ diameter due to the affect of departure from MMC. Therefore, the location of hole #1 is acceptable. Plotting the location errors of all four holes shows the given part to be good.

Inspection Calculations

It is possible to calculate the produced position errors without plotting coordinate errors on the concentric circles as described in the previous section. The coordinate errors can be used to calculate the amount of needed position tolerance to accept the measured coordinate errors

See Figure 8-19. To determine if a part is good, it is necessary to know the measured position error for every hole. The measured position error must be compared to the allowable

Allowable Error	=	Specified Tolerance	+	Any Applicable Bonus Tolerance

Produced Hole Size	.191		Specified Tolerance	.012
MMC Size	.188		Bonus Tolerance	.003
Bonus Tolerance	.003		Allowable Tolerance	.015

$$\text{POSITION ERROR} = 2\sqrt{\Delta X^2 + \Delta Y^2}$$

	X	Y
Measured Location	.506	.374
Specified Location	−.500	−.375
Coordinate Error	.006	−.001

$$\text{Position Error} = 2\sqrt{.006^2 + (-.001^2)}$$
$$2\sqrt{.000036 + .000001}$$
$$2\sqrt{.000037}$$
$$2(.00608)$$
$$\text{Position Error} = .0122$$

Measured Position Error Must Be ≤ Allowable Error

$$\text{Actual Position Error} \leq \text{Allowable Position Error}$$

$$.0122 \leq .0150$$

Figure 8-19. Allowable and actual position errors can be calculated.

position error. Remember, the allowable position error is the specified tolerance plus any applicable bonus tolerance.

The given figure shows how to complete the required calculations using data from hole #1 in the previous figure. The coordinate errors are used to calculate the measured position error. This is done by squaring the two coordinate errors, then solving for the square root of the sum of the two squares. This number is then multiplied by two. The result for this example is a position tolerance value of .0122″ diameter.

The allowable position tolerance is determined by adding the specified tolerance of .012″ to the bonus tolerance. The bonus tolerance is .003″ since the hole is produced at a diameter .003″ larger than the MMC size. The total allowable position tolerance is calculated to be .015″, which makes the measured error of .0122″ acceptable.

Functional gages can be produced for simple verification of hole positions. When many parts must be inspected, it is often less expensive to fabricate a functional gage than to measure each part and analyze the data. Explanations regarding the usage of functional gages for position tolerances are given in the next chapter.

ADVANTAGES OF POSITION TOLERANCES

Position tolerances have advantages when compared to the use of coordinate tolerances for defining location requirements. One advantage is the clarity of the requirement that a position tolerance indicates. When the basic dimensions, datums, and position tolerance are properly applied, there is only one correct interpretation of the requirements. Coordinate tolerances are subject to interpretation.

Position tolerances take advantage of the full amount of tolerance that is acceptable in the design. This is done through the application of circular zones on round features and bonus tolerances. Coordinate tolerances create square or rectangular zones and can't take advantage of bonus tolerances.

All the freedom and flexibility needed to meet any design application is available when using position tolerances. Either small or large tolerances can be specified, and they can be applied on simple or complex parts.

CLARITY OF POSITION TOLERANCES

Application of position tolerances results in a clearer definition of requirements than a location tolerance defined by coordinate tolerances. There is a well-defined meaning for any properly specified position tolerance. Coordinate tolerances are subject to interpretation and can result in confusion.

The clarity of a position tolerance is attributed to several factors. One significant factor is the documented guidelines provided in the national dimensioning and tolerancing standard. Position tolerancing is well-defined by this standard. Written standard guidelines are important since they eliminate the need for people to assume a set of rules by which to interpret dimensions. Using a dimensioning system that has no standardized interpretation leaves a drawing subject to varying interpretations. There is no current standard that defines the meaning of coordinate location tolerances; therefore, the use of coordinate location tolerances leaves a drawing subject to varying interpretations.

Another contributor to the clarity of position tolerances is the fact that tolerance zones are located relative to a datum reference frame. A datum reference frame provides a clear reference from which to manufacture and inspect the part. Coordinate location tolerances are not as clearly defined since they originate from surfaces that may be produced with some amount of error on them. The datum reference frame is not affected by surface errors.

Clarity of the position tolerance requirement is further enhanced by the usage of basic dimensions to define the nominal feature locations. There is no tolerance accumulation on basic dimensions. Since there is no tolerance accumulation, the position tolerance zones are always centered on the true positions. This does not require that the part be perfect. Only the tolerance zone must be located in the true position. The toleranced features can be anywhere within the defined tolerance zone.

Freedom exists to express the required level of control, ranging from very small to very large tolerances, when using position tolerances. One misconception about position tolerancing is that it should only be used to achieve a high degree of accuracy. Of course it can be used to express small tolerances, but it is just as useful for expressing large tolerances. The only limits on the tolerance values placed in a feature control frame are the functional requirements of the part and the available manufacturing capabilities.

Flexibility to meet a wide range of design applications is built into the position tolerancing methods. The methods are not restricted to complex parts, or limited to simple ones. Only the level of control required by the function of the part needs to be specified when using position tolerances.

Position tolerancing methods are well-defined by the current dimensioning and tolerancing standard. They are widely accepted by major industries throughout the United States. The military and defense contractors use these methods extensively. Many small companies are recognizing the benefits of this system and are adopting them for utilization in their design drawings. It is necessary to understand and properly utilize these methods to be compatible with a large segment of our industry.

Ambiguity of Coordinate Location Tolerances

Coordinate location tolerances have been used for many years, and a great number of parts have been successfully produced. However, the past successful fabrication of parts is not adequate reason for continued use of a system that is ambiguous.

There currently isn't a standard that defines the meaning of coordinate location tolerances. A simple plate with four holes in it can be used to show the ambiguity of coordinate tolerances. See Figure 8-20.

There are four holes in the given plate. The drawing shows hole #1 dimensioned from the lower-left corner of the plate. Holes #2, #3, and #4 are dimensioned from hole #1. Edges of the plate are drawn straight and at a 90° angle to one another. The dimensions to the holes are drawn parallel to the edges. Tolerances on all the hole location dimensions are ± .010″.

One possible produced part is shown below the drawing. The edges of the part have been saw cut and are rough. The edges don't form a perfect 90° angle. This is acceptable since the drawing permits a ± .5° tolerance on angles.

Immediate questions start to arise as it becomes necessary to locate hole #1. Since the edges are not perpendicular, from where should measurements be made and from which edge should measurements be oriented? This is a difficult question since no datums are identified or referenced.

If some assumptions are made, it is possible to make measurements for holes #1, #2, and #3. Locations for these holes could be made relative to the edges nearest the hole. Of course, surface errors will cause some location error for the holes. The given figure shows one possible position for the first three holes if we assume that measurements can be made from the edges. The hole locations result in centerlines spread to an angle greater than 90°.

A considerable problem now exists for locating hole #4. Should measurements continue from the edges of the part, or should measurements now be made from the other three holes? If they are made from the holes, how can this be done with the centerlines forming an angle greater than 90°? There are no defined answers to these questions. There are no answers because there are no guidelines defining how the problem is to be addressed.

The ambiguity in coordinate location tolerances is caused by part variations that can exist and have been shown in the given figure. Each person that attempts to determine how the part is to be measured must assume a set of guidelines. Each person that assumes different guidelines can produce parts that are different from anyone else's parts.

Position tolerances have not always been shown on drawings, so how is it that good parts were ever produced to coordinate location tolerances? In many situations, manufacturing processes were set up to produce good parts based on the obvious function of the part. This was done in spite of the incomplete requirements defined by the drawings.

In some situations, parts have been produced that the manufacturer thinks meets the drawing requirements, yet the part will not fit into the required assembly. When the fabricator and the customer disagree about the requirements on a drawing, the disagreement can often lead to a lawsuit.

It is much better to completely and clearly define requirements on the drawing than to argue about drawing interpretation after parts have been manufactured.

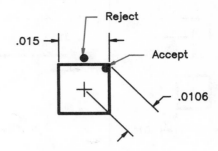

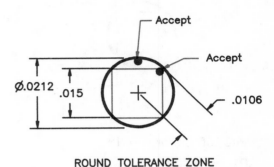

Figure 8-21. A round tolerance zone permits the same amount of error in all directions. The square tolerance zone does not.

INCREASED TOLERANCE ZONE AREA

Round tolerance zones specified by positional tolerances provide a significantly increased amount of area when compared to the square tolerance zone created by a coordinate tolerance. See Figure 8-21. The top half of the given figure shows a square tolerance zone. The square is .015″ across the flats. Locations for the centers of two holes are indicated by small solid circles. Both of the center points are the same distance (.0106″) from the center of the tolerance zone. One center point is in the corner of the tolerance zone, and is therefore acceptable. The other center point is on the vertical centerline. It is located outside the square tolerance zone; therefore, it is a reject location.

The illustrated square zone forces the rejection of one of the two parts, although the amount of error in the two parts is identical. Fortunately, position tolerances can be used to avoid rejecting functionally good parts.

The bottom half of the figure shows a circular tolerance zone circumscribed on the square. The tolerance zone has a .0212″ diameter. This size of round tolerance zone permits a .0106″ location error in any direction, therefore the two shown center point locations are acceptable.

Initial impressions of the improvement from a square to a circular zone may not seem significant, but calculation of the areas for the two shapes show there is actually a *57% increase in area*. See Figure 8-22. The area of a .015″ square is .000225 in^2. The area of a .0212″ diameter circle is .000346 in^2.

The increased tolerance area realized with the round zones can be used to increase the producibility of parts. An increase in producibility can often reduce part cost. Ignoring the advantages of the larger zone, and continuing to use square tolerance zones, reduces the functionally acceptable

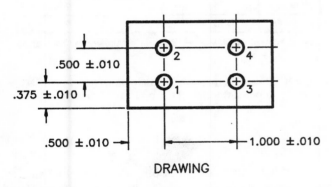

DRAWING

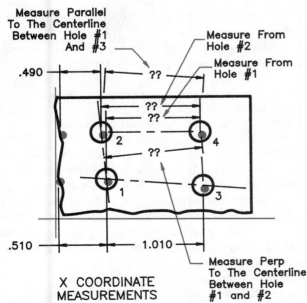

X COORDINATE
MEASUREMENTS

Figure 8-20. Coordinate location tolerances are ambiguous. The origin and direction for measurements are not defined. All measurements taken must be done on the basis of assumptions since no standard defines the coordinate tolerancing system.

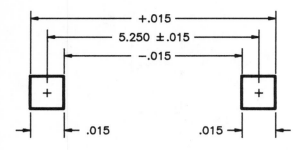

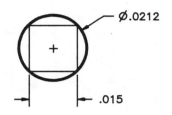

$$\text{Area Of Square} = L^2$$
$$.015^2 = .225 \times 10^{-3} \, in^2$$

$$\text{Area Of Circle} = \pi R^2$$
$$\pi\, .0106^2 = .346 \times 10^{-3} \, in^2$$

Figure 8-22. The area of a round tolerance zone is 57% greater than the area of a square zone.

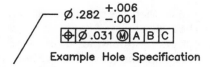

Example Hole Specification

Tabulated Hole Location Data

		SPECIFIED LOCATION	PRODUCED LOCATION	POSITION ERROR	
Hole #1	X	1.000	1.009	+.009	Ø .0241
	Y	2.500	2.492	−.008	
Hole #2	X	1.375	1.390	+.015	Ø .0316
	Y	2.500	2.505	+.005	

	SPECIFIED MMC	MEASURED DIAMETER	BONUS POSITION TOLERANCE
Hole #1	.281	.284	.003
Hole #2	.281	.285	.004

Measured Size − Specified MMC = Bonus Position Tolerance

	ALLOWABLE TOLERANCE	POSITION ERROR	ACCEPTABLE PART
Hole #1	.034	.0241	YES
Hole #2	.035	.0316	YES

Specified Tol At MMC + Bonus = Allowable Tolerance

Figure 8-23. Bonus tolerances due to departure from MMC further increase the size of the position tolerance zone.

tolerance by 33%. Unnecessarily reducing the available tolerance can cause rejection or rework of functionally good parts. It is a bad practice to create drawings that force the rejection or rework of functionally acceptable parts.

BONUS TOLERANCES

It has been shown that using a circular tolerance zone increases the tolerance zone area by 57% over the area of a square zone. The same tolerance zone can be further increased through bonus tolerances if the MMC or LMC modifier is applied.

Figure 8-23 shows how the bonus tolerance can be used to increase the acceptable amount of location error. An example hole specification is given. Just below it, a table compiling inspection data from two holes on a produced part is shown. The resulting position error is tabulated in the right column.

A second table shows how the allowable bonus tolerance on each of the holes is determined. The bonus tolerance on hole #1 is .003″. This value added to the specified .031″ results in an allowable tolerance of .034″ on hole #1. The bonus tolerance on hole #2 is .004″. This value added to the specified .031″ results in an allowable tolerance of .035″ on hole #2.

The added bonus tolerance on holes #1 and #2 significantly increases the allowable variation on the part. However, the variation will not prevent the part from functioning properly.

The advantage of a bonus tolerance is only achieved by using position tolerances and the MMC or LMC material condition modifiers. There is no way of taking advantage of bonus tolerances when using coordinate tolerances. Coordinate tolerances provide a fixed tolerance value. It remains unchanged regardless of how large the holes are produced.

There are several ways in which bonus tolerances can be used to increase producibility. Careful planning can actually let the machine shop take advantage of the bonus before any holes are produced. For the example holes in Figure 8-23, the holes could purposely be drilled large to increase the allowable position tolerance. If holes are drilled at a diameter of .285″, the resulting bonus tolerance of .004″ can be added to the specified position tolerance to obtain an allowable position tolerance of .035″ diameter.

Bonus tolerance can also be used when reworking holes that are slightly outside the specified position tolerance. All that needs to be done to correct the out-of-tolerance condition is to open the hole diameter. This will result in the needed bonus tolerance.

Using the hole specification in Figure 8-23, consider a

part in which one hole is produced too far out-of-position. Consider a hole that has a .281″ diameter and a position error of .032″ diameter. For this hole, a bonus of only .001″ is needed. This means the hole only needs to be increased to a diameter of .282″.

SPECIAL APPLICATIONS

Position tolerances are commonly applied to features of size other than simple clearance holes. Some common applications include counterbored holes, threaded holes, and noncylindrical features of size. Some application requirements also affect where the tolerance zone is located. Although the default location of a tolerance zone is along the length of the controlled feature, the zone is sometimes specified to lie outside the feature. This is done with a projected tolerance zone.

COUNTERBORED HOLES

A common use for a counterbored hole is to permit the installation of a flush-head screw or bolt. It is only possible to install the screw or bolt if the clearance hole and the counterbore are the correct size and accurately located. Counterbores may or may not be produced in a single machining operation, so the two features may not lie on the same centerline.

Positional tolerance specifications can be applied that will ensure assembly of the screw or bolt into the counterbored hole. There are three methods of achieving the desired results. See Figure 8-24. The first method that may be used is to place a single position tolerance specification next to the hole and counterbore callout. The position tolerance applies to both features. The two features act independently, and can float opposite directions within the specified zone.

If different tolerance zones are desired for the two features, then a position tolerance specification can be placed by the hole callout and another by the counterbore callout. The features act independently, and can travel to opposite sides of their respective tolerance zone.

It is possible to reference the counterbore to the hole if, for some reason, there is a need to do this. In this case, a position tolerance is applied to the clearance hole. If there are four holes in the hole pattern, then the position tolerance applies to all four holes. A datum feature identifier is applied to one of the holes, with a note underneath stating, "4X INDIVIDUALLY". This indicates that each hole is going to serve as a datum for its counterbore. Omission of the notation would indicate that all four holes act simultaneously to establish a datum.

A position tolerance is applied to one of the counterbores to show the counterbore position relative to the clearance hole. The note of "4X INDIVIDUALLY" is placed under the position tolerance for the counterbore. This allows each of the counterbores to act as a single entity that is located only relative to its clearance hole.

Counterbore diameter calculations are completed using the same formulas as are used for calculating the clearance hole diameter. The maximum material condition for the

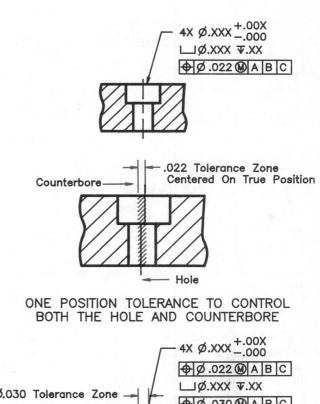

ONE POSITION TOLERANCE TO CONTROL BOTH THE HOLE AND COUNTERBORE

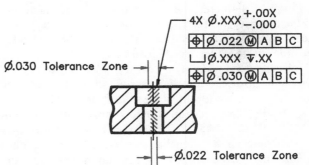

SEPARATE POSITION TOLERANCE FOR THE HOLE AND COUNTERBORE

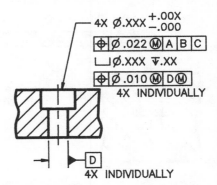

COUNTERBORE POSITION TOLERANCE REFERENCED TO THE HOLE LOCATION

Figure 8-24. Three methods for controlling the location of a counterbore are shown.

screw head is used in calculations to determine the counterbore diameter.

APPLICATION TO THREADS

Position tolerances applied to a threaded feature are assumed to be applicable to the location of the pitch diameter. This is because threads are self-centering on the pitch diam-

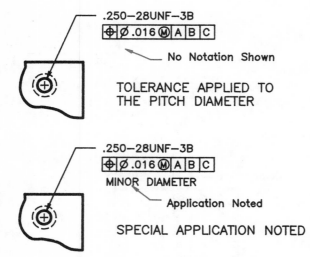

.250—28UNF—3B

⊕ | ⌀.016 Ⓜ | A | B | C

No Notation Shown

TOLERANCE APPLIED TO
THE PITCH DIAMETER

.250—28UNF—3B

⊕ | ⌀.016 Ⓜ | A | B | C

MINOR DIAMETER

Application Noted

SPECIAL APPLICATION NOTED

Figure 8-25. A position tolerance applied to a thread is
understood to control the location of the pitch diameter
unless noted otherwise.

eter. The pitch diameter, therefore, serves as the logical controlling feature. The feature control frame can be placed adjacent to the thread specification, or it can be connected to the hole with a leader. See Figure 8-25.

If it is necessary to apply a position tolerance to control either the minor of major diameter location of a threaded feature, then a notation must be placed under the feature control frame. The notation should simply state "MINOR DIAMETER" or "MAJOR DIAMETER". Control of these features in place of the pitch diameter should be approached with caution since there is no certainty that the various diameters will actually be coaxial when the thread is produced.

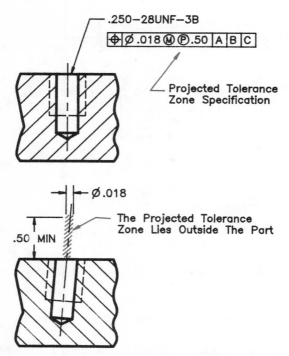

.250—28UNF—3B

⊕ | ⌀.018 Ⓜ Ⓟ .50 | A | B | C

Projected Tolerance
Zone Specification

⌀.018

The Projected Tolerance
Zone Lies Outside The Part

.50 MIN

Figure 8-26. Projected tolerance zones can be specified with
the projection distance shown in the feature control frame.

PROJECTED TOLERANCE ZONES

Position tolerances applied to threaded holes often carry the requirement for a projected tolerance zone. See Figure 8-26. *A projected tolerance zone lies outside the controlled feature*. This is appropriate for a feature such as a threaded hole, since the thread actually locates a screw that extends outside the hole. The location of the screw where it passes through the adjacent part is what is important, so the tolerance zone should extend a distance at least equal to the mating part thickness.

One method for indicating a projected tolerance zone is to show the projection distance and a circled "P" following the position tolerance specification. The given example shows a requirement for a projected zone to extend .50″ outside the part.

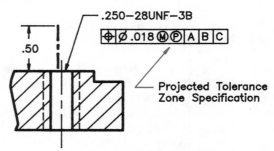

.250—28UNF—3B

⊕ | ⌀.018 Ⓜ Ⓟ | A | B | C

.50

Projected Tolerance
Zone Specification

Figure 8-27. Projected tolerance zones can be specified with
the projection distance dimensioned on the feature.

The projected tolerance zone lies totally on the outside of the feature. On the shown part, the axis of the thread must be within the specified .018″ diameter tolerance zone. The tolerance zone extends from the part surface to a distance .50″ outside the surface. There is no control of the thread location on the inside of the part.

A projected tolerance zone can also be indicated with only the circled "P" inside the feature control frame and a dimension applied directly to the controlled feature. See Figure 8-27. A heavy chain line is drawn adjacent to the controlled axis and a dimension applied to it. This is especially useful

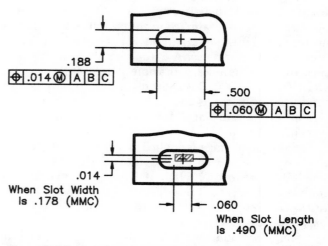

.188

⊕ | .014 Ⓜ | A | B | C

.500

⊕ | .060 Ⓜ | A | B | C

.014

When Slot Width
Is .178 (MMC)

.060

When Slot Length
Is .490 (MMC)

Figure 8-28. Two feature control frames are used to show
a bidirectional position tolerance.

when the controlled feature is a through hole. The dimension and chain line clearly indicate to which side of the part the zone is to project.

BIDIRECTIONAL CONTROL

It is sometimes desirable to permit more position tolerance in one direction than the other. This is referred to as *bidirectional control.*

A common application of bidirectional control is on a slotted hole. See Figure 8-28. Two feature control frames are required to show the amount of tolerance in each direction. The given figure shows a position tolerance of .014″ applied to the .188″ slot width dimension. A position tolerance of .060″ is applied to the .500″ slot length dimension. Neither of the two position tolerances include a diameter symbol.

The tolerance zone created by the bidirectional specification is rectangular. This tolerance zone can't take full advantage of the functionally acceptable zone, but it does provide advantages over coordinate tolerances. The bidirectional zone is specified relative to a datum reference frame. This gives a well-defined true position. Also, the bidirectional zone can be specified with the MMC modifier. This provides the advantage of having bonus tolerances.

NONCYLINDRICAL FEATURES OF SIZE

Location requirements for slots, tabs, rails, and other noncylindrical features of size can be specified using position tolerances. See Figure 8-29. The tolerance specification is placed adjacent to the size dimension of the feature to be controlled, or it may be attached to the dimension line.

The feature control frame must not be placed on one of the extension lines. Placement on an extension line would incorrectly indicate a control on one surface. This would be wrong since position tolerances are only used to control features of size.

The first example shows a position tolerance of .016″ on a slot. The tolerance zone is referenced to datum A, which is the center plane established by the width of the part. Since the controlled slot is centered on the referenced datum, the slot is understood to be symmetrically located. For this reason, no dimension is needed to locate the slot.

The tolerance zone for the slot is a rectangular zone .016″ wide. The tolerance zone is centered on datum plane A. The tolerance zone has a height equal to the slot, and extends the full length of the slot. The center plane of the slot must fall inside the tolerance zone. Sides of the slot may be in any condition provided the limits of size are met, and the center plane falls inside the position tolerance zone.

The second example shows a keyseat in a shaft. A position tolerance of .010″ is applied to the width of the keyseat. It is referenced to datum axis B. Datum axis B is established by the diameter of the shaft. The .010″ tolerance zone is centered on the datum axis.

Neither of the position tolerances applied to the shown slots has any affect on the bottoms of the slots. The position tolerances are placed adjacent to the slot width dimensions, and therefore only control the position of the slot width.

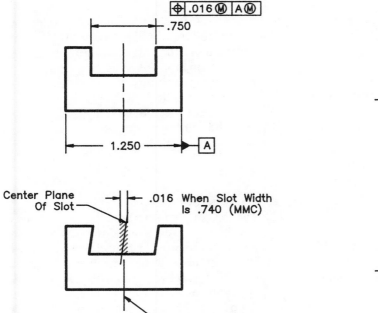

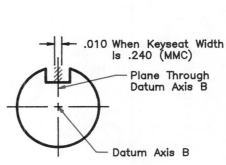

Figure 8-29. A position tolerance applied to a rectangular feature of size creates a rectangular tolerance zone that controls the location of the feature's center plane.

Position tolerances are use to specify the required location accuracy for features of size.

Datum references are required on position tolerance specifications.

RFS modifiers are assumed on position tolerances and on any references to datum features of size.

The position tolerance zone for a hole is typically cylindrical in shape.

True positions for holes are defined with basic dimensions when position tolerances are applied to the holes.

The MMC and LMC modifiers can be used to permit tolerance zones to increase, with the amount of increase dependent on the produced size of the toleranced feature.

Two simple formulas permit the calculation of position tolerances for fixed and floating fastener conditions.

Bonus tolerances are a result of applying the MMC or LMC modifier. Bonus tolerances permit significant increase in the tolerance zone size, depending on the actual produced size of the controlled feature.

Inspection data from produced parts can be plotted on grid paper to determine if position tolerance requirements have been met.

Position tolerances are not limited in application to simple through holes. They can be applied to counterbored holes, threaded holes, blind bottom holes, and non-cylindrical features of size.

REVIEW QUESTIONS

Answer the following questions on a separate sheet of paper. Do not write in this book. Accurately complete any required sketches.

MULTIPLE CHOICE

1. One position tolerance specification may be used to control _____ at a time.
   A. one feature
   B. one or more features
   C. two or more features
   D. None of the above.

2. A feature control frame that specifies a position tolerance must include _____.
   A. a tolerance value
   B. a diameter symbol
   C. implied datums
   D. All of the above.

3. Prior to _____, implied datums were permitted.
   A. 1989
   B. 1982
   C. 1973
   D. 1966

4. An RFS material condition modifier is assumed on _____.
   A. a reference to a datum feature of size
   B. a reference to a datum feature
   C. all datum references
   D. none of the datum references

5. The optimum position tolerance zone shape for a round hole is _____.
   A. square
   B. rectangular
   C. cylindrical
   D. spherical

6. Datum references in a position tolerance specification result in measurements being made from _____.
   A. datum feature surfaces
   B. assumed locations
   C. reference points
   D. a datum reference frame

7. The tolerance zone is unchanged by produced feature size when the _____ modifier is applicable.
   A. LMC
   B. RFS
   C. MMC
   D. All the above.

8. A tolerance specification on a hole has an MMC modifier. For every .002″ departure of the hole diameter from MMC size, the tolerance zone increases _____″.
   A. .001
   B. .002
   C. .004
   D. None of the above.

9. A dowel pin pressed into one part and passing through a clearance hole in another part is an example of a(n) _____ fastener application.
   A. fixed
   B. floating
   C. standard
   D. unusual

10. If a hole is .001″ out of the acceptable tolerance zone and the tolerance is specified at MMC, it might be possible to make the hole acceptable by _____.
    A. reducing the hole diameter
    B. moving the hole
    C. increasing the hole diameter
    D. None of the above.

11. A round tolerance zone has _____% more area than a square zone.
    A. 100
    B. 57
    C. 25
    D. None of the above.

12. The axis of a counterbore _____ the axis of the hole in their respective position tolerance zones.
    A. is free to move in a direction opposite of
    B. must move in the same direction as
    C. must remain coaxial with
    D. None of the above.

13. A position tolerance applied to a thread controls the location of the _____ diameter unless indicated otherwise.
    A. minor
    B. pitch

C. major

D. root

14. A position tolerance applied to a slot controls _____ of the slot.

   A. the center plane

   B. the axis

   C. one side

   D. both sides

15. Advantages of position tolerances include _____.

   A. clarity

   B. flexibility in usage

   C. MMC bonus tolerances

   D. All the above.

TRUE/FALSE

16. Position tolerances are used to specify the required location accuracy for features of size. (A)True or (B)False?

17. Implied datums require that the machinist assume which features should be used to locate datums. (A)True or (B)False?

18. A floating fastener condition exists when a bolt passes through a clearance hole in one part and threads into a hole in the second part. (A)True or (B)False?

19. A round tolerance zone has a smaller area than a square tolerance zone. (A)True or (B)False?

20. Bonus tolerances can't be achieved when using the square tolerance zone created by coordinate tolerances. (A)True or (B)False?

21. If one position tolerance is specified for a hole and counterbore, then the counterbore has the same size tolerance zone as the hole. (A)True or (B)False?

FILL IN THE BLANK

22. _____ dimensions must be used to define the true positions for features that have position tolerances applied.

23. The _____ and _____ modifiers permit the tolerance zone size to change in relationship to the produced feature size.

24. Rule #_____ of the current standard requires that the RFS material condition modifier be assumed on position tolerances.

25. A position tolerance zone for a hole is centered on the _____ position of the hole.

26. A _____ fastener condition exists when a bolt passes through clearance holes in two parts.

27. A _____ tolerance zone extends outside the feature that is controlled.

28. A position tolerance applied to a noncylindrical feature of size should not include a _____ in the feature control frame.

SHORT ANSWER

29. Create a feature control frame to control position within a diameter of .026" at MMC relative to datum A primary, B secondary, and C tertiary.

30. Create a feature control frame to control position

within a diameter of .018" at RFS relative to datum B primary, E secondary, and F tertiary.

31. Explain why implied datums are no longer permitted.

32. Explain what is meant by "true position."

33. Show the formula used to calculate a position tolerance for a fixed fastener condition.

34. Show the formula used to calculate a position tolerance for a floating fastener condition.

35. Sketch the symbol for a projected tolerance zone.

APPLICATION PROBLEMS

Each of the following problems require calculations related to position tolerances. All of the problems can be solved using the information contained in this chapter.

36. A hole specification and position tolerance are given. What is the allowable position tolerance for a hole produced at a .502" diameter?

$$\varnothing \ .500 \ {}^{+.006}_{-.000}$$

$$\boxed{\oplus \ | \ \varnothing .018 \ \text{\textcircled{M}} \ | \ A \ | \ B \ | \ C}$$

37. A hole specification and position tolerance are given. What is the allowable position tolerance for a hole produced at a .384" diameter?

$$\varnothing \ .381 \ {}^{+.005}_{-.001}$$

$$\boxed{\oplus \ | \ \varnothing .036 \ \text{\textcircled{M}} \ | \ A \ | \ B \ | \ C}$$

38. Calculate the position tolerance to be applied to the two holes in a floating fastener application. Use the given information.

   Bolt MMC: .250"

   Clearance Hole MMC: .281"

39. Calculate the position tolerance to be applied to the two holes in a floating fastener application. Use the given information.

   Bolt MMC: .190"

   Clearance Hole MMC: .221"

40. Calculate the position tolerance to be applied to the two holes in a fixed fastener application. Use the given information.

   Bolt MMC: .164"

   Clearance Hole MMC: .188"

41. Calculate the position tolerance to be applied to the two holes in a fixed fastener application. Use the given information.

   Bolt MMC: .190"

   Clearance Hole MMC: .220"

42. Calculate the required MMC clearance hole size for the given floating fastener application.

   Bolt MMC: .250"

   Position Tolerance: .016"

43. Calculate the required MMC clearance hole size for the given fixed fastener application.

   Bolt MMC: .312"

   Position Tolerance: .018"

44. A hole is dimensioned at X = 1.500" and Y = 1.250". It

is produced at X = 1.505″ and Y = 1.246″. Calculate the position tolerance error for the produced hole.

45. A drawing of a simple plate with four holes is given. Inspection data showing produced hole sizes and locations is also provided. The data has already been analyzed to determine delta X and delta Y errors. Calculate the position errors for each of the four holes.

46. Use the inspection data from problem 45 to answer this question. Determine the amount of bonus tolerance on each of the four holes in the produced part. Using the bonus tolerance, calculate the total allowable position tolerance for each hole? Use the data from your answer to complete the given table.

Hole Number	Bonus Tolerance	Allowable Position Tolerance
1		
2		
3		
4		

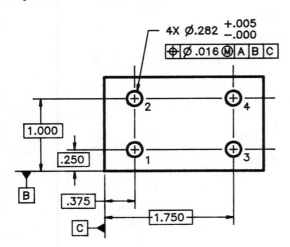

Hole	DIA	Location		Errors		Position Error
		X	Y	ΔX	ΔY	
1	.284	.378	.242	.003	−.008	
2	.285	.379	1.002	.004	.002	
3	.284	1.754	.248	.004	−.002	
4	.285	1.755	1.003	.005	.003	

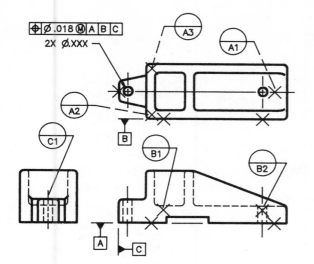

Chapter 9

Position Tolerancing-Expanded Principles, Symmetry and Concentricity

---
CHAPTER TOPICS

☐ The tolerance zones created by composite position tolerances and the effect of repeated datum references in the second line of a composite tolerance specification.

☐ Functional gaging of position tolerances.

☐ Control of multiple groups of features as a single pattern.

☐ Control of multiple groups of features as separate patterns.

☐ Specifying the position requirements for in-line holes, such as on a hinge plate.

☐ Tolerances on symmetrical features.

☐ Controlling coaxial relationships between the axis of two or more features.

☐ A brief overview of past practices related to position tolerances.

INTRODUCTION

Location tolerance applications include position, symmetry, and concentricity. Position tolerancing methods permit a great deal of flexibility in the level of control that is specified on a drawing. The single line position tolerance specification presented in the previous chapter is adequate for many situations, but falls short of meeting all the levels of control necessary to meet complex design requirements. The addition of composite position tolerances provides a significant increase in the flexibility of what can be controlled.

A special functional application of location tolerances is to control symmetry. Depending on the application, this may be done by a position or symmetry tolerance specification — each being specified with a different symbol. Position tolerances are commonly used to control the symmetrical location of features of size at MMC. Symmetry tolerances are intended for applications where features of size are symmetrically located at RFS. The practice of using a symmetry symbol was not included in the 1982 standard, but is now acceptable for RFS applications.

The coaxiality of features is commonly controlled by position or runout tolerances. Some situations occur that require the control of one feature axis relative to another feature axis. These situations can be met through the use of a concentricity tolerance.

Relatively simple paper gaging practices for position tolerances were presented in the last chapter. Those methods require data to be gathered and analyzed. Although paper gaging is a valuable process for many situations, gathering

and analyzing data can be a significant expense if it is necessary to inspect a large quantity of parts.

This chapter shows how functional gages can be used to verify position tolerances. Functional gages provide a means for verifying location accuracy without any need for location measurements or data analysis.

Position tolerancing methods are intended to provide a method of specifying tolerances that reflect the functional requirements of the parts. Expressing functional requirements through properly applied tolerances not only maximize the available tolerance and improve the producibility of the parts, but they can also make it possible to check the parts with functional gages.

Although composite tolerance specifications were included prior to the 1994 standard, complex applications were not explained in detail. This resulted in varying interpretations of composite tolerances. To minimize the potential for variation in interpretation, an extensive effort was made by the ASME Y14.5 committee to expand and clarify the desired interpretation.

In the process of writing the current standard, many viewpoints regarding composite tolerances were discussed. When differing viewpoints were presented, it was necessary to ultimately select only one interpretation or formulate a compromise that would meet the needs of industry.

Much time was spent over a ten year period to define composite position tolerances as described in this chapter. There is some possibility of further definition or reversal of this material by a future standard. For now, consistency can be achieved using composite position tolerances as described here.

COMPOSITE POSITION TOLERANCES

A single line position tolerance specification creates tolerance zones that are in fixed locations relative to a datum reference frame. This results in one tolerance zone for each controlled feature. The tolerance zone on each feature has the same affect on the true position distance between features as it has on the distance of the features from the datum reference frame.

Consider a single line position tolerance of .010″ applied to two holes and referenced to datum reference frame A, B, and C. Tolerance zones measuring .010″ diameter are located at true positions relative to one another and to the datum reference frame. The tolerance zones are not permitted to move relative to each other nor relative to the datum reference frame.

The single line position tolerance creates the same level of control on the location of a hole pattern as it does for the location of holes within a pattern. There are many applications for which this level of control is acceptable. There are other applications for which the location of holes relative to the edge of a part is less critical than the location of holes within the hole pattern.

In many design applications, there is more importance placed in the alignment of holes on mating parts than there is in the alignment of edges on those parts. This situation calls for a means of specifying two levels of control on a hole pattern. One level of control would define a *pattern locating tolerance*. The pattern locating tolerance would allow a relatively large tolerance relative to the edges of the part. The second level of control would define a *feature (hole) relating tolerance*. This control would define a relatively small tolerance to control locations of holes relative to one another. A feature relating tolerance is sometimes referred to as a feature locating tolerance.

Composite position tolerances provide a means to specify the two needed levels of control. One of the controls creates a tolerance for the locations relative to the edges, and the other creates a tolerance for the distance between holes. In other words, one tolerance is for the location of the pattern, and the second is a tolerance on the dimensions between features in the pattern.

A composite position tolerance specification has a distinct appearance that makes it easy to recognize. See Figure 9-1. The feature control frame has two lines, and *there is only one position tolerance symbol*. The feature control frame for a composite position tolerance should not have two tolerance characteristic symbols. This is a requirement that is clearly illustrated in the standard. The single tolerance symbol is the feature that identifies the composite tolerance.

Composite position tolerances are always formatted the same. The first line specifies the *pattern locating tolerance*. The second line specifies the *feature relating tolerance*. The pattern locating tolerance is always a larger value than the feature relating tolerance.

Figure 9-2 shows a simple plate with four holes in it. A composite position tolerance is applied to the four hole pat-

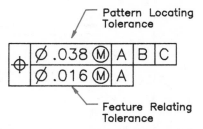

Figure 9-1. The feature control frame for a composite position tolerance includes a pattern locating tolerance and a feature relating tolerance.

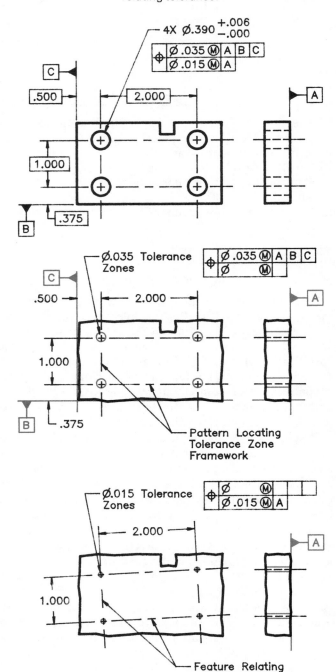

Figure 9-2. Each line in a composite tolerance creates a tolerance zone framework that is sized by the basic dimensions that define the true positions for the holes.

tern. The first line specifies a .035″ diameter pattern locating tolerance. The second line specifies a .015″ diameter feature relating tolerance.

The pattern locating tolerance includes references to three datums–datum A is primary, B is secondary, and C is tertiary. The feature relating tolerance shown in the second line only references primary datum A. *Whenever datum references are shown in the second line of a composite tolerance, they must be repeated from the first line, in the same order of precedence, and with the same material condition modifiers.*

The pattern locating tolerance of the first line has the same effect as a one line position tolerance specification. Tolerance zones are located relative to the datum reference frame, at the true positions indicated by the basic dimensions. In the given figure, the tolerance zones have a diameter of .035″, assuming the holes are produced at MMC. The four tolerance zones are perfectly located relative to the datum reference frame, and also have perfect orientation to it. The four pattern locating tolerance zones at the proper relationship to the datum reference frame can be referred to as the *pattern locating tolerance zone framework.*

The feature relating tolerance of the second line has a different effect than the pattern locating tolerance. The second line always controls the feature-to-feature locations and the orientations between the features. Orientation relative to any datum specified in the second line is also controlled. In compliance with the current standard, the second line does not control the location of the features relative to the datums.

On the given part, the second line establishes tolerance zones located on a 1.000″ by 2.000″ rectangle. The tolerance zones are perpendicular to datum A since that datum is referenced. The four tolerance zones in their proper relative positions to each other and in the proper orientation to the specified datums can be referred to as the *feature relating tolerance zone framework.*

The feature relating tolerance zone framework must be properly oriented to any datums referenced in the second line. The second line does not indicate any location requirement relative to those datums. Location relative to the datums is only included in the pattern locating tolerance of the first line.

Composite tolerances provide two levels of control on toleranced features. The two levels of control permit the freedom to specify a relatively small tolerance to match hole patterns and still permit relatively large tolerances on edge distances. See Figure 9-3. The two requirements in the composite tolerance are not completely separated. Both requirements must be met by the same set of holes.

The pattern locating tolerance zone framework has a fixed location and orientation relative to a datum reference frame. Relatively large tolerance zones are located at the true positions defined by this framework.

Locations and orientations within the feature relating tolerance zone framework are perfect. However, there is no required position of this framework relative to any of the specified datums. Only the orientation of the framework is controlled relative to the datums specified in the second line of the feature control frame. The relatively small tolerance zones are centered on the true positions defined by this framework.

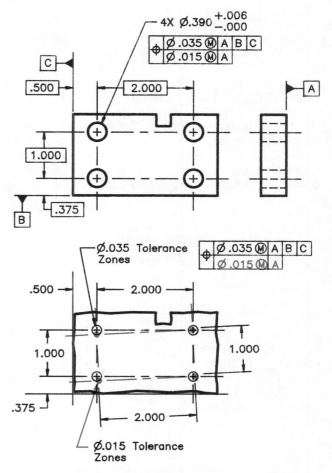

Figure 9-3. The feature relating tolerances can move within the pattern locating tolerance.

The feature relating tolerance zone framework is free to move. The feature relating tolerances move with the framework. It is permissible for the feature relating tolerance zones to overlap the pattern locating tolerance zones. There is a limit to the allowable movement of the feature relating tolerance zone framework and the associated tolerance zones, because the toleranced features must meet the requirements of both lines in the tolerance specification.

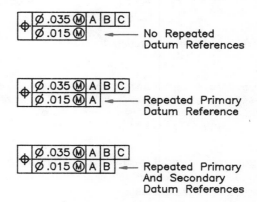

Figure 9-4. The number of datum references repeated in the second line of a composite tolerance is determined by the desired level of orientation control on the feature relating tolerances.

Datum references are repeated in the second line of the composite feature control frame on the basis of the required level of control. See Figure 9-4. Any shown datum reference must be made in the same order of precedence as in the first line.

NO REPEATED DATUM REFERENCES

Absence of any datum reference in the second line of a composite tolerance creates a feature relating tolerance zone framework that is free to float relative to the datum reference frame. See Figure 9-5. A composite position tolerance is shown. The first line specifies a .025″ diameter pattern

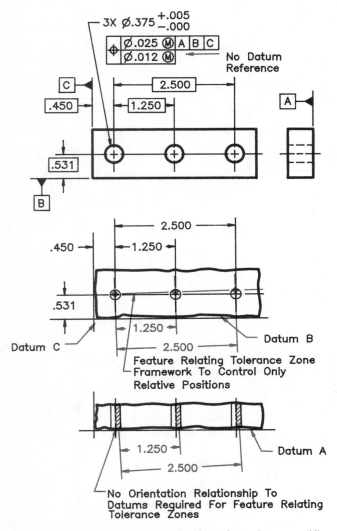

Figure 9-5. Omitting all datum references from the second line releases all orientation requirements from the feature relating tolerance. The second line of the composite tolerance only controls feature-to-feature locations.

locating tolerance, and it is referenced to datums A, B, and C. The feature relating tolerance is .012″ diameter, and it is not referenced to any datums.

The pattern locating tolerance zone framework creates .025″ diameter tolerance zones located on the true positions dimensioned from the datum reference frame. The feature relating tolerance zone framework creates .012″ diameter

tolerance zones that are located relative to one another at the basic dimensions defined on the drawing. The feature relating tolerance zone framework is free to float at any orientation relative to the datum reference frame since no datums are referenced in the second line.

Holes produced in the part must be able to meet both tolerance requirements.

REPEATED PRIMARY DATUM REFERENCE

Repeating the primary datum reference in the second line of a composite tolerance creates a feature relating tolerance zone framework that is oriented to the primary datum. See Figure 9-6. The feature relating tolerance zone framework is free to move relative to the datum reference frame so long as the proper orientation to the primary datum is maintained. A composite position tolerance is shown in the given figure. The first line requires a .025″ diameter tolerance on the pattern locating tolerance zone framework, which is referenced to datums A, B, and C. The feature relating tolerance is .012″ diameter and is referenced to datum A.

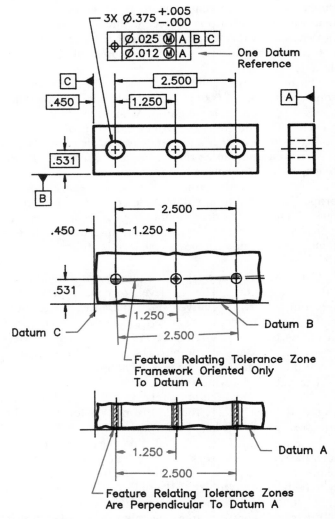

Figure 9-6. The primary datum reference requires the feature relating tolerance zones to be properly oriented to the primary datum.

The pattern locating tolerance zone framework creates .025″ diameter tolerance zones located on the true positions dimensioned from the datum reference frame. The feature relating tolerance zone framework creates .012″ diameter tolerance zones that are located relative to one another at the basic dimensions defined on the drawing. The orientation of this framework is controlled relative to datum A, but its location is not fixed.

Holes produced in the part must be able to meet both tolerance requirements.

REPEATED PRIMARY AND SECONDARY DATUM REFERENCE

The current standard clearly illustrates composite position tolerances that include two datums in the second line. When working to the current standard, the explanation of composite tolerances in this book is applicable and no explanatory notes are required on the drawing.

If a drawing is being completed in compliance with 1982 or earlier standards, it is advisable to include a note to define the interpretation of any shown composite tolerances. ANSI Y14.5M-1982 and earlier standards did not clearly illustrate the interpretation of multiple datums in the second line, so some controversy regarding the correct interpretation may exist if no note is shown. The following is a suggested note.

THE SECOND LINE OF A COMPOSITE POSITION TOLERANCE CONTROLS RELATIONSHIPS OF FEATURES WITHIN A PATTERN. DATUM REFERENCES IN THE SECOND LINE OF A COMPOSITE POSITION TOLERANCE REQUIRE ORIENTATION TO THE REFERENCED DATUMS AND RELEASES THE LOCATION DEFINED BY THE BASIC DIMENSIONS FROM THOSE DATUMS.

Repeating the primary and secondary datum reference in the second line of a composite tolerance creates a feature relating tolerance zone framework that is oriented to the two referenced datums. See Figure 9-7. The datum references in the second line do not imply any location requirement relative to the referenced datums. The feature relating tolerance zone framework is free to move relative to the datum reference frame so long as the proper orientation to the primary and secondary datums is maintained.

A composite position tolerance is shown in the given figure. The first line requires a .025″ diameter tolerance on the pattern locating tolerance zone framework, which is referenced to datums A, B, and C. The feature relating tolerance is .012″ diameter and is referenced to datums A and B.

The pattern locating tolerance zone framework creates .025″ diameter tolerance zones located on the true positions dimensioned from the datum reference frame. The feature relating tolerance zone framework creates .012″ diameter tolerance zones that are located relative to one another at the basic dimensions defined on the drawing. The orientation of this framework is controlled relative to datums A and B. The figure shows that the framework is permitted to move relative to datum B, but it remains parallel to that datum.

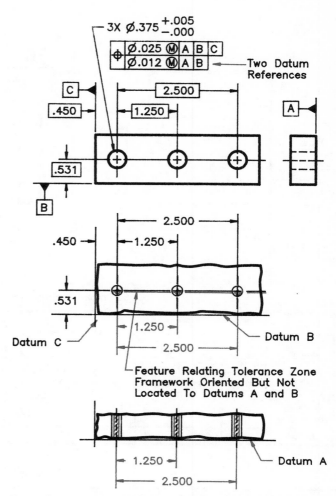

Figure 9-7. Including both the primary and secondary datum references requires the feature relating tolerance zone framework to be oriented to the two referenced datums.

PAPER GAGING OF COMPOSITE TOLERANCES

One method for inspection of hole positions to verify compliance with a composite position tolerance requires two sets of measurements if using manual inspection methods. The first set of measurements determines the locations of the holes relative to the datums. Another set determines the locations of the holes relative to each other.

Figure 9-8 shows a four hole pattern with a composite tolerance applied to it. Data from a produced part is entered into the shown table.

Data in the first table shows measured hole locations relative to the datum reference frame. The measured coordinate location errors are used to determine the amount of positional error that exists. This can be done either by calculation or by plotting the coordinate errors on a grid with concentric circles representing the position tolerance zones. One segment of the figure shows the errors plotted on a grid.

Data in the second table shows the measured hole locations relative to one another. Measurement of the relative hole locations requires that two holes be used to establish a coordinate system. Any two holes can be used for this purpose. Hole #1 in the figure was used as the origin, and hole #3 was used to establish the orientation of the X axis.

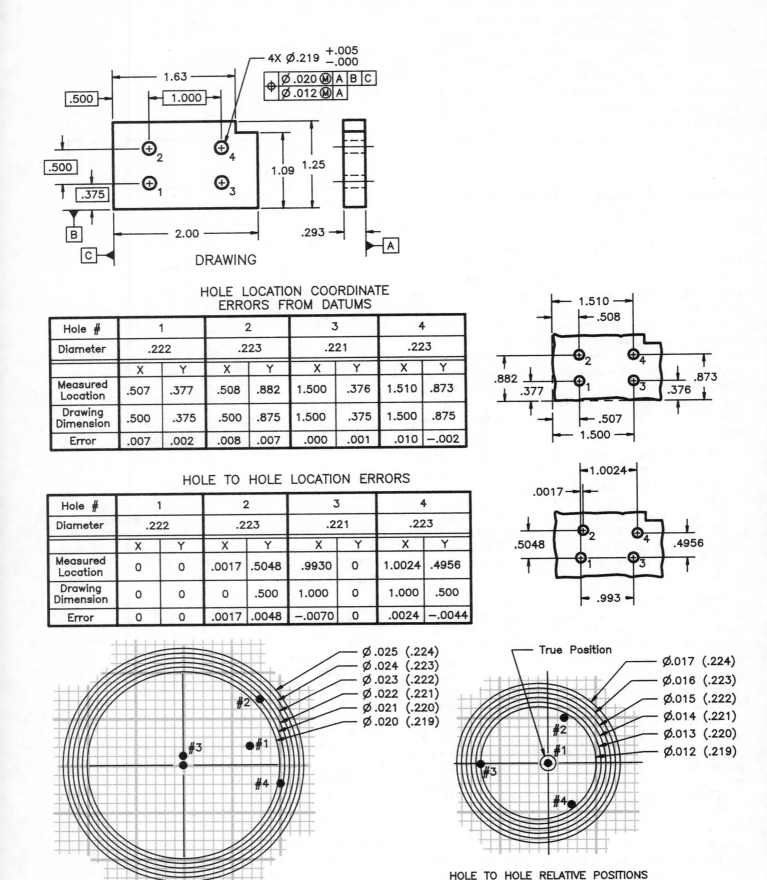

HOLE LOCATION COORDINATE ERRORS FROM DATUMS

Hole #	1		2		3		4	
Diameter	.222		.223		.221		.223	
	X	Y	X	Y	X	Y	X	Y
Measured Location	.507	.377	.508	.882	1.500	.376	1.510	.873
Drawing Dimension	.500	.375	.500	.875	1.500	.375	1.500	.875
Error	.007	.002	.008	.007	.000	.001	.010	−.002

HOLE TO HOLE LOCATION ERRORS

Hole #	1		2		3		4	
Diameter	.222		.223		.221		.223	
	X	Y	X	Y	X	Y	X	Y
Measured Location	0	0	.0017	.5048	.9930	0	1.0024	.4956
Drawing Dimension	0	0	0	.500	1.000	0	1.000	.500
Error	0	0	.0017	.0048	−.0070	0	.0024	−.0044

Ø.025 (.224)
Ø.024 (.223)
Ø.023 (.222)
Ø.022 (.221)
Ø.021 (.220)
Ø.020 (.219)

POSITION RELATIVE TO DATUMS

True Position

Ø.017 (.224)
Ø.016 (.223)
Ø.015 (.222)
Ø.014 (.221)
Ø.013 (.220)
Ø.012 (.219)

HOLE TO HOLE RELATIVE POSITIONS

Figure 9-8. Coordinate hole location errors can be plotted on graph paper and an overlay of concentric circles can be used to determine positional tolerance errors for the holes.

It is not important which holes are selected to establish the coordinate system. If measurements indicate one or more holes is too far out-of-position, then another set of holes may be used to establish a new coordinate system. Any two holes can be used that will result in the optimum set of measurements.

The coordinate errors for each hole location is plotted on a grid. The hole that acted as the origin for the measurements is plotted at the 0,0 coordinate of the grid. In this figure, hole #1 is placed at the 0,0 coordinate. The coordinate errors of the other holes are plotted relative to the location of hole #1.

A set of concentric circles representing tolerance zone diameters is placed over the plotted hole locations. There is no requirement for the circles to be centered on the origin. Since location relative to the datums is not required by the second line of the composite tolerance, it would be incorrect to force the tolerance zones to be in a fixed location.

The location of the concentric circles for the feature relating tolerance zones can be optimized when needed. Each concentric circle represents a specific diameter tolerance zone, and that diameter is related to a produced hole diameter. The concentric circles can be moved into any position, provided the point for each hole is contained within the circle that corresponds to the produced hole diameter. Movement of the concentric circles is representative of the feature relating tolerance zone framework movement relative to the datum reference frame.

TWO LINE FEATURE CONTROL FRAME

A two line position tolerance containing two position tolerance symbols has a different meaning than a composite position tolerance. See Figure 9-9. The two line specifications establishes two tolerance zone frameworks, and each is located from the referenced datums.

Each tolerance zone framework is related to the identified datums. In the given example, line one requires a position tolerance of .020″ diameter relative to three datums–A, B, and C. The tolerance zones are located by basic dimensions from the referenced datums. The tolerance zones are also oriented relative to the datums.

The second line requires a position tolerance zone of .012″ diameter relative to two datums–A and B. The location and orientation of the tolerance zones is established relative to the two referenced datums. There is no location or orientation requirement relative to the omitted datum.

FUNCTIONAL GAGING OF POSITION TOLERANCES

Efficiency in production can often be improved by using functional gages to inspect parts. Function gages are tools used to verify that produced parts meet tolerance requirements. Gages for some applications are very simple. Simple gages are inexpensive to design and produce. The cost of these gages can quickly be recovered because of the reduced amount of time spent performing labor intensive manual measurements.

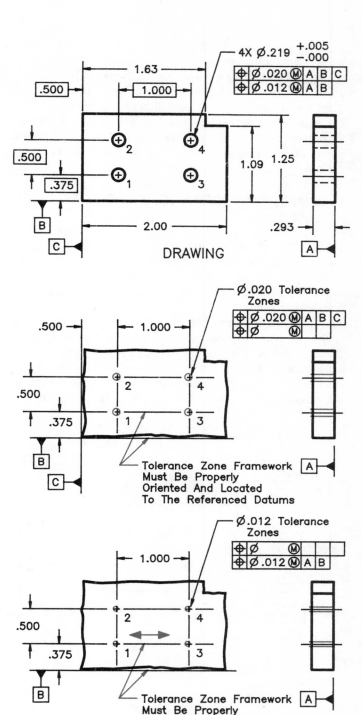

Figure 9-9. Using two position tolerance symbols creates a two line position tolerance specification. Each of the lines creates an independent tolerance zone framework.

Complex parts and tolerances impact the complexity of functional gages. Complex gages require more design and fabrication time than simple gages. The cost of a complex gage must be balanced against the cost of manual measurements, or the use of computer-driven measurement machines. If a part is complex and difficult to check manually, the functional gage may pay for itself very quickly, even if the gage is expensive to design and produce.

The material contained in this section is provided to show what functional gages are, and to demonstrate through example gages what the workpiece tolerance specifications require. Gage dimensions in the given examples are determined from the workpiece dimensions and tolerances. No gage design tolerances are calculated or shown. The inclusion of gage tolerance calculations in the examples would complicate the explanations of what the functional gages are designed to verify.

GAGING FOR TOLERANCES SPECIFIED AT MMC

Position tolerances can be functionally gaged when the tolerances include an MMC modifier. See Figure 9-10. The given figure includes a drawing of a plate that contains four holes. A positional tolerance including an MMC modifier is applied to the holes.

A functional gage for checking hole locations produced to the given drawing must include features to properly locate the datum reference frame, and to check the locations of the holes relative to that datum reference frame. The shown position tolerance on the four holes includes a reference to datums A primary, B secondary, and C tertiary. To properly

locate the part in the gage, there must be datum simulators for all three datums. The gage shown in the figure has three plates that correctly simulate the datums.

Four pins are installed in the gage to check the locations of the holes. The pins are located at the true positions specified on the drawing. Assuming the gage can be made perfect, the pins are sized at the virtual condition of the holes. The virtual condition is equal to the hole MMC minus the position tolerance.

A part can be inspected for hole positions by placing the part over the four pins. In addition to fitting over the four pins, several conditions must be met to verify that the holes are in acceptable positions. The part must rest against the three datum simulators, properly establishing the datum reference frame, with all four pins extending through the holes. To properly establish the datum reference frame, at least three points must be in contact with the primary datum simulator, two with the secondary datum simulator, and one with the tertiary datum simulator.

The shown gage is designed to check the hole positions based on the virtual condition of the holes. Sizing the pins at the virtual condition results in an adequate check of position for any hole that is produced within size tolerance. The fixed diameter pins automatically permit the allowable bonus tolerance when a hole is produced at any size larger than MMC.

The shown gage does not verify hole size. Hole diameter must be checked separately.

GAGING WHEN A PRIMARY DATUM IS REFERENCED AT MMC

References to a datum feature of size in a position tolerance must include a material condition modifier. Material condition modifiers applied to datum references have a significant impact on how the datum reference frame is established. Generally, the application of an MMC modifier makes functional gaging easier to accomplish.

Figure 9-11 includes a drawing of a simple part with a .500″ diameter shaft identified as datum A. Three holes in the part include a primary reference to datum A, and the MMC modifier is applied to the reference.

Since primary datum A is a cylinder, two planes of the datum reference frame are established on the datum axis. The third plane is established by secondary datum surface B. Datum surface C is referenced only to establish a rotational position for the hole pattern.

There are special rules about how a datum feature of size must be established. One of the rules applies to a primary reference to a datum feature of size. *When a primary datum reference to a datum feature of size includes the MMC modifier, and no axis straightness tolerance is applied to the feature, the datum feature must be simulated at the MMC size.*

A functional gage for the given part includes a hole to simulate the datum. The hole is sized at the .503″ MMC size of the shaft. The shaft will fit snugly into the hole when it is at the MMC, but any pin produced at a diameter smaller than MMC will be free to float within the gage.

With the primary datum feature in the .503″ hole in the gage, secondary datum feature B is moved into contact with

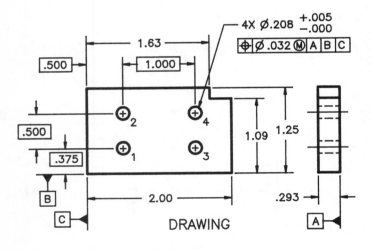

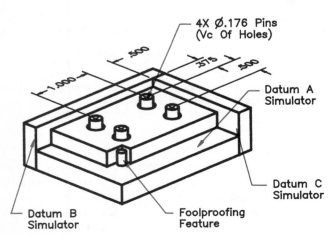

Figure 9-10. Functional gages can be used to check the location requirements defined by position tolerances.

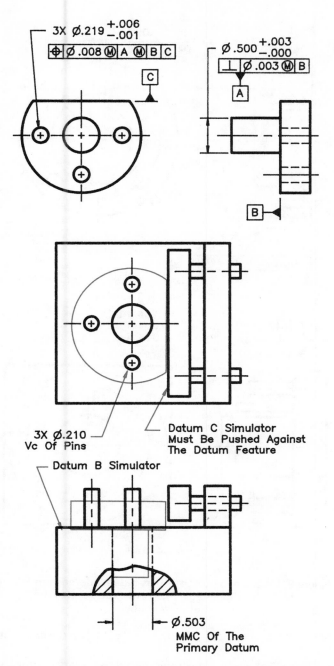

Figure 9-11. A primary datum referenced at MMC is simulated at the MMC size of the datum feature.

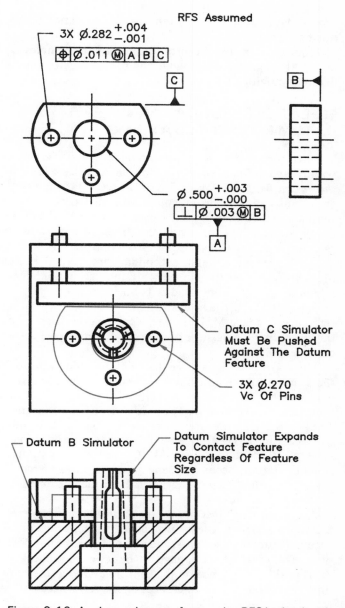

Figure 9-12. A primary datum referenced at RFS is simulated by a gage feature that makes contact with the datum feature.

the top surface of the gage. An adjustable plate is designed to press up against datum feature C to establish the correct rotational position of the part.

The three pins in the gage are sized at the virtual condition of the holes. The pins must extend through the holes while datum features A, B, and C are properly located in the gage.

GAGING WHEN A PRIMARY DATUM IS REFERENCED AT RFS

References to datum features of size at RFS can complicate gage design, but the requirement can often be accurately simulated. See Figure 9-12. The given figure includes

a drawing of a circular plate with four holes in it. The center hole is identified as datum feature A. The three hole pattern has a position tolerance that references datum A as a primary datum. No modifier is shown, so RFS is applicable.

Since datum A is a primary datum, and referenced at RFS, a datum simulator must be used that actually makes contact with the part surface. Contact is required regardless of the produced diameter of the hole. Adequate contact must be made to align the axis of the datum simulator and the axis of the produced hole.

The gage for this part has an expanding pin that makes contact with the primary datum. It is expanded enough to align the axis of the gage pin with the produced hole. With the pin properly expanded, the part is moved down the pin until it makes contact with the datum B simulator. With the relationship between the part and two datum simulators already established, the tertiary datum simulator is pressed against datum feature C to establish a fixed rotational position of the part.

Three pins sized at the virtual condition of the holes are passed through the holes while the part is located on the datum simulators.

EFFECT OF SECONDARY AND TERTIARY DATUM REFERENCES AT MMC

Secondary and tertiary datum references made at MMC require simulation of the datums at the virtual condition of the datum features. Simulation of the secondary and tertiary datums at their virtual condition is different than the simulation requirements for a primary datum. Sizing the datum simulator at the virtual condition of the feature is essential to achieve the proper level of control.

It is incorrect to simulate a secondary or tertiary datum at the MMC size, even though the MMC modifier is used. Only a primary datum is simulated at its MMC size.

Figure 9-13 illustrates how secondary and tertiary datum references made at MMC impact a gage. Several aspects of the given drawing must be observed before the gage design can be explained.

The given part has a large diameter hole bored through the center, and the hole has a keyseat. A perpendicularity tolerance of .003″ diameter at MMC is permitted on this hole, relative to datum D. The hole is identified as datum J. The keyseat has a position tolerance of .004″ at MMC relative to datum J. The four small holes have a position tolerance of .024″ diameter at MMC relative to datum D primary, datum J secondary, and datum K tertiary. Secondary datum J and tertiary datum K both include an MMC modifier.

Application of MMC modifiers on the datum references makes it possible to use a single gage to check some of the datum feature requirements, and also to check the hole locations. This means that one gage can be designed to check the .024″ position tolerance on the four holes and at the same time verify some of the tolerances on the referenced datum features of size.

The given figure shows a gage for checking the hole locations and the datum features. Primary datum feature D is a flat surface. It is simulated with a simple flat plate. Secondary datum feature J is a hole. A pin sized to the virtual condition of the datum hole is used to locate datum J. In addition to establishing the location of datum J, this pin will verify the perpendicularity requirement on the hole. Tertiary datum K is a keyseat. A key sized to the virtual condiiton of the keyseat is used as the tertiary datum simulator.

Primary datum feature D is a simple flat surface. It locates a primary datum plane.

Secondary datum feature J is a hole. It establishes the location of a datum axis. This datum axis must be perpendicular to the primary datum plane. Two of the planes in the datum reference frame intersect on the secondary datum axis. These two planes must be perpendicular to the primary datum plane since the datum axis is perpendicular to the primary plane.

Although the two planes located by datum feature J must be perpendicular to the primary datum plane, the hole itself

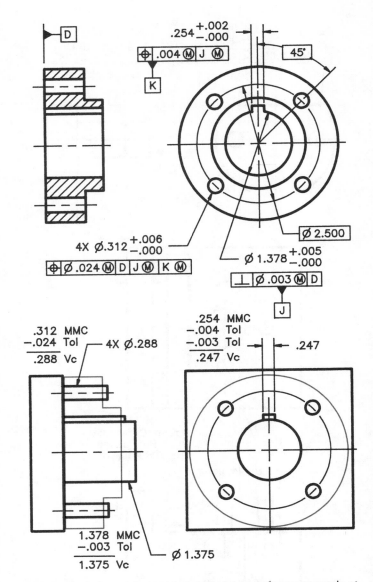

Figure 9-13. Secondary and tertiary datum references made at MMC are simulated at the virtual condition of the datum features.

can have a perpendicularity error of .003″ when the hole is at MMC.

The datum simulator for referenced secondary datum J at MMC must be sized to the virtual condition of the datum feature. This permits the hole to be produced with the specified amount of error, and still have the datum reference frame be perfect. Since the datum simulator is smaller than the MMC of the feature, the feature can have some error and still fit into the gage. Proper installation of the workpiece in the gage requires that the primary datum maintain at least a three point contact with its simulator when the secondary datum simulator is inserted into the secondary datum feature.

For the given example, the pin in the gage must have a 1.375″ diameter. This is the virtual condition for datum feature J. The 1.375″ diameter virtual condition is determined by calculating the combined affect of the MMC size and the tolerance applied to the feature. In the given example, the virtual condition is equal to:

1.378″ - .003″ = 1.375″ diameter

It is incorrect to size the pin in the given gage to the MMC size of the secondary datum. The result of such an incorrectly designed gage would be incorrect simulation of the datums. If the pin is equal to the MMC size of the hole, it could incorrectly act as the primary datum. A gage pin size at MMC will act as a primary datum simulator if the hole in the given part is produced at MMC and has a .003″ perpendicularity error.

Tertiary datums referenced at MMC must also be properly simulated to properly establish the datum reference frame. The tertiary datum reference in the given figure includes an MMC modifier. The gage must therefore simulate the tertiary datum at its virtual condition.

Tertiary datum K is the keyseat on the part. The key in the gage that simulates datum K is sized at the virtual condition of the keyseat. This permits the key to enter the slot if all features are produced within the specified tolerance.

Incorrectly producing a gage with a key sized to the MMC size of the slot might incorrectly establish the datum reference frame. A key sized to the MMC size of the keyseat would not enter a keyseat that is produced at MMC, unless the keyseat is perfectly located.

Position of the four .312″ diameter holes can only be verified if the datum reference frame is properly established. The hole locations are checked by pins that are located at the true positions relative to the datum reference frame created by the datum simulators.

Pins in the gage, for checking hole locations, are sized at the virtual condition of the holes. Pins used to check hole locations are always sized to the virtual condition if the position tolerance includes the MMC modifier.

GAGING A COMPOSITE POSITION TOLERANCE

Composite position tolerances are gaged in a similar manner to the single line position tolerance specification. See Figure 9-14. The given example shows a drawing with a composite position tolerance and the two gages used to check the produced hole locations. It is possible to design a single gage that will check both lines of a composite tolerance, but two separate gages better illustrate how the hole positions are checked.

The given composite tolerance specification requires a pattern locating tolerance of .032″ diameter at MMC relative to datum A primary, B secondary, and C tertiary. The gage used to check this line of the composite tolerance is the same as it would be for a single line position tolerance specification. Plates are used to simulate the three datum planes. Pins are located in the positions for the holes, and the pins are sized to the virtual condition created by the .032″ position tolerance.

Pins in the gage must pass through the holes and proper contact between the datum features and the three datum simulators must be made. This gage will verify that the hole positions relative to one another and to the datum planes are within the .032″ tolerance at MMC.

The second line of the composite tolerance shows a .016″ diameter feature relating tolerance at MMC relative to datum A. The gage for this tolerance has only one plate, and

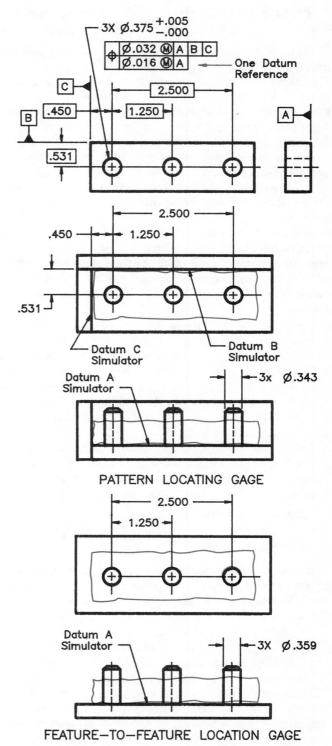

Figure 9-14. Separate gages can be used to check the requirements for the two lines of a composite position tolerance.

it simulates primary datum A. Pins are located in the plate at the true positions of the holes. The pin diameters are sized to the virtual condition created by the .016″ diameter position tolerance.

Pins in the gage must pass through the holes and permit the plate to rest flat against datum feature A. This gage checks the relative positions of the holes and also verifies the orientation of the holes relative to datum A.

POSITION CONTROL OF MULTIPLE AND SINGLE PATTERNS

Holes serve many functional purposes. A very common purpose for holes is to provide clearance for installing the fasteners that hold parts together. In the case of an engine block, there are many holes for attaching the various parts that complete the engine; for example, the pattern of holes for attaching the engine head, or another pattern of holes for mounting the oil pan. Since there is no relationship between the engine head and oil pan mounting positions, the groups of holes used for mounting these parts should act as separate patterns. The drawing of the engine block should clearly show that the two groups of holes act as separate patterns.

Locations of holes, apparent grouping, and hole size often provide clues as to which holes form a pattern. However, these characteristics can be misleading. For purposes of position tolerances, the three listed characteristics are not considered in determining which holes fall into the same pattern.

According to the current dimensioning and tolerancing standard, *features act as a single pattern if the tolerances applied to them reference the same datums in the same order of precedence and with the same material condition modifiers.* This requirement is applicable to only single line feature control frames and the first line of a composite tolerance. It is not applicable to the second or third line of a composite tolerance.

When it is desired that a specific single line position tolerance act as a separate requirement from the other tolerances containing the same datum references, it is necessary to apply a note. The note simply states:

SEPARATE REQUIREMENT
  or
SEP REQT

It is also possible to apply the same note to the first line of a composite tolerance. There is no need to apply this note to the second line of a composite tolerance since the second line is understood to create a separate requirement.

The following sections describe how features act as single and multiple patterns. Holes are used in the examples, but the single or multiple patterns could be any type of feature.

The designer must determine the functional purpose of the holes and other features in a part and determine whether there is one or more patterns of features. The function of the features determine how position tolerances are applied on a drawing, and the position tolerances indicate whether the features belong to one or more patterns.

MULTIPLE GROUPS RELATED TO ONE DATUM REFERENCE FRAME

When single line feature control frames contain the same datum references in the same order of precedence with the same material condition modifiers, all the controlled features form a single pattern. See Figure 9-15. The single pattern requirement for a single line feature control frame is most significant when the referenced datums are features of size.

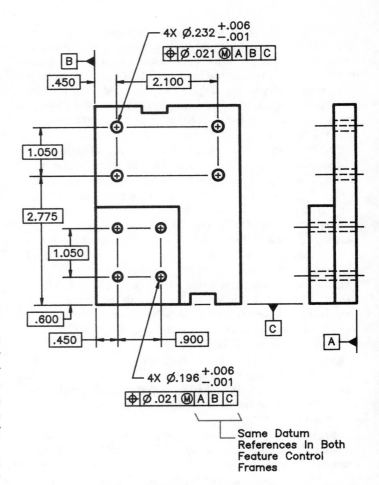

Figure 9-15. The eight holes in the given part create a single pattern since the position tolerances reference the same datums, in the same order of precedence, and with the same material condition modifiers.

Simultaneous Requirement Referenced to Flat Datum Surfaces

Some design applications occur for which all holes need to form a single pattern. See Figure 9-15. The given part has two groups of holes. Each group has four holes. The group near the top of the part has four .232″ diameter holes. A position tolerance of .021″ diameter is referenced to datum A primary, datum B secondary, and datum C tertiary. The group of holes near the bottom of the part has four .196″ diameter holes. A position tolerance of .021″ diameter is referenced to datum A primary, datum B secondary, and datum C tertiary.

Both groups of holes have position tolerances with identical datum references; therefore, all eight holes act as one pattern. The given example shows the same tolerance value in both feature control frames. This has no affect on whether the holes act as one or two patterns.

One functional gage can be used to verify the locations of the holes in a single pattern. See Figure 9-16. The shown functional gage will check the eight hole locations of the part in the previous figure.

The shown gage is made of plates that act as datum simulators. Pins located at the true positions are sized at the virtual condition of the holes. A part inserted in the **gage**

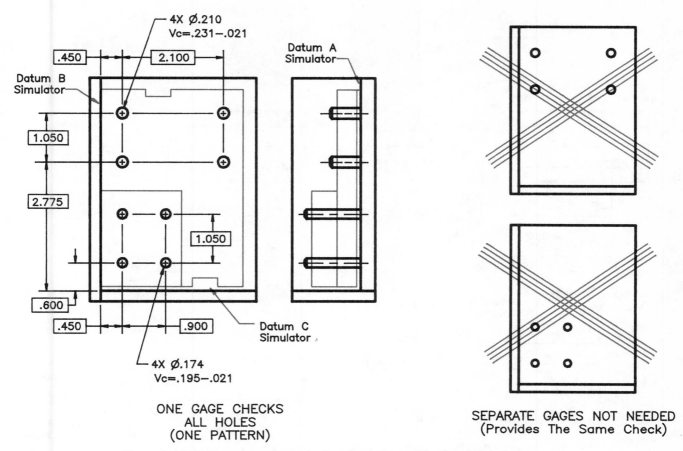

**ONE GAGE CHECKS
ALL HOLES
(ONE PATTERN)**

**SEPARATE GAGES NOT NEEDED
(Provides The Same Check)**

Figure 9-16. Only one gage is required to check the position for all eight holes.

must properly contact the three datum simulators with the eight pins passing through the holes.

It is unnecessary to make two gages to check the positions of the holes, although it would not be wrong for this particular part. The two gages would result in the same check as a single gage. If two gages are used, they must each have the three datum simulators, with pins located at the true positions of the holes and sized to the virtual conditions. Placing the part in each of the two gages provides the same check as placing the part in the single gage that simultaneously checks all eight holes.

Simultaneous Requirement Referenced to Datum Features of Size

Simultaneous requirements are significant when the referenced datums are features of size. See Figure 9-17. The figure shows two groups of holes. Position tolerances for these groups of holes are referenced to the same datums, in the same order of precedence, and with the same material condition modifiers. Because of the identical datum references, all four holes create one pattern.

The effect of the simultaneous requirement can be seen by observing the functional gage that can be used for inspection of the hole locations. Only one tool is used to check all holes simultaneously. The datum A simulator is sized to the MMC of primary datum A. The datum B simulator is a plate to contact datum feature B. A key on the side of the tool simulates datum C. The key is sized to the virtual

condition of the datum C slot. Pins for checking hole location are sized to the virtual condition of the holes. These pins are slipped into the tool after the tool is placed inside the workpiece.

If the datum C slot is produced at LMC and at perfect position, the tool is free to rotate on datum axis A by an amount equal to the clearance between the datum C slot and the datum C simulator. This allowable rotation will permit the four holes in the part to rotate out of position from the slot, but only by the amount of the clearance between the slot and key. Also, it is necessary for all four holes to rotate around the axis in the same direction.

Separate Requirement Noted for References to Datum Features of Size

Groups of features can be specified to move independently even though datum references are identical. Exception to the simultaneous requirement is taken by placing the SEPARATE REQUIREMENT notation under the feature control frames of those items that are to act separately. See Figure 9-18. Although this figure contains the same part shown in Figure 9-17, the notation of separate requirement has been added beneath the two position tolerances.

Both sets of holes are separately controlled due to the notation. The effect of this type control is shown by the functional tools for inspection of the hole patterns. Two tools are required to check the two separate requirements. Each tool has datum simulators that are sized in the same

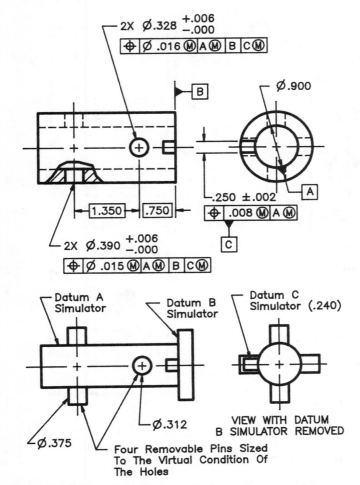

Figure 9-17. Identical references to datum features of size require that the features be considered as a single pattern (simultaneous requirement).

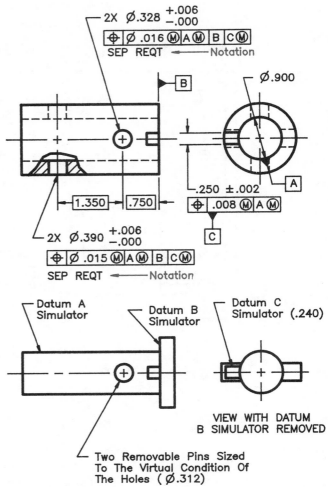

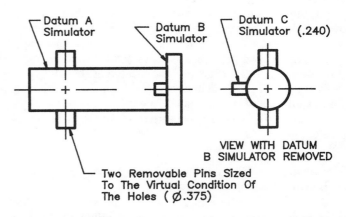

Figure 9-18. A notation of separate requirement under a position tolerance specification requires that the controlled features be verified separate from any other groups of features that are toleranced to the same datums.

manner as described for Figure 9-17. Datum A is simulated at its MMC size since it is primary. Datum C is simulated at its virtual condition since it is tertiary.

The functional gages are free to rotate on datum axis A by an amount equal to the clearance between the datum feature C slot and the datum C simulator key. The tools will permit one group of holes to rotate in one direction and the other group to rotate in the opposite direction.

Tolerance specifications should always permit the maximum amount of variation that can be accommodated by the function of the part. If the function of the part does not require that the holes act as a single pattern, the tolerances should be specified as separate requirements.

Default Simultaneous and Separate Requirements for Composite Tolerances

The first line of each composite position tolerance specification creates a simultaneous requirement if the datum references are identical. However, subsequent lines in composite position tolerances always create a separate requirement unless noted otherwise. See Figure 9-19. This figure contains the same part as Figure 9-15, but composite position tolerances have been added.

The first line in each composite position tolerance contains identical datum references. The first line, therefore, creates a single pattern requirement for the pattern locating tolerance. The tool shown in Figure 9-16 can be used to check the hole locations for this requirement.

The second line of each composite position tolerance creates a separate requirement, regardless of the datum references. In effect, the two sets of holes can float in opposite directions within the pattern locating tolerance. This

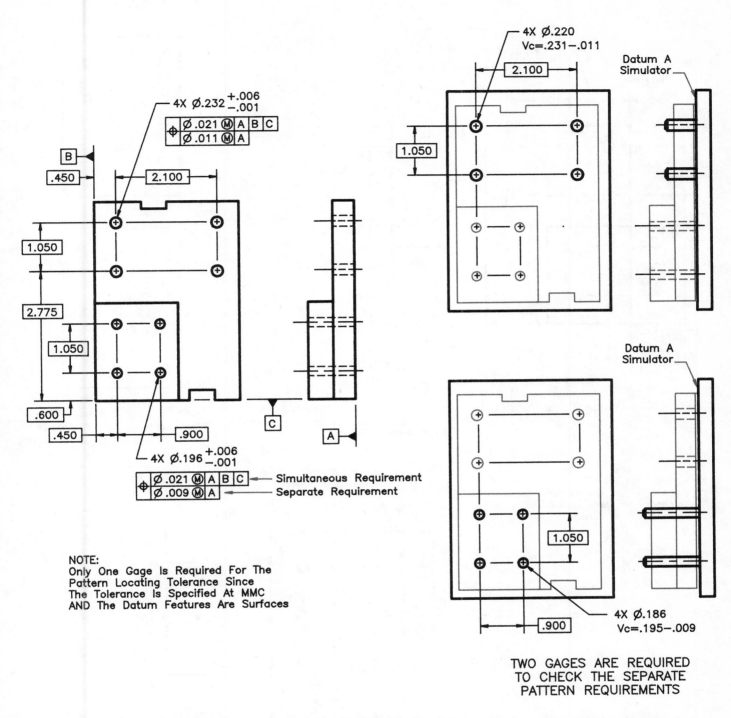

Figure 9-19. The first line of multiple composite tolerance specifications creates simultaneous requirements when datum references are identical. The second line creates separate requirements unless noted otherwise.

requires that two functional gages be created to check the two patterns of four holes. The pins in each tool are sized to the virtual condition of the hole that is created by the feature relating tolerance. Pins in the gages are located at the true positions for the holes.

The separate requirement of the second line of a composite tolerance is useful when mounting two or more parts that can vary in their relative locations to one another. There is no need to hold a close relationship between groups of holes if there is not a functional need to control them relative to one another.

Noted Simultaneous Requirement for Composite Position Tolerances

Design requirements may require that two sizes of holes be used when mounting a single part to an assembly. Although there are two sizes of holes, the functional requirements make it necessary for all holes to act as a single

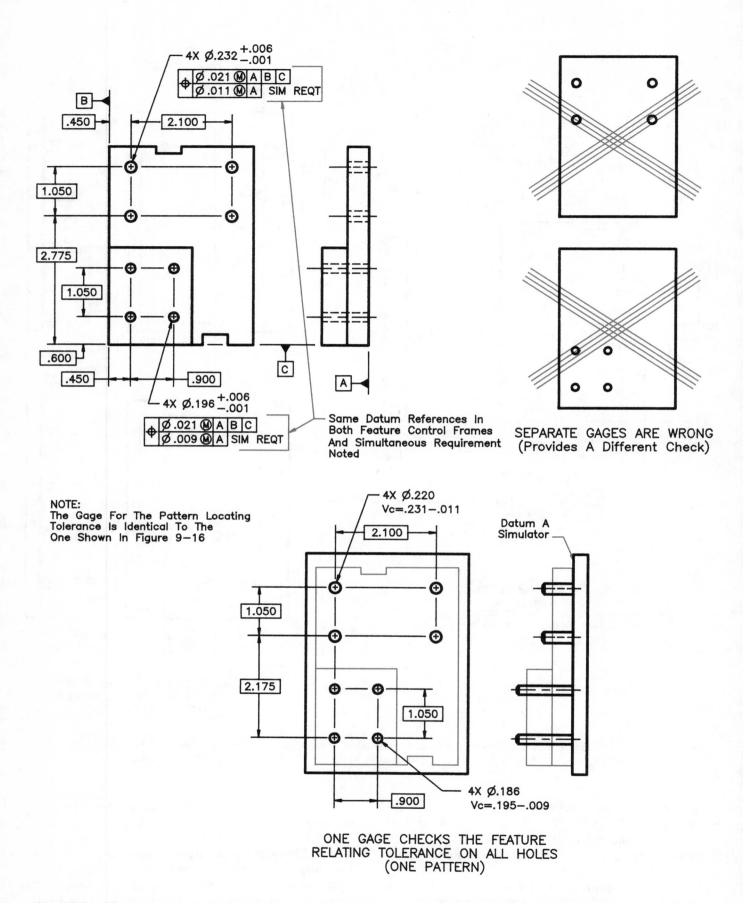

4X Ø.232 $^{+.006}_{-.001}$

⊕ | Ø.021 Ⓜ | A | B | C |
Ø.011 Ⓜ | A | SIM REQT |

B

.450

2.100

1.050

2.775

1.050

.600

.450   .900

4X Ø.196 $^{+.006}_{-.001}$

⊕ | Ø.021 Ⓜ A | B | C |
Ø.009 Ⓜ | A | SIM REQT |

C

A

Same Datum References In Both Feature Control Frames And Simultaneous Requirement Noted

SEPARATE GAGES ARE WRONG
(Provides A Different Check)

NOTE:
The Gage For The Pattern Locating Tolerance Is Identical To The One Shown In Figure 9—16

4X Ø.220
Vc=.231-.011

2.100

1.050

2.175

1.050

.900

4X Ø.186
Vc=.195-.009

Datum A Simulator

ONE GAGE CHECKS THE FEATURE
RELATING TOLERANCE ON ALL HOLES
(ONE PATTERN)

Fig. 9-20. When one or more datums are shown in the second line of a composite tolerance specification, and if the datums in the second line are pulled from the first line in the same order of precedence as in the first line, then only one gage is needed to check the feature relating tolerance on all the features when the simultaneous requirement is noted.

pattern. If composite position tolerances are applied on two groups of holes that must act as a single pattern, a notation must be added to the composite tolerance specifications. The recommended notation is:

SIMULTANEOUS REQUIREMENT

or

SIM REQT

Placement of the note in line with the affected line of the composite position tolerance is recommended. See Figure 9-20. Placement under the entire feature control frame might cause confusion since the first line is already understood to create a simultaneous requirement.

One functional gage can be used to check the second line of the shown composite position tolerances. The gage is made of one plate with eight pins sized to the virtual condition of the holes. Eight pins in a single plate have fixed locations to one another and simulate one feature-relating tolerance zone framework. Since the pins are sized to their respective virtual conditions, the gage checks the position tolerance that is centered on the true positions created by the feature-relating tolerance zone framework.

It would be incorrect to create a separate gage to check each group of holes. Separately checking the two groups of holes is not acceptable on the basis of the given tolerance specifications. Checking them separately would permit the groups of holes to float independently; this is not what the drawing specifies. The given tolerance specification requires that the holes act as one pattern.

COAXIAL HOLE PATTERNS

Coaxial holes are commonly used on parts such as hinges. Coaxial holes often require a closer position relationship to one another than to other features on the part. See Figure 9-21. The alignment of coaxial holes must be accurate enough to insert the shaft that passes through them. Location of the holes relative to other features on the part can usually be permitted to vary by a larger tolerance.

The given figure shows a part containing two coaxial holes. A composite tolerance is applied to the holes. It shows a pattern locating tolerance of .016″ diameter relative to datums A, B, and C. This defines the location requirement for the pattern of holes relative to the referenced datums.

The shown feature relating tolerance is .003″ diameter. It is not referenced to any datums. This establishes a .003″ diameter tolerance zone for each hole. The two .003″ tolerance zones are located on the feature relating tolerance zone framework, which is a single axis for coaxial holes. Movement of the feature relating tolerance zone framework is permitted. It can move in any direction and change orientation. Its position is only restricted by the requirement for the axis of the two holes to be within the .016″ diameter pattern locating tolerance.

Coaxial holes can have different diameters. See Figure 9-22. A composite tolerance can be applied to coaxial holes of different diameters. To ensure the composite tolerance applies to all holes, the current standard requires that a notation be placed under the tolerance specification. The nota-

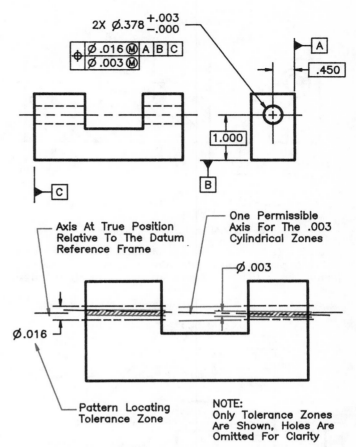

Figure 9-21. A composite tolerance can be applied to coaxial holes.

tion indicates the number of coaxial holes to which the tolerance applies.

The control of the hole axes is identical to the explanation given for holes of the same diameter. The pattern locating tolerance locates the holes relative to the referenced datums, and the feature relating tolerance controls the coaxiality of the holes.

When no datums are referenced in the feature relating tolerance, the feature relating tolerance zone framework may be in any orientation. The only restriction on the pattern orientation is invoked by the pattern locating tolerance.

It is possible to control the orientation of the feature relating tolerance zone framework. To establish an orientation requirement, datum references are included in the feature relating tolerance. See Figure 9-23. The number of datum references shown depends on the level of control desired.

References to datums A and B are made in the feature relating tolerance of the given figure. The tolerance requires that both holes be located within .006″ diameter tolerance zones. Those zones must be located on a feature relating tolerance zone framework that is properly oriented relative to the referenced datums. Proper orientation of the framework for the given figure must be parallel to datums A and B.

The reference to datums A and B in the second line of the composite tolerance does not invoke any location requirement to those datums. The datum references in the second line only invoke an orientation requirement.

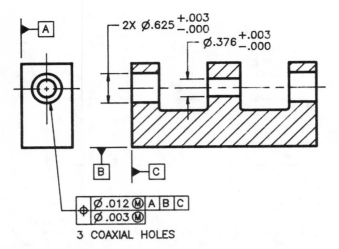

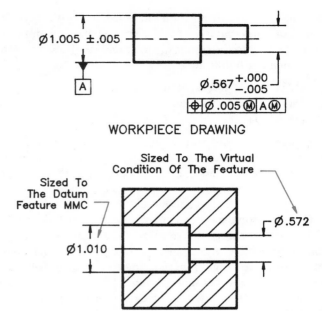

Figure 9-22. When a composite tolerance is applied to coaxial holes of different diameters, a note is placed beneath the feature control frame. The note indicates the number of coaxial holes.

Figure 9-24. Coaxiality of features such as a stepped shaft can be controlled with a position tolerance.

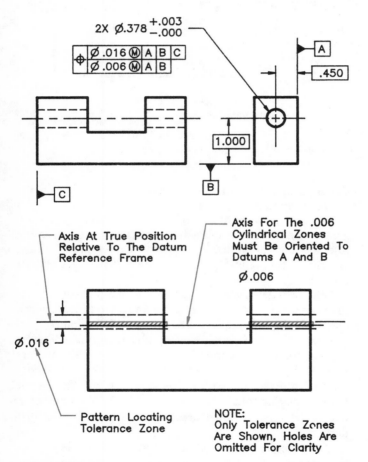

Figure 9-23. Datum references can be included in the second line of a composite tolerance that is applied to coaxial holes.

OTHER COAXIAL CONTROLS

Two or more coaxial features can be toleranced relative to one another. This is done by identifying one of the coaxial features as a datum feature, and applying a position tolerance on the other features to reference the identified datum. See Figure 9-24.

Position tolerances are typically applied to coaxial features when the main concern is assembly of parts. A properly calculated and applied position tolerance with the MMC modifier will ensure that the parts can be assembled.

The given figure shows a simple step shaft. The larger diameter is identified as datum feature A. The other diameter has a position tolerance of .005″ diameter at MMC referenced to datum A.

A gage for checking the position tolerance is shown. Primary datum A is simulated at the MMC size. A hole to check the position tolerance on the .567″ diameter is sized to the .572″ diameter virtual condition. If the stepped shaft fits into the gage, the position tolerance requirement has been met.

RADIAL HOLE PATTERNS

Radial holes have axes that pass through the center of a circular shape. See Figure 9-25. Four radial holes are shown near the left end of the given part. There is a position tolerance applied to the holes. It specifies a position tolerance of .020″ diameter at MMC relative to datum A-B primary and datum C secondary. The two datum references completely define a datum reference frame. Tolerance zones created by the tolerance specification are located on a framework that is located and oriented relative to the datum reference frame.

Datum features A and B simultaneously establish an axis through the given part. This axis is the primary datum for the specified tolerance. The tolerance zone framework is centered on and perpendicular to the primary datum axis.

Two of the three planes in the datum reference frame intersect on the primary axis. The third plane of the datum reference frame is perpendicular to the axis and is located relative to datum feature C. Datum feature C does not affect

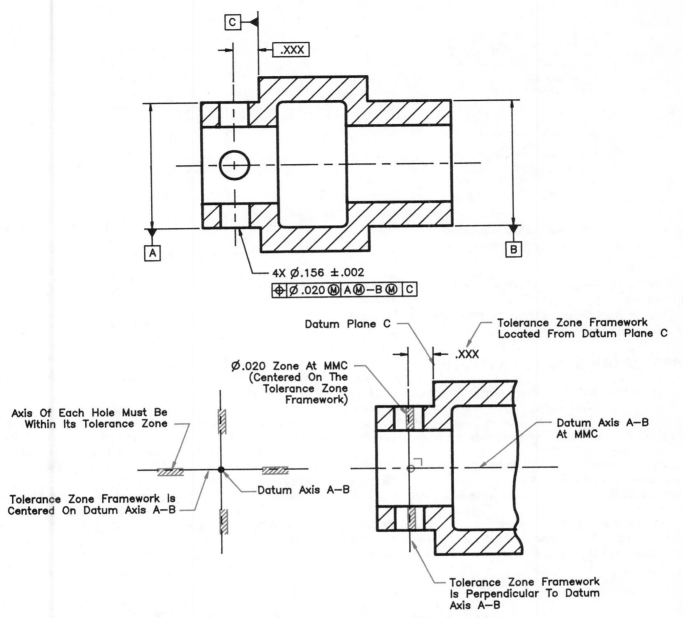

Figure 9-25. Radial holes can move independently within the tolerance zones.

the orientation of the third plane since the plane must be perpendicular to the two planes located on datum axis A-B. It must be perpendicular since the planes of a datum reference frame must always be mutually perpendicular.

The tolerance zone framework for the four radial holes is located and oriented relative to the primary datum axis. Since it is oriented by the primary datum, the secondary datum only serves to provide a location for the framework.

Tolerance zones for the four holes are located on the tolerance zone framework. The axis of each hole may lie anywhere within its tolerance zone. There is no requirement for holes on opposite sides of the pattern to lie on a common axis.

SYMMETRY

Symmetry is a condition where features on each side of a centerline or center plane are equal. Symmetrical features can be hole patterns or any other type of feature.

There are two types of tolerances that can be applied to control the location of symmetrical features. One control is a symmetry tolerance; the other is a position tolerance.

In 1982 standard did not include the symmetry tolerance symbol. When working to the 1982 standard, all symmetry location requirements are controlled by position tolerances.

FEATURE CONTROL FRAME

Feature control frames used to indicate symmetrical position requirements are formatted the same as for any other tolerance specification. See Figure 9-26. Prior to 1982, some symmetrical tolerances were specified using a feature control frame that included a symmetry tolerance symbol.

It was perceived that position tolerances could meet most, if not all symmetry specification needs, and the symmetry symbol was removed from the 1982 standard in an

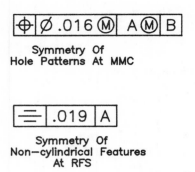

Symmetry Of
Hole Patterns At MMC

Symmetry Of
Non—cylindrical Features
At RFS

Figure 9-26. Two symbols can be used for symmetry applications.

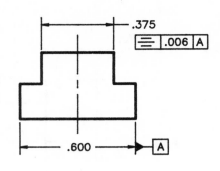

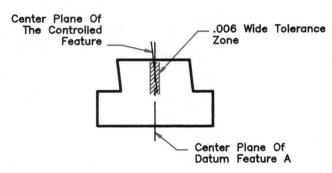

Figure 9-28. No location dimension is required for a feature of size when symmetry is specified.

attempt to reduce errors in tolerance application. There has been some concern in American industry about the removal of the symmetry symbol from the standard. That concern has resulted in the symbol being reinstated in the standard.

SYMMETRY OF HOLE PATTERNS

One common application of symmetry control is on hole patterns. When functional requirements require that a hole pattern be located symmetrically to some feature on a part, position tolerances can be applied to express the required control. See Figure 9-27.

Dimensions within the hole pattern are applied as they would be for any hole pattern. In the given figure, all holes are dimensioned to show their locations relative to one another.

The location of the hole pattern is defined with the position tolerance. There is no need for a dimension that shows the location of the holes relative to the referenced datums.

The position tolerance specification in the given figure is .008″ diameter at MMC relative to datum A primary, datum B secondary at MMC, and datum C tertiary at MMC. Datum features B and C are features of size. These datum features both establish center planes.

The position tolerance zones for the eight holes are symmetrically located relative to the datum planes. The requirement for a symmetrical location is defined by the position tolerance, and does not require any dimensions to show the hole locations relative to the center planes. The given

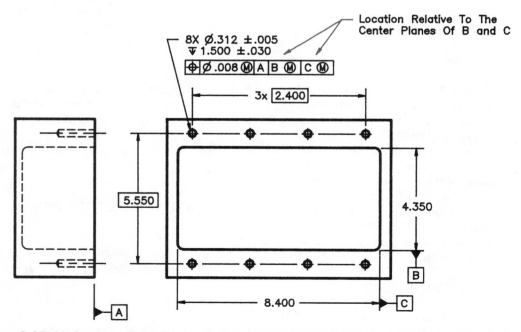

Figure 9-27. No location dimension is required for a hole pattern when position is used to specify symmetry.

drawing does not include any dimensions that show the symmetrical location.

SYMMETRY FOR FEATURES OF SIZE

Features of size such as tabs, rails, and slots can be toleranced for location by indicating a symmetry tolerance requirement.

Dimensions are applied to the features to indicate size. See Figure 9-28. No location dimension is needed, and symmetry tolerances are applied if the tolerance is to apply RFS.

The given figure shows a rail that is symmetrically located to the base of the part. The base is dimensioned and is identified as datum feature A. The rail is dimensioned and has a position tolerance of .006″ at RFS referenced to datum A at RFS. There is not a dimension that shows the location of the rail.

Datum A is the center plane of datum feature A. The tolerance zone for the rail is .006″ wide and is centered on datum plane A, regardless of feature sizes. The tolerance zone controls the location of the center plane for the controlled rail, but does not directly control the sides. Control of the center plane only has an indirect effect on the location of the sides of the rail.

CONCENTRICITY

Concentricity is a control of one axis to another. It is always applied on an RFS basis. The use of any other modifier is incorrect.

A very small number of tolerancing applications call for a concentricity tolerance. The tolerance should only be used when the location of one axis to another needs to be accurately controlled. Concentricity should only be applied when it is certain that axis relationships are the only means of producing a functionally acceptable part. Verifying concentricity tolerances is very difficult when using manual inspection equipment.

Most coaxial requirements can be met with either a position tolerance at MMC or a runout tolerance at RFS. These tolerance types can both be easily checked on the basis of surface conditions. They are preferable to concentricity requirements because of the ease in verifying position and runout tolerances. Concentricity should only be used when absolutely necessary.

There are many applications where concentricity is incorrectly applied, and runout is the control that should have been applied.

FEATURE CONTROL FRAME

Concentricity tolerances always include a diameter symbol in front of the tolerance value. See Figure 9-29. There must be a datum reference. The tolerance and datum reference must be at RFS.

CONCENTRICITY TOLERANCE APPLICATION

Concentricity can be applied to any feature that creates an axis. This means that a concentricity tolerance could be

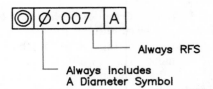

Figure 9-29. Concentricity always includes a diameter symbol and applies RFS.

applied to a cylinder, square bar, hexagonal bar, or any other symmetrical shape. Figure 9-30 shows a concentricity tolerance applied to a cylinder.

The tolerance specification in the given figure requires the axis of the cylinder to be concentric within .006″ diameter relative to datum axis A. The concentricity tolerance does not directly affect the surface condition of the cylinder.

Two possible conditions for parts produced to the given drawing are shown in the figure. The first one shows a cylinder produced at an angle to the datum axis. The size of the cylinder varies between .435″ and .438″ diameter, which is within the specified limits of size. Measuring the surface of the part results in a full indicator movement of .006″, which means there is a total runout error of .006″. *The full indicator movement by itself does not always indicate the amount of concentricity error.* On this particular part, the shape and

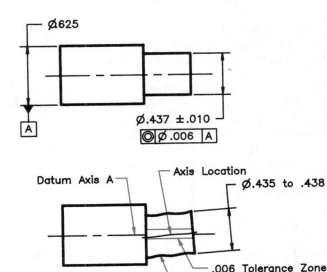

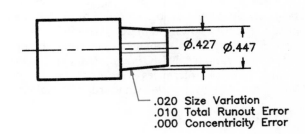

Fig. 9-30. Concentricity controls axis location, and can be difficult to evaluate on the basis of surface conditions.

orientation of the cylinder does result in a concentricity error that is equal to the runout error.

The second part shows a tapered feature that ranges from .427″ to .447″ diameter. This is within the limits of size. The tapered feature is aligned with the datum A axis. Although the tapered feature is aligned with the axis, the full indicator movement is .010″. This indicates a total runout error of .010″. Since the axis of the tapered feature is aligned with the datum axis, there is .000″ concentricity error.

The two parts show that measuring surface errors with one dial indicator is not adequate to determine the concentricity errors. One of the given parts showed readings that equalled the concentricity error; the other part showed readings that far exceeded the concentricity error.

Verification of concentricity requires that the axis location be determined to lie within the allowance tolerance zone. This must be done through a series of measurements. The measurements can show if the axis is within the required boundary. However, defining the exact axis location for an object can be very difficult, is not impossible.

It is advisable to specify part requirements through the application of position and runout tolerances when these can meet the functional application for the part. Concentricity tolerances should be avoided whenever possible.

PAST PRACTICES

The dimensioning standard continues to evolve as input is received from American industry. Changes to the standard are sometimes necessary to clarify existing guidelines, to expand the guidelines to cover a wider range of applications, and to improve the methods previously created. Methods in the current standard should be used when working on new drawings unless an existing contract requires compliance with a previous standard.

It is necessary to recognize some of the old practices since many existing drawings were created prior to the current standard. These existing drawings are easier to understand if past practices are known.

IMPLIED DATUMS

ANSI Y14.5M-1982 made it a requirement that position tolerances include any needed datum references, and does not permit implied datums. Prior to the 1982 standard, the use of implied datums was permitted. Figure 9-31 shows a position tolerance specification that implies datums.

The hole locations in the given figure are defined with basic dimensions. A position tolerance of .032″ diameter is applied to the four holes. No material condition modifier is shown. Since the given drawing is created in compliance with the 1973 standard, MMC is assumed. Prior to the 1982 standard, position tolerances were assumed to apply at MMC. According to the 1982 standard, material condition

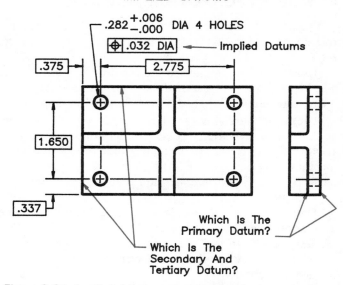

Figure 9-31. Implied datums were permitted prior to the issue of ANSI Y14.5M-1982.

modifiers must be shown on all position tolerances. Currently, all tolerances are assumed to apply RFS.

No datum references are included in the shown position tolerance specification. There are no identified datum features on the part. Datums must be assumed since they aren't identified or referenced.

This part appears relatively simple, and therefore it may seem easy to identify the features that should act as datum features. However, caution should be used because all of the datum features aren't obvious.

It is relatively easy to identify two of the datums for this part since basic dimensions for the holes originate at the bottom and left surfaces of the front view. Selecting the third datum feature is not easy.

The mounting surface of a part is typically used as the primary datum. Observing the given drawing, there are three possible planes that could be the mounting surface. One of these three must be assumed to act as the primary datum. After assuming a primary datum, it is necessary to select the order of precedence for the secondary and tertiary datums. Although selection of two of the datum features was easy, determining the optimum order of precedence is left to chance unless the function of the part is known.

Although implied datums were permitted by early versions of the dimensioning and tolerancing standard, they are no longer permitted because of the ambiguity that can result. It is recommended that datum references and datum feature symbols be added to old drawings that include implied datums if those drawings are resulting in parts that don't always function as intended.

Composite position tolerances create a pattern locating and a feature relating tolerance.

The pattern locating tolerance of a composite position tolerance specification is always larger than the feature relating tolerance.

Pattern locating tolerance zones are located on a pattern locating tolerance zone framework. This framework is

Paper gaging of the first line in a composite tolerance is completed in the same manner as for a single line position tolerance.

The second line of a composite position tolerance is paper gaged by plotting the relative position errors of the holes and then placing a set of concentric circles over the plotted errors. There is no requirement to center the concentric circles on the origin for the plotted errors.

A functional gage for the first line of a composite tolerance is the same as a functional gage for a single line position tolerance.

A functional gage for the second line of a composite tolerance must check the relative locations of features in a pattern.

defined by the basic dimensions that locate the pattern relative to datums and basic dimensions that give the relative positions of features in the pattern.

Feature relating tolerance zones are located on a feature relating tolerance zone framework. This framework is defined by the basic dimensions that give the relative positions of features in the pattern.

All features that have tolerances specified with the exact same datum references create a single pattern. All features in one pattern can be checked with a single functional gage.

Coaxial holes can be controlled with a composite position tolerance. The first line controls hole locations relative to datums, and the second line controls the relative location of the holes.

Symmetrical patterns of features can be position toleranced to control symmetry relative to specified datums.

Concentricity tolerances are only used when the axes of two features must be controlled. Position or runout tolerances should be used for coaxial features when they provide adequate control.

REVIEW QUESTIONS

Answer the following questions on a separate sheet of paper. Do not write in this book. Accurately complete any required sketches.

MULTIPLE CHOICE

1. A composite position tolerance has _____ tolerance symbol(s).
   A. only one
   B. two
   C. either one or two
   D. None of the above.

2. Datum references in the first line of a composite position tolerance invoke _____ control of the tolerance zones relative to the referenced datums.
   A. location and orientation
   B. location
   C. orientation
   D. None of the above.

3. The pattern locating tolerance zones must be located and oriented relative to _____.
   A. any referenced datums
   B. the feature relating tolerance zones
   C. the production equipment
   D. None of the above.

4. A tolerance zone _____ is the set of centerlines that define the true positions on which tolerance zones are located.
   A. specification
   B. cylinder
   C. plane
   D. framework

5. Datum references in the second line of a composite tolerance specification _____ order of precedence than/as shown in the first line.
   A. may be in a different
   B. must be in the same
   C. must be in a different
   D. None of the above.

6. Paper gaging to verify the feature relating tolerance requires that measurements be made _____.
   A. between features
   B. from datums
   C. in polar coordinates
   D. None of the above.

7. _____ can be used as datum simulators when the datum features are flat surfaces.
   A. Flat tool posts
   B. Spherical-ended tool posts
   C. Flat plates
   D. All of the above.

8. Several groups of holes form _____ if the position tolerances on the holes all have identical datum references.
   A. multiple patterns
   B. one pattern
   C. one or more patterns
   D. a radial pattern

9. If two groups of holes are noted as separate patterns, then _____ required to check the holes.
   A. one gage is
   B. two gages are
   C. Either A or B.
   D. Neither A nor B.

10. If the second line of a composite position tolerance applied to coaxial holes doesn't include any datum refer-

ences, then the tolerance controls _____ of the holes.
- A. only coaxiality
- B. coaxiality and orientation
- C. coaxiality and location
- D. coaxiality, orientation, and location

11. _____ needed to locate a symmetrical pattern when a position tolerance is used to locate the pattern relative to a centerline or center plane.
- A. No dimension is
- B. One dimension is
- C. Two dimensions are
- D. Some dimensions are

12. A primary datum reference including an RFS modifier requires that the datum simulator be _____ the datum feature.
- A. adjusted to make contact with
- B. fixed in size at the LMC limit of
- C. fixed in size at the MMC limit of
- D. fixed in size at the virtual condition of

13. A _____ datum reference including an MMC modifier must be simulated at its MMC size if no form tolerance is applied to the datum feature.
- A. primary
- B. secondary
- C. tertiary
- D. All the above.

14. Concentricity tolerance zones control the _____.
- A. surface of a feature relative to a datum axis
- B. axis of a feature relative to a datum axis
- C. surface of a feature relative to a datum feature surface
- D. None of the above.

TRUE/FALSE

15. A composite position tolerance requires a pattern locating tolerance that is smaller than the feature relating tolerance. (A)True or (B)False?

16. The second line of a composite tolerance specification is not required to include datum references. (A)True or (B)False?

17. The pattern locating and feature relating tolerance zones may overlap, provided the feature meets all tolerance zone requirements. (A)True or (B)False?

18. Hole location errors cannot be plotted on grid paper with sufficient accuracy to determine if composite tolerance requirements are met. (A)True or (B)False?

19. Pins in a functional gage for checking hole locations are made equal in size to the MMC of holes if the holes include a position tolerance of .001″ or more. (A)True or (B)False?

20. Hole size alone is not a good indicator of which holes form a complete pattern. (A)True or (B)False?

21. Two groups of holes can be checked with one gage if the holes act as a single pattern. (A)True or (B)False?

22. Coaxiality of features, such as the cylinders on a step shaft can be controlled with a position tolerance. (A)True or (B)False?

23. Symmetry is when a feature or group of features are offset to one side of a centerline or center plane. (A)True or (B)False?

24. Concentricity can be perfect when the feature surfaces include large runout (surface) errors. (A)True or (B)False?

25. Implied datums are permitted on position tolerances. (A)True or (B)False?

26. The datum simulator for a secondary datum reference including an RFS modifier must pull the workpiece into alignment with the simulator, even if this pulls the workpiece off the primary datum simulator. (A)True or (B)False?

FILL IN THE BLANK

27. The _____ line of a composite position tolerance specifies the feature-to-feature location requirements.

28. When paper gaging, _____ circles are used to represent tolerance zone diameters.

29. A functional gage for checking hole locations should include pins that are sized to the _____ of the holes.

30. A notation that states _____ can be placed under the feature control frame if it is necessary to have a group of features act as a separate pattern.

31. _____ hole patterns exit on parts such as hinges.

32. Drawings completed in compliance with the _____ issue of the standard may not include the symmetry symbol.

33. A position or symmetry tolerance used to locate a tab, rail, slot, or other rectangular feature actually creates a tolerance zone that controls the _____ instead of the feature surfaces.

34. Concentricity is difficult to verify since the _____ of the controlled feature must be derived from surface errors.

35. A secondary or tertiary datum reference including an MMC modifier requires the datum simulator be sized to the _____ of the datum feature.

SHORT ANSWER

36. Two tolerance zone frameworks are created by a composite position tolerance. What is the name of the framework created by the first line of a composite position tolerance?

37. Describe the requirements of the first line in a composite position tolerance.

38. What is the effect of referencing a datum in the second line of a composite position tolerance that is applied to coaxial holes?

39. Describe a problem that is caused by implied datums.

40. How is a datum feature of size simulated if it is referenced as primary and the datum reference includes an MMC modifier?

APPLICATION PROBLEMS

Some of the following problems require that a sketch be made. All sketches should be neat and accurate. Each problem description requires the addition of some dimensions for completion of the problem. Apply all required dimensions in compliance with dimensioning and tolerancing requirements. Show any required calculations.

41. Draw a feature control frame that requires a .027″ diameter pattern locating tolerance that is positioned and oriented to primary datum A, secondary datum D, and tertiary datum F. It must require a .012″ diameter feature relating tolerance that is oriented to datum A.

42. Make a sketch that shows the pattern locating tolerance zones for the given drawing. Show datums, true position requirements, and tolerance zones. Do not show the holes. Superimpose feature relating tolerance zones. These must be some amount of permissible offset shown from the true positions for the pattern locating tolerances. See Figure 9-3. Include dimensions and notes to clarify requirements.

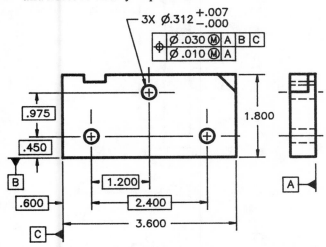

43. Sketch a functional gage that checks hole locations for the pattern locating tolerance in problem 42. Include dimensions, but assume that a gage needs no tolerance.

44. Sketch a functional gage that checks hole locations for the feature relating tolerance in problem 42. Include dimensions, but assume that a gage needs no tolerance.

45. Sketch a functional gage for the given drawing. Include dimensions, but assume that a gage needs no tolerance.

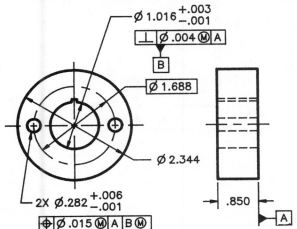

46. Sketch a functional gage for the given drawing. Include dimensions, but assume that a gage needs no tolerance.

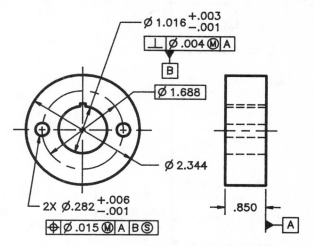

47. Sketch the necessary functional gages to check the feature relating tolerances in the given drawing. Include dimensions, but assume that a gage needs no tolerance.

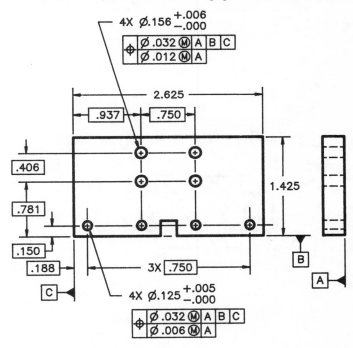

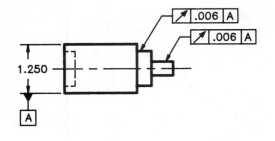

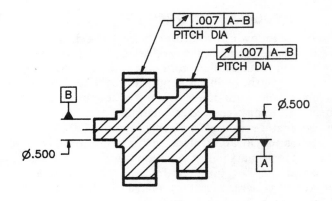

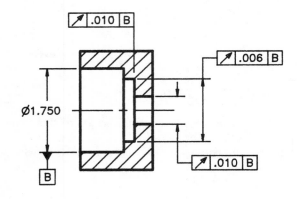

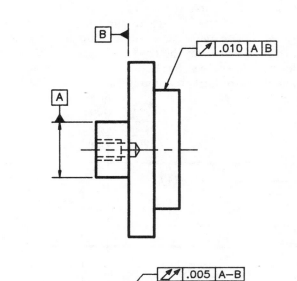

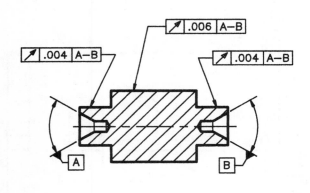

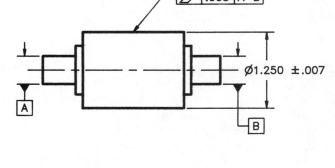

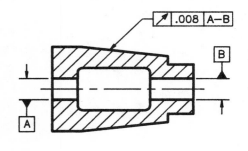

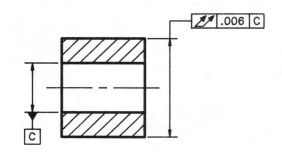

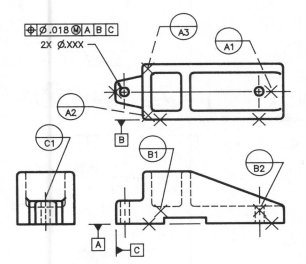

Chapter 10

Runout

CHAPTER TOPICS

□ Two types of runout and how they are measured.
□ The feature control frame construction for specifying each type of runout tolerance.
□ Application of runout tolerances on circular features and face surfaces.

□ Runout tolerances referenced to simultaneous datums.
□ How runout tolerances can be referenced to one datum or to two datums.
□ How to limit the area on which a runout tolerance applies.

INTRODUCTION

Runout error is the surface variation that occurs relative to an axis of rotation. These variations may occur on a cylindrical surface that is parallel to the axis of rotation, or on a face surface that is perpendicular to the axis of rotation.

There are two types of runout error–circular runout and total runout. *Circular runout is the error on a single circular element.* Circular runout can be measured on any object that has circular elements. *Total runout is the error across an entire surface.* Total runout only applies to cylindrical surfaces and surfaces perpendicular to an axis of rotation.

Runout tolerance specifications define the amount of runout error that is permitted on a feature. The tolerance symbol in the feature control frame indicates whether the permitted variations are circular or total runout. Both types of runout tolerance specifications must always include datum references.

CIRCULAR RUNOUT

Circular runout is the surface variation at each circular element on a part. Each circular element is evaluated separate from all others. See Figure 10-1. The surface variations are measured relative to an axis of rotation.

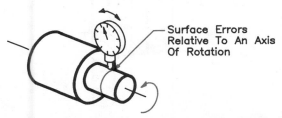

Figure 10-1. Individual circular elements are measured relative to an axis of rotation to verify circular runout tolerances.

The given part is a simple step shaft. The circular runout of a single circular element can be measured by placing a dial indicator in a fixed position and rotating the part on an axis of rotation. The full indicator movement caused by one complete revolution of the part is the runout error at the measured location.

Multiple circular elements on the part must be measured, but each measurement is evaluated separately from all others. The largest reading obtained for all the measured circular elements is the circular runout error for the part. This worst-case measurement must be less than the specified circular runout tolerance. If the circular runout error at any circular element exceeds the specified tolerance, the part is not within the allowable tolerance.

FEATURE CONTROL FRAME

A circular runout tolerance specification has a required format. See Figure 10-2. The circular runout tolerance

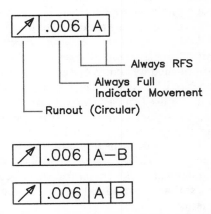

Figure 10-2. Runout tolerances are always specified at RFS and must include one or more datum references.

symbol is shown first. It is a single arrow. The tolerance value follows the symbol. A diameter symbol is never used since the runout tolerance indicates surface variations. A diameter symbol would incorrectly indicate a cylindrical tolerance zone.

Runout tolerances are always applied regardless of feature size. The tolerances must be specified at RFS since the measurements are made on the surface of the part.

An axis of rotation must be established by a reference to one or more datums. A single datum reference to a feature of size, such as a cylinder, is often adequate to establish an axis. It is sometimes necessary to use more than one feature.

The number of datum references and selection of datum features is usually determined from how the part is supported in its assembly. If one feature supports the part, then only one datum reference is needed. If two bearing surfaces support the part, then two simultaneous datums are referenced. If a primary bearing surface and a secondary surface support the part, then a primary and secondary datum are referenced.

It is not acceptable to reference datum features in a way that establishes more than one datum axis. A part can only be rotated on one axis at a time.

CIRCULAR RUNOUT APPLICATIONS

Surfaces with circular cross sections can be controlled relative to an axis of rotation by using a circular runout tolerance. Typical surfaces that include circular elements are cylinders, cones, spherical surfaces, and flat surfaces.

Circular Runout Applied To A Cylinder

The feature control frame for circular runout may be applied in any of three ways. All three indicate the same control on the indicated feature. See Figure 10-3. The given step shaft has three cylindrical segments. The .562″ diameter is identified as datum A. Circular runout tolerances on the shaft reference this datum feature to establish an axis of rotation.

Three methods of specifying a circular runout tolerance are shown. The feature control frame may be applied to an extension line, to the size dimension, or connected by a leader. Each of the three methods has exactly the same interpretation. Any of the shown application methods can be used when completing a drawing.

Circular runout on a part produced to the given drawing can be checked through the following procedure. Datum feature A must be used to establish an axis of rotation. This can be accomplished by clamping the feature in a collet or chuck. Of course, it is assumed the collet or chuck has an accuracy adequate for use in measuring the specified tolerance.

Multiple, but separate, dial indicator readings must be taken along the length of each toleranced feature. Each check must be made with the dial indicator in a fixed location, and the part must be rotated one full revolution. Provided the full indicator movement for each circular check is equal to or less than the specified circular runout tolerance, the feature is acceptable.

The full indicator movement at each cross section on the .375″ diameter of the shown part must be less than or equal

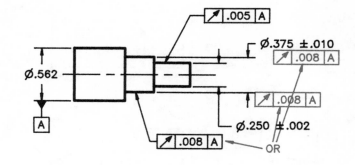

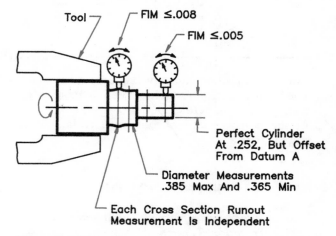

Figure 10-3. The datum reference in a runout tolerance determines which feature is used to establish an axis of rotation for verification of the tolerance.

to .008″. The circular runout variations on each cross section of the .250″ diameter must be equal to or less than .005″.

Applications of runout tolerance are most commonly seen on cylindrical features or on surfaces perpendicular to the axis of rotation. See Figure 10-4. Cylindrical features may be internal features such as holes or counterbores. Regardless of whether the runout tolerance is applied to an internal or external feature, the tolerance is a surface control

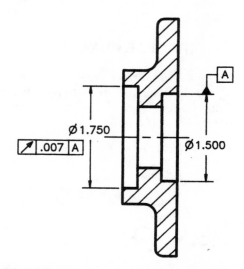

Figure 10-4. Runout tolerances can be used to control internal features, and they can also reference internal datum features.

relative to an axis of rotation. The tolerance zones apply along the full length of the feature.

Circular Runout Applied To Noncylindrical Features

Circular runout can be applied to noncylindrical features if the features include circular elements. See Figure 10-5. A circular runout tolerance of .010″ is applied to a conical surface. This requires that each circular element along the length of the cone be controlled. Surface errors on the cone are measured with the dial indicator positioned *normal* (perpendicular) to the cone surface.

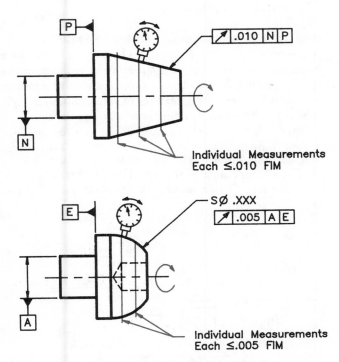

Figure 10-5. Circular runout can be controlled on any feature having circular elements that can be related to an axis of rotation.

The given figure shows a circular runout tolerance of .005″ applied to a spherical surface. Each circular element perpendicular to the datum axis must be controlled to within the .005″ tolerance. The runout errors must be measured with the dial indicator positioned normal to the surface.

Circular Runout Applied To Face Surfaces

Circular elements exist on the face surfaces of a shaft or counterbore. The face surfaces may be flat, a large-angle conical surface, or a large-radius spherical surface. See the examples in Figure 10-6.

Circular runout tolerances may be applied on extension lines from these surfaces or connected to the surface with a leader. Both application methods result in the same control.

The given figure shows a .008″ circular runout tolerance applied to a flat face surface. This requires that each circular element on the surface be flat and properly oriented within the .008″ tolerance when the part is rotated on the datum

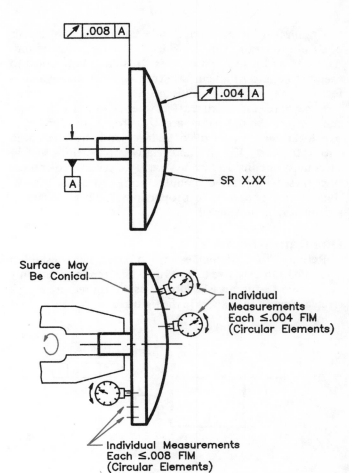

Figure 10-6. Flat, conical, and spherical face surface variations can be controlled with circular runout tolerances.

axis. This is checked by positioning a dial indicator perpendicular to the surface and rotating the part. If the full indicator movement is less than or equal to .008″, then the circular element is acceptable. Multiple circular elements must be checked.

Each circular element on the surface is checked separately. This would permit the surface to be wavy or conical and still meet the requirement for each circular element to be within the runout tolerance. Circular runout applied to a face surface does not control the overall shape of the surface; it only controls individual elements.

A circular runout tolerance of .004″ is applied to a large-radius spherical face on the given part. Measurements of the circular elements are completed with the dial indicator positioned normal to the surface.

Circular runout applied to a face surface controls the amount of allowable wobble that can be present when the part rotates. It is not meant to control the shape of the overall surface.

DATUM AXIS

All runout tolerances must include one or more datum references that can be used to establish a datum axis. Referenced datums may be established by entire surfaces or by datum targets.

Datum feature selection must be completed on the basis of design function, with consideration given to the manufacturing and inspection process. It isn't very practical to specify tolerances that can't be verified by available manufacturing and inspection equipment.

Some of the factors that affect whether or not a tolerance can be produced or verified are size, proportions, and relative locations of controlled features and the referenced datum features. The size and relative proportions of features must be considered when specifying runout tolerances since these factors affect manufacturing processes. In addition to size and relative proportions, relative locations should also be considered.

One Datum Feature

Reference to one datum feature is often enough to establish a good datum axis. See Figure 10-7. The step shaft has only one datum feature; it is identified as datum feature A. All circular runout tolerances in the figure reference datum A.

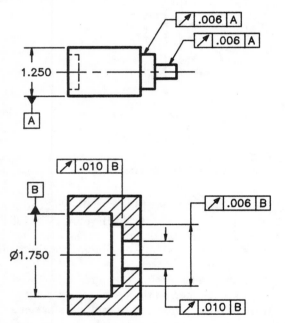

Figure 10-7. The size and proportions of features should be considered when assigning runout tolerances and identifying datum features.

Datum feature A is large enough to provide a good location for the datum axis. When the part is produced, a chuck or collet can clamp on datum feature A while cutting the two small diameters. The part will be stable in the chuck, and the forces applied while cutting the small diameters will not move the part within the chuck or collet.

The runout tolerances can be measured with the part properly located within the chuck or collet. There is little or no chance for the part to be improperly located or oriented relative to the datum axis.

Improper selection of a datum feature on this part could make it difficult to produce and inspect. Selecting the wrong feature to establish the datum axis can increase product cost

or even make the tolerances impossible to verify. Consider selection of the smallest diameter as datum A. The datum would be difficult to establish accurately because of the large size of the part relative to the small diameter. The mass of the part alone would tend to cause the part to move in the chuck or collet.

Holes and counterbores can be used to establish a datum axis. The given figure shows a counterbore that is identified as datum B. On this particular part, a bearing is installed in the counterbore. Since the bearing establishes the axis of rotation for the part, it is logical to use the counterbore to establish the datum axis. The size of the counterbore relative to the size of the part is also large enough to support the part during fabrication and inspection.

Compound (Simultaneous) Datum Features

Many rotating parts are supported by two bearings. Since two features are used to establish the axis of rotation of the part, the drawing should identify those features to establish the datum axis. Two datum features used to establish a single datum are referred to as *compound datum features*.

Compound datum features referenced in a feature control frame indicate one datum. See Figure 10-8. Two datum letters are placed in a single cell within the feature control frame. The letters must be separated by a hyphen. The two letters identify the two datum features that establish the datum.

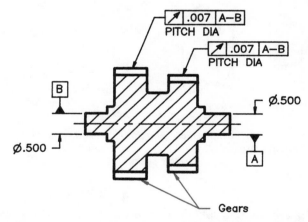

Figure 10-8. A notation is placed under a runout tolerance to indicate which feature is to be controlled on a gear.

The given figure shows a small gear shaft. Bearing surfaces at each end of the shaft are identified as datum features. One is identified as datum feature A, and the other is identified as datum feature B.

When installed in the functioning assembly, the two datum features are inserted in bearings. The axis of rotation is therefore established by both of these features. To establish a datum axis that simulates the functional application of the part, both datum features must be used simultaneously to establish the datum axis.

The runout tolerances in the given figure reference compound datum features A and B. This is done by showing datum letters A and B in a single cell. The two letters are separated by a hyphen. This means that the part must be rotated on both diameters to establish a single datum axis.

It would be incorrect to reference only datum A or only datum B for this part. The function of the part does not use only one of the features. The proportions of the part would make it unlikely that proper location of the datum axis could be established from only one of the datum features.

The given figure includes runout tolerances applied to gears. A notation of pitch diameter is placed under each of the tolerance specifications. When applying a feature control frame to control gear teeth, it is required that a notation be applied to indicate what feature is controlled. This type notation is also required when a feature control frame is applied to features such as a spline.

Prior to ANSI Y14.5M-1982, it was generally assumed that tolerances applied to gears and splines were for control of the pitch diameter. A notation was only required if the tolerance applied to some feature other than the pitch diameter.

Runout tolerances applied to the pitch diameter of a gear can't be measured with a dial indicator that travels along the surface of the teeth. Measurements are made using a master gear. A master gear is considered, for practical purposes, to be perfect.

The gear being inspected for runout error (the workpiece) is located on its datum features to establish the datum axis. A master gear is located on a shaft that is parallel to the datum axis. The inspection setup is designed to permit the center distance between the master gear shaft and the datum axis to vary. With the teeth of the two gears fully meshed, the workpiece is rotated. Runout errors in the workpiece cause movement in the center distance between the gears. The variations in center distance are equal to the runout error. (See AGMA 390.03).

Compound datum features can be internal features such as the countersinks of centerdrilled holes or coaxial holes. See Figure 10-9. Centerdrilled holes as datum features are good for production and inspection. However, centerdrilled holes are seldom functional features in a design.

It is possible to use nonfunctional features such as centerdrilled holes as datum features. All that needs to be done is to take into account all tolerance accumulations between functional features. The given part has runout tolerances of .004" on the two bearing surfaces at the end of the shaft, and a runout tolerance of .006" on the large diameter. These are all referenced to the datum axis established by the two centerdrilled holes. The worst-case runout error between the two bearing diameters and the large diameter is equal to the sum of the tolerance on the features, in this case .010".

The datum axis for this part is established using two machine centers. A machine center is placed in each centerdrilled hole, and the part is rotated.

Coaxial holes in a part can be used to establish a datum axis. The second illustration in Figure 10-9 shows a runout tolerance referenced to a datum axis that is established from two coaxial holes. Each of the shown coaxial holes has a datum feature symbol attached. The runout tolerance specification references the two datum features.

An expanding mandrel can be used to locate the datum axis. A fixed-diameter shaft must not be used (unless it fits tightly), since datum references for runout tolerances are

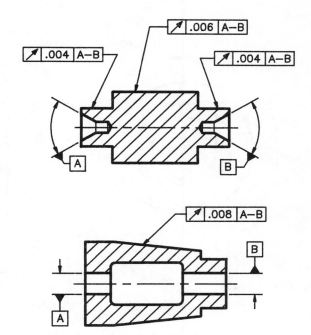

Figure 10-9. Compound datum references can be used to establish a single axis of rotation through two datum features.

always RFS. Dial indicator readings are taken while the part is rotated on the datum axis established by the coaxial holes.

Datum Targets

Datum targets are used whenever specific locations on a feature are used to establish a datum. There are many plausible reasons for using datum targets. One reason is that a functional datum feature may be large enough to make contact with the whole feature impractical. Another reason is that a feature may be irregular, and therefore require that specific locations be selected.

Whatever the reason, datum targets as defined in a previous chapter can be used with runout tolerances. See Figure 10-10. Datum target lines are used to define datum axis A-B on the given part. The runout tolerance is referenced to datum axis A-B.

Since the target lines are equally spaced on each end, two machine chucks of adequate precision can be used to establish the datum axis. Three-jaw chucks must be used since three line contact is required.

LIMITED TOLERANCE ZONE APPLICATION

Tolerances should only be applied to establish the amount of control required for the parts to properly function. Specifying only the needed level of control prevents the cost increases associated with overspecification of tolerances, and thus overall production costs.

Runout tolerances control all of the surface to which they are applied unless indicated otherwise. When only a portion of a surface requires a runout tolerance, the application of the tolerance can be limited. See Figure 10-11.

A chain line is used to show the area of application for a profile tolerance. The chain line is drawn along the outside

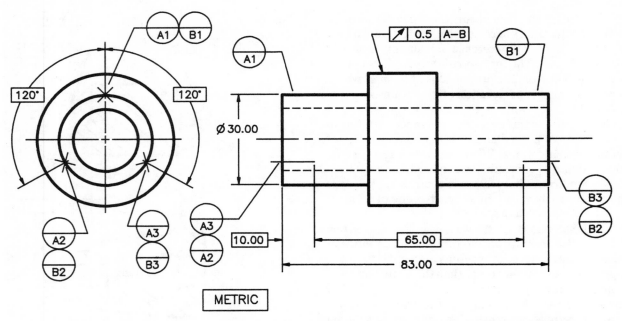

Figure 10-10. Runout tolerances may be referenced to a datum axis that is established from datum targets.

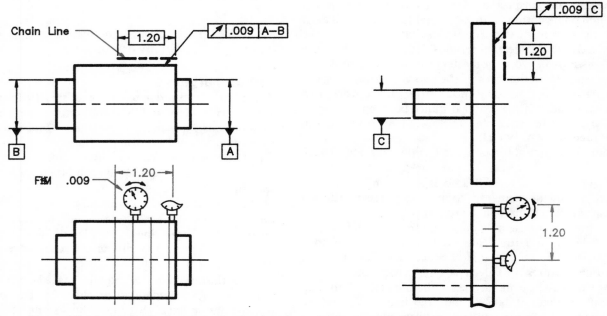

Figure 10-11. A limited zone of application for a runout tolerance is indicated by a heavy chain line.

of a view that shows the profile of the controlled surface. The chain line length is dimensioned with a basic dimension.

The chain line only indicates a limit of application in one direction. The runout tolerance still applies all the way around the circular elements that lie within the controlled length.

RUNOUT REFERENCED TO TWO DATUMS

One datum reference is adequate for many applications, but some applications require a second datum to stabilize the part. Runout tolerances having a primary and secondary datum reference must include one datum axis and one

datum plane. Only the design function determines whether the axis or the plane is referenced as primary. It is permissible for either the axis or the plane to be primary.

Selection of the correct primary datum is important. The primary datum will establish the orientation of the part and will have a significant impact on the produced part.

Primary Axis, Secondary Plane

An axis is usually chosen as the primary datum for runout tolerances when the dimensioned part is mounted on the diameter of a shaft, or inserted in a hole. The cylindrical datum feature should be primary in these situations since the diameter of the feature serves to orient and locate the part.

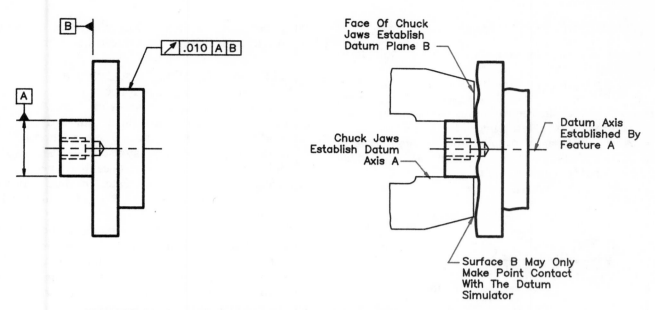

Figure 10-12. A secondary datum can be used to stabilize a part on the primary datum axis.

See Figure 10-12. The primary datum reference on the given part establishes an axis. The secondary datum reference establishes a datum plane that is perpendicular to the axis.

Primary datum axis A, for the given part, is established from datum feature A. Datum feature A is a cylinder. Datum feature A has a length-to-diameter ratio that may not adequately stabilize the part when clamped in a chuck or collet. Secondary datum feature B is used to stabilize the part on datum axis A.

The datum axis is both located and oriented by primary datum A. Datum feature B is only used to maintain the axis.

Primary Plane, Secondary Axis

Some types of rotating parts are mounted on a face surface. The face surface therefore establishes the orientation for the axis of rotation. Another feature must locate the axis of rotation.

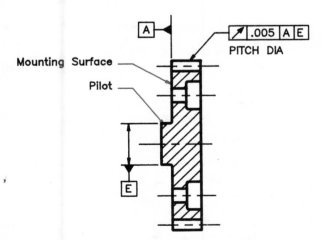

Figure 10-13. The mounting surface for a part sometimes requires that the primary datum be a plane. The datum axis is located by a secondary datum reference when runout is controlled relative to a primary datum plane.

See Figure 10-13. A gear is shown in the given figure. The face of the shown gear bolts against the face of another gear. A pilot on the shown gear fits into a hole in the other gear.

The gear face is identified as datum feature A, and the pilot is identified as datum feature E. The runout tolerance applied to the gear teeth references datum plane A primary and datum axis E secondary.

The face of the given gear establishes orientation relative to the axis of rotation. The pilot serves to locate the axis. Proper simulation of the functional application for this part requires the flat surface be used as the primary datum.

TOTAL RUNOUT

Total runout is the variation across the entire surface of a cylindrical feature or a perpendicular face surface. All elements on the surface are evaluated together. See Figure 10-14. The surface variations are measured relative to an axis of rotation.

The given part is a simple step shaft. The total runout of the large cylinder is measured by moving a dial indicator parallel to the axis of rotation while rotating the part on an axis of rotation. The full indicator movement caused by the surface variations along the feature is the total runout error.

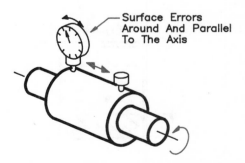

Figure 10-14. The combined effect of all surface variations on a feature, relative to an axis of rotation, is the total runout.

FEATURE CONTROL FRAME

A total runout tolerance specification has a required format. See Figure 10-15. The total runout tolerance symbol is shown first; it is a double arrow. The tolerance value follows the symbol. A diameter symbol is never used since runout tolerances indicate surface variations. A diameter symbol would incorrectly indicate a cylindrical tolerance zone.

As explained for circular runout, total runout tolerances are always applied RFS. Datum references are made in the same manner as for circular runout.

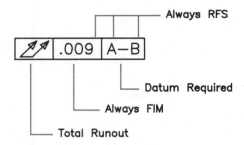

Figure 10-15. Present practice requires a double arrow be used to indicate total runout.

TOTAL RUNOUT APPLICATIONS

Cylindrical surfaces and flat surfaces perpendicular to the datum axis may be controlled with total runout tolerances. Total runout can't be applied to conical or curved surfaces as can circular runout. Total runout controls the form, orientation, and location (coaxiality) of the controlled surface relative to a referenced datum or datums.

Total Runout Applied To A Cylinder

The feature control frame for total runout may be applied to cylindrical features in any of three ways, and all three indicate the same control on the indicated feature. See Figure 10-16. The feature control frame may be attached to the surface using a leader. It may also be associated with the size dimension through placement near the dimension value or attachment to the dimension line. The third option is to attach the feature control frame to an extension line from the surface.

Two parts are shown in the given figure. They show that total runout tolerances, like circular runout tolerances, can be controlled relative to internal or external datum features.

One of the parts is a shaft that has a bearing diameter at each end. These are identified as datum features A and B. The total runout tolerance on the shaft is referenced to compound datum A-B.

Total runout on a part produced to the given drawing can be checked through the following procedure. Datum features A and B must be used together to establish an axis of rotation. This can be accomplished by clamping the two features in collets or chucks. Continuous dial indicator readings must be taken along the length of the of the part as it is rotated. The feature is acceptable provided the full indicator movement is equal to or less than .005".

Although the figure doesn't show an example, total runout can be applied to internal features. The interpretation is

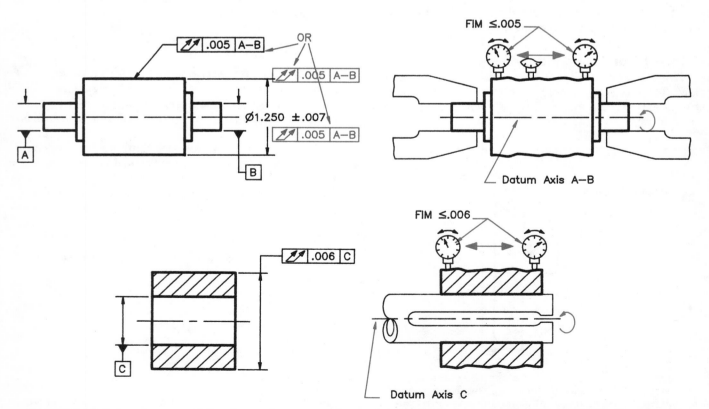

Figure 10-16. A feature control frame for total runout may be applied in any of several methods and have the same meaning.

the same as when applied to external features. Total runout applied to an internal cylinder requires that the full indicator movement across the entire surface be equal to or less than the specified tolerance value.

Total Runout Applied To Face Surfaces

Total runout tolerances may be applied on extension lines from face surfaces or connected to them with a leader. Both application methods result in the same control on the surface.

See Figure 10-17. The given figure shows a .009″ total runout tolerance applied to a flat face surface. This requires that the surface be flat and properly oriented within the .009″ tolerance when the part is rotated on the datum axis. This is checked by positioning a dial indicator perpendicular to the surface and moving the dial indicator across the surface as the part is rotated. The dial indicator is moved across the face surface in a direction perpendicular to the axis of rotation. If the indicator readings show .009″ or less variation, then the total runout tolerance specification has been met.

The shown total runout tolerance only applies to one surface. Datum feature B is not affected by the runout tolerance.

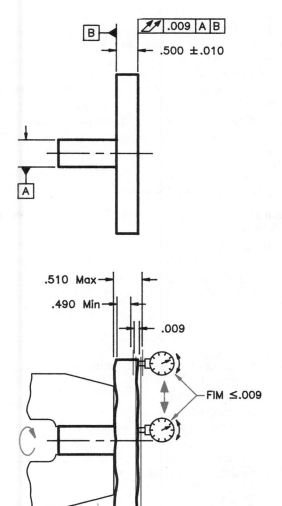

Figure 10-17. Total runout applied to a face surface should be applied on an extension line from the surface or attached to the surface using a leader.

COMBINED EFFECTS OF SIZE AND RUNOUT TOLERANCES

Runout tolerances applied to a feature may be larger than the size tolerance applied to that feature. See Figure 10-18. It is also acceptable to apply a runout tolerance that is smaller than the size tolerance. Neither the runout tolerance nor the size tolerance is constrained by the other.

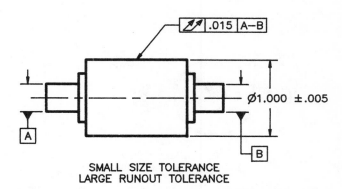

SMALL SIZE TOLERANCE
LARGE RUNOUT TOLERANCE

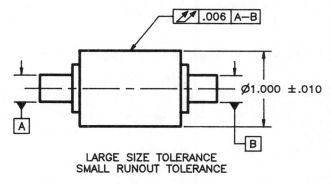

LARGE SIZE TOLERANCE
SMALL RUNOUT TOLERANCE

Figure 10-18. A runout tolerance applied to a feature may be larger or smaller than the size tolerance on the feature.

Runout error is possible even when a feature has perfect form. See Figure 10-19. The given figure has a dimension of 1.000″ ±.005″ on one feature. When this feature is at the MMC size of 1.005″ diameter, it must have perfect form. The perfect form requirement at MMC does not control the location of the feature relative to other features on the part.

The given figure shows a permissible total runout tolerance of .015″ relative to datum axis A-B. This allowable error may be caused by a combination of size, orientation, and location variations. A feature produced at MMC has no size variations, but runout errors may still exist due to orientation and location errors.

One acceptable condition for a part produced to the given drawing is shown. It has a perfect form cylinder at the 1.005″ MMC diameter. The cylinder is offset from the axis by a distance that results in a full indicator movement of .015″. This variation of the surface relative to the datum axis is permitted.

Size limits are not controlled through the application of total runout tolerances. See Figure 10-20. The given figure has a size tolerance of .020″. It also has a total runout tolerance of .006″. This shows that the limits of size can be

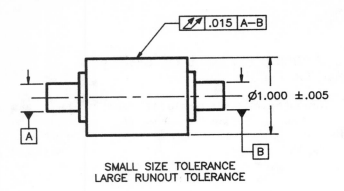

SMALL SIZE TOLERANCE
LARGE RUNOUT TOLERANCE

FIM ≤.015

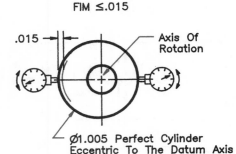

.015

Axis Of
Rotation

Ø1.005 Perfect Cylinder
Eccentric To The Datum Axis

Figure 10-19. A feature may be at MMC, have perfect form,
and still include runout errors relative to the referenced
datum axis.

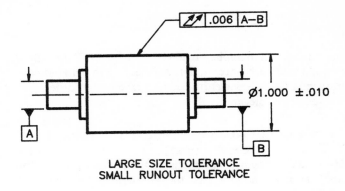

LARGE SIZE TOLERANCE
SMALL RUNOUT TOLERANCE

FIM ≤.006

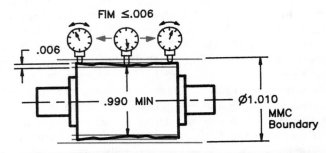

.006

.990 MIN

Ø1.010
MMC
Boundary

Figure 10-20. Runout tolerances are not meant to control the
limits of size; they do control the amount of permissible

assigned a larger permissible tolerance than the total runout tolerance.

The 1.000″ cylinder on the given part may be produced at any diameter between .990″ and 1.010″. The cylinder can't exceed the 1.010″ MMC envelope. Whatever the produced diameter of the cylinder, it must not have surface variations more than .006″ relative to the datum axis.

It is possible to produce a part that has at least one cross-sectional measurement that is at the .990″ minimum diameter. A part produced with one .990″ diameter cross section is not permitted to have any other cross section that is larger than 1.002″ diameter. This is because the total runout tolerance only permits a .006″ full indicator movement. The .006″ full indicator movement permits a maximum of .012″ variation on the diameter of a produced part. If a part is produced with a .994″ minimum cross-sectional measurement, then its maximum permitted cross-sectional measurement is 1.006″ diameter.

CHAPTER SUMMARY

Circular runout tolerances control each of the circular elements on an entire surface.

Total runout tolerances simultaneously control all elements on a surface.

Runout tolerances apply to an entire surface unless a limited application zone is shown.

Runout tolerances control surface errors relative to an axis of rotation.

Both the runout tolerance and datum references are always RFS.

The tolerance value for a runout tolerance is the maximum allowable full indicator movement.

The amount of runout tolerance applied to a part is not directly affected by the size tolerances on the controlled features.

REVIEW QUESTIONS

Answer the following questions on a separate sheet of paper. Do not write in this book. Accurately complete any required sketches.

MULTIPLE CHOICE

1. _____ requires that only individual circular elements meet the specified tolerance value.
   A. Total runout
   B. Circular runout
   C. Face runout
   D. None of the above.

2. A runout tolerance always includes _____ datum reference(s).
   A. one or more
   B. one
   C. two
   D. compound

3. Circular runout is measured at _____ location(s) on the controlled surface.
   A. one
   B. multiple
   C. five
   D. None of the above.

4. A cylinder may include acceptable errors that result in a tapered (conical) shape if _____ runout is applied. The amount of taper can exceed the specified runout value.
   A. total
   B. circular
   C. Either A or B.
   D. Neither A nor B.

5. One method of checking runout tolerance requires the part to be _____ with a dial indicator in contact with the controlled feature.
   A. held stationary
   B. moved parallel to its axis
   C. rotated on its axis
   D. None of the above.

6. Circular runout tolerances can be applied to any feature that has _____.
   A. a cylindrical shape
   B. a flat shape perpendicular to the axis
   C. circular elements related to an axis of rotation
   D. All the above.

7. A _____ zone is created by drawing a chain line along a portion of the surface to which a runout tolerance is applied.
   A. limited application tolerance
   B. tolerance
   C. projected tolerance
   D. None of the above.

8. Runout tolerances applied to gear teeth are usually applied to the _____ diameter.
   A. outside
   B. base
   C. pitch
   D. minor

9. Runout applied to a face surface controls _____.
   A. wobble
   B. radial movement
   C. eccentric cam motion
   D. None of the above.

TRUE/FALSE

10. A dial indicator does not indicate the diameter of the object being measured. (A)True or (B)False?

11. The primary datum feature for a runout tolerance must never be a flat surface. (A)True or (B)False?

12. Each measurement for circular runout is considered separately. (A)True or (B)False?

13. The surface variations permitted by a runout tolerance does not affect the size limits on a controlled surface. (A)True or (B)False?

14. Runout tolerance may be applied to internal and external surfaces. (A)True or (B)False?

15. Datum targets may be used to establish a datum axis. (A)True or (B)False?

16. When a runout tolerance references two datums, the primary datum must establish an axis. (A)True or (B)False?

17. Application of runout to gear teeth requires a notation to indicate which feature is controlled. (A)True or (B)False?

FILL IN THE BLANK

18. The symbol for total runout has _____ arrowheads.

19. Runout variations are always measured relative to a datum _____.

20. One continuous measurement is taken to check _____ runout of a feature.

21. Prior to the 1982 standard, total runout was indicated by a single arrow and the word _____ placed under the feature control frame.

22. _____ runout can be applied to conical surfaces, but _____ runout cannot.

23. Total runout tolerances can only be applied to flat surfaces only when the surface is _____ to the axis of rotation.

24. Total runout applied to a cylinder creates a tolerance zone bounded by two _____ cylinders.

SHORT ANSWER

25. What is meant by the phrase "Full Indicator Movement"?

26. Explain the effect of referencing compound datum features.

27. Is it permissible to specify a circular runout tolerance on a conical feature? Explain your answer.

28. Why isn't the application of total runout on a conical feature permitted?

29. How can runout of gear teeth be checked?

APPLICATION PROBLEMS

Each of the following problems require that a sketch be made of the figure. All sketches should be neat and accurate. Each problem description requires the addition of some dimensions for completion of the problem. Apply all required dimensions in compliance with dimensioning and tolerancing requirements. Show any required calculations.

30. Sketch a feature control frame that controls total runout to .009" relative to a datum axis that runs through datum features A and B.

31. Sketch a feature control frame that controls circular runout to .005" relative to datum axis D.

32. Control diameters B and C with a total runout tolerance of .005" relative to datum A. Control face surfaces D and E with a circular runout tolerance of .006" relative to datum A.

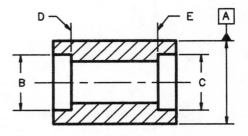

33. Sketch an inspection setup that includes a means for establishing the datum axis and measuring runout

errors. Also show some permitted surface variation and the tolerance zone.

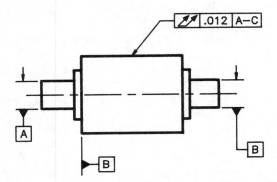

34. Sketch the shown produced part in an inspection setup for measuring the specified runout tolerance. Include datum simulators and also show how measurements would be made.

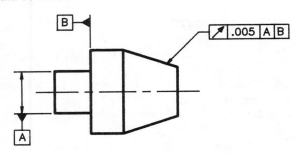

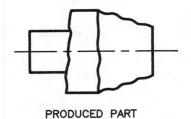

PRODUCED PART

35. Apply a total runout tolerance of .006″ to surfaces A, B, D, and E. Also apply a circular runout tolerance of .005″ to surface C. Identify the two datum features and reference them as compound datum features in the runout tolerance specifications.

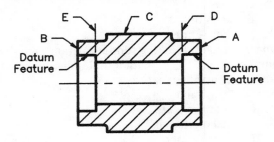

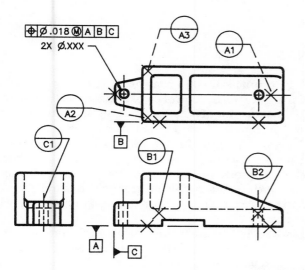

Chapter 11

Profile

CHAPTER TOPICS

☐ Line profile and surface profile definitions.
☐ Application of profile tolerances to a limited area on a feature, to all of a feature, and all around the profile of a part.
☐ Bidirectional and unidirectional tolerance zones specified with profile tolerances.

☐ Three levels of control achievable with profile tolerances: form only; form and orientation; or form, orientation, and location.
☐ Profile tolerances for control of coplanar features such as a series of flat bosses or rails.
☐ Profile tolerances applied to conical features.
☐ Composite profile tolerances.

INTRODUCTION

Profile tolerances may be used for a wide variety of controls. They may be used to control form, orientation, and location. The manner in which the tolerance is applied, how the toleranced feature is dimensioned, and how datums are utilized all impact the level of control achieved through the use of profile tolerances.

Because of the flexibility in the levels of control that can be achieved with profile tolerances, they are often used when no other tolerance control is appropriate. The flexibility that is provided by profile tolerances make specification of the desired control possible, but it also makes mastery of profile utilization more complex.

Subtle differences in how profile tolerances and the associated dimensions are applied to a drawing can make significant differences in the required tolerance zone. Care should be used in the application of profile tolerances to ensure specification of the desired control.

PROFILE SPECIFICATION

There are two distinctly different types of profile tolerances. Line profile is specified using a single curved line as the tolerance symbol. See Figure 11-1. Surface profile is specified using a curved line with a line drawn across the bottom to close the symbol.

Feature control frames for the two profile tolerance types have the same format, except for the tolerance symbol. The tolerance value never includes a diameter symbol, and the tolerance value is applied RFS. Depending on the situation, datum references may be omitted. It is also possible that one or more datums may be referenced. Datum references to features of size are typically RFS, but they may be MMC or LMC.

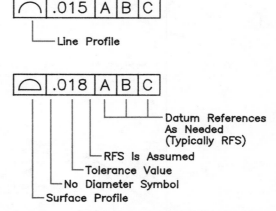

Figure 11-1. Tolerance values for line profile and surface profile are assumed to apply at RFS.

Several variables are involved in determining the correct application of a profile tolerance to a feature. If the profile tolerance is applied to a feature that is anything other than flat, then the form of the feature must be defined with basic dimensions. See Figure 11-2. The curved end on the shown forming punch is controlled by a profile tolerance. The curve must therefore be defined with basic dimensions. The two radii that form the punch are both shown basic. The relative locations of the two radii centers is also given with a basic dimension. The four basic dimensions completely define the form of the curved surface.

A surface profile tolerance is attached to the curved surface with a leader. The tolerance specification does not include any datum references, therefore the tolerance only controls the form and size of the surface. This control is established by a tolerance boundary that is defined relative

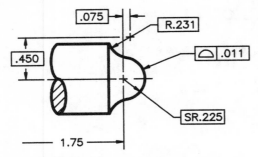

Figure 11-2. Profile tolerances control the shape (form) of the feature to which they are attached. Shapes controlled by profile tolerances are defined by basic dimensions.

to a theoretically perfect shape created by the basic dimensions on the drawing. Orientation and location of the surface is not controlled by the profile tolerance since no datum references are shown.

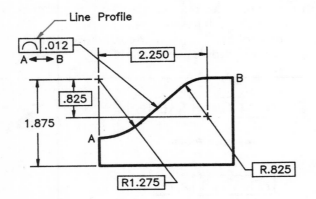

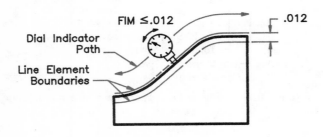

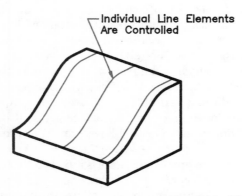

Figure 11-3. Line profile tolerances control individual line elements that are parallel to the view in which the tolerance is specified.

LINE PROFILE

Line profile is used to control single line elements on a feature. The level of control invoked on the feature depends on a combination of what is shown in the feature control frame and how the feature is dimensioned.

The simplest level of control that can be specified with a line profile tolerance is to control only form. See Figure 11-3. A line profile tolerance of .012″ is applied to the shown curved surface. Each line element that is parallel to the shown profile must have a form that lies within a .012″ wide tolerance zone.

Since no datums are referenced in the tolerance specification, the .012″ wide boundary has no required orientation or location relative to any datum reference frame. In fact, no datum reference frame is invoked.

Tolerances on location dimensions for the curved surface provide control of the orientation and location for the curved surface. The .012″ wide profile tolerance zone only controls the form of individual elements on the surface.

Each surface element parallel to the profile on which the tolerance is applied has a .012″ wide zone. Each element is independent of all others. There is no control of form in any direction other than parallel to the profile to which the tolerance is applied.

SURFACE PROFILE

Surface profile is used to simultaneously control all elements on a feature. Whether form, form and orientation, or form, orientation, and location are controlled depends on what is shown in the feature control frame and how the feature is dimensioned.

The simplest level of control that can be specified with a surface profile tolerance is to control only form. See Figure 11-4. A surface profile tolerance of .014″ is applied to the shown curved surface. All points on the surface defined by the shown profile must lie within a .014″ wide boundary. The boundary is centered on a perfect form profile defined by the dimensions on the curved surface.

Since no datums are referenced in the tolerance specification, the .014″ wide boundary has no required orientation or location relative to any datum reference frame. In fact, no datum reference frame is invoked.

Tolerances on location dimensions for the curved surface provide control of the orientation and location for the curved surface. The .014″ wide profile tolerance zone only controls the form of the surface.

The difference between surface profile and line profile can be compared to the difference between flatness and straightness. One controls the entire feature, and the other one only controls individual elements.

LIMITS OF APPLICATION

Profile tolerances only control the individual features to which they are applied unless indicated otherwise. Special applications include an all around profile requirement or a constrained limit over which a profile tolerance is to apply. It is possible to indicate by the use of an all around symbol

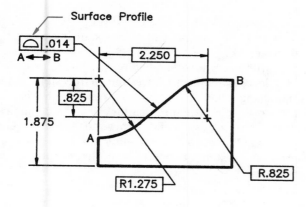

Surface Profile

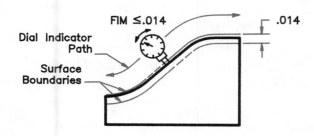

FIM ≤.014

Dial Indicator Path

Surface Boundaries

.014

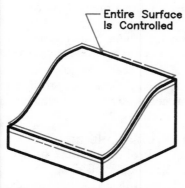

Entire Surface Is Controlled

Figure 11-4. Surface profile tolerances simultaneously control all points on the surface to which the tolerance is applied.

that a profile tolerance should apply all the way around a profile. It is also possible to limit the tolerance application to a segment of a feature by showing limits and referencing them.

Application of a profile tolerance specification to a feature is generally understood to create a tolerance zone across the entire feature. See Figure 11-5. Also, the tolerance zone created by the specification is understood to extend only to where there is an abrupt change in direction. An edge created by the intersection of two surfaces is an example of an abrupt change in direction.

The given figure shows a profile tolerance on a spherical surface that is on the end of a cylinder. There is an obvious abrupt change in direction where the spherical surface intersects the cylinder. The shown surface profile tolerance only applies to the spherical surface.

Abrupt changes in direction don't always exist at the desired limits of application for a profile tolerance. In these situations, limits of application must be shown.

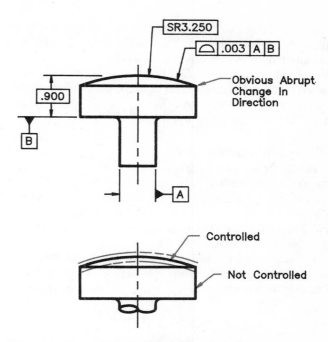

SR3.250

Obvious Abrupt Change In Direction

Controlled

Not Controlled

Figure 11-5. Profile tolerances apply to the entire surface to which the specification is attached.

Limits of application may be located anywhere on a part. See Figure 11-6. Two limits of application are shown in the given figure. The limits are identified by labelling points. Point A in the given figure is the point of tangency between a curved line and a straight line segment on the surface. Point B is the intersection of two straight segments that are at slight angles to one another.

Point A is definitely not an abrupt change in direction. It must be identified and referenced if the tolerance zone is not to continue past this point.

Point B may be considered an abrupt change by some people. Others would say that it is not an abrupt change. To

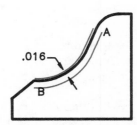

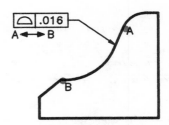

Figure 11-6. Limits of application can be labelled and referenced under the feature control frame.

prevent different interpretations of the tolerance requirements, points of application should be defined when there isn't an obvious abrupt change in direction.

In addition to labelling limits of application on the controlled surface, the limits must be referenced below the profile tolerance specification. The "between" symbol is used to indicate the limits of application.

All Around Application

Design requirements often require a profile tolerance that extends beyond abrupt changes in direction. These requirements can be met through the use of an all around symbol. The all around symbol indicates that the profile tolerance is to extend all the way around the feature to which the tolerance is applied.

The all around symbol is similar to the all around symbol used in welding specification. See Figure 11-7. It is a circle placed at the corner in the tolerance specification leader.

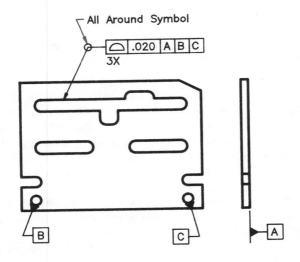

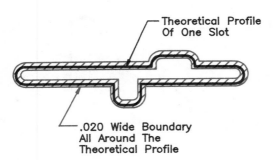

Figure 11-7. An all around symbol can be used to indicate that profile tolerances extend beyond abrupt changes in direction.

The shown profile tolerance has a "3X" notation. This indicates that the tolerance applies to all three slots. For simplicity, the figure only shows the effect of the tolerance specification on one slot.

The given drawing is a thin sheet metal plate with three slots cut in it. This part serves as a heat sink on a printed wiring board. The slots must provide clearance for electronic component leads (wires). Size, form, orientation, and

location of all slot surfaces must be properly controlled to prevent the component leads from shorting against the metal heat sink.

A profile tolerance is applied to the slots to provide the desired level of control. Since one of the slots includes abrupt changes in direction, an all around symbol is placed on the profile tolerance specification. This symbol indicates that the tolerance zone extends all around the feature. Omission of the all around symbol would result in a tolerance zone that only extends between two of the abrupt changes in direction.

The all around symbol only indicates a requirement for the tolerance to apply all around the feature profile as it is seen in the view to which the tolerance is applied. It does not indicate a requirement for the tolerance to apply over all surfaces of the part.

BIDIRECTIONAL AND UNIDIRECTIONAL CONTROL

Profile tolerances are always assumed to create a tolerance zone that is centered on the theoretical profile defined by basic dimensions unless indicated otherwise. When profile tolerances are applied to a feature of size, this has the effect of creating a bidirectional tolerance zone.

Exceptions can be indicated to create the affect of a unidirectional tolerance zone. Unidirectional zones can be created as a plus tolerance or a minus tolerance.

Through careful application of tolerances, it is also possible to indicate unequally distributed profile tolerance zones. This practice was not illustrated in the standard prior to 1994.

Bidirectional Profile Tolerance

Bidirectional control of a feature of size is established when a profile tolerance is applied and no special indications are placed on the tolerance specification. See Figure 11-8. Size and location dimensions on the given slot are all known to be basic because a note in the figure specifies this requirement. The basic dimensions establish a theoretically perfect slot profile and location.

The profile tolerance of .020″ is referenced to datums A primary, B secondary, and C tertiary. Since nothing indicates otherwise, a .020″ wide tolerance zone is centered on the theoretically perfect shape and location defined for the slot. The slot surface may fall anywhere within the tolerance zone.

Size, form, orientation, and location of the slot is controlled by the profile tolerance. Size is, in effect, controlled to a bilateral tolerance of ± .020 from the basic dimensions shown. The location of the slot center is, in effect, controlled to a bilateral tolerance of ± .010 relative to the referenced datums.

Unidirectional Profile Tolerance

A phantom line is used to indicate that a profile tolerance applies to only one side of the theoretical profile of a controlled feature. The phantom line is placed to one side, and parallel to the feature on which variations are allowed.

NOTES:
1. ALL UNTOLERANCED DIMENSIONS ARE BASIC.
2. ALL UNDIMENSIONED RADII ARE R.125

NOTES:
1. ALL UNTOLERANCED DIMENSIONS ARE BASIC.
2. ALL UNDIMENSIONED RADII ARE R.125

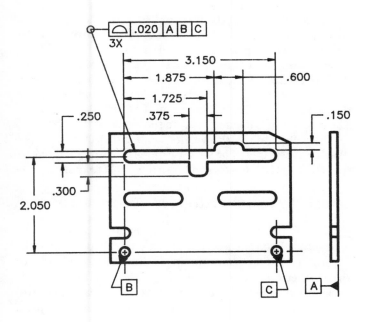

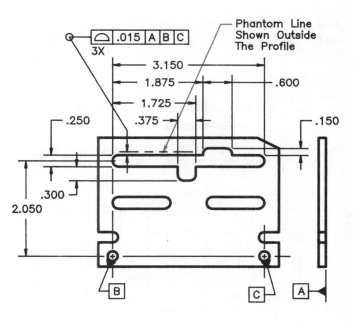

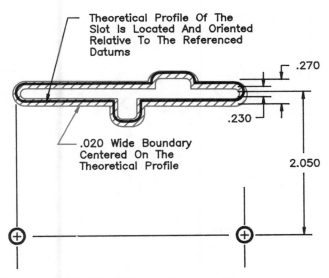

Figure 11-8. Profile tolerances are assumed to be bilateral unless indicated otherwise.

Figure 11-9. A unidirectional tolerance to the outside of the profile can be indicated by placing a phantom line to the outside of the toleranced feature.

The phantom line is only made a length that makes the direction of application obvious. It is not necessary or recommended that a phantom line be shown along the full length of the controlled feature. The distance between the phantom line and the controlled feature must be adequate to ensure clear separation of the lines when the drawing is reproduced. The distance between the lines is not determined by the profile tolerance value.

One possible application of a unidirectional profile tolerance results in a plus tolerance on the controlled feature. See

Figure 11-9. Size and form dimensions on the feature must be basic. Location dimensions may be basic, depending on the desired level of control.

Size and location dimensions in the given figure are basic. A profile tolerance of .015" is applied to the slot and referenced to three datums. The tolerance specification is modified by showing a phantom line on the outside of the slot.

The phantom line indicates the direction in which the specified profile tolerance is to apply. With the phantom line located to the outside of the feature, the entire .015" tolerance zone is to be located to the outside of the theoretical profile. A perfect profile sized and located by the basic dimensions forms one limit of the tolerance zone. The other limit is located .015″ to the outside of this one.

The effect on the shown slot is a plus .030″ and minus .000″ size tolerance. The location of the slot center relative to the referenced datums is similar to a ± .0075″ tolerance.

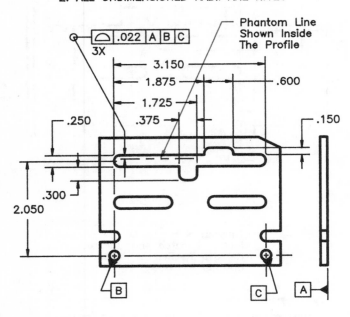

NOTES:
  1. ALL UNTOLERANCED DIMENSIONS ARE BASIC.
  2. ALL UNDIMENSIONED RADII ARE R.125

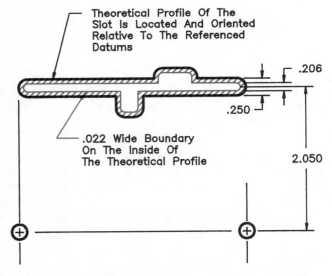

Figure 11-10. A unidirectional tolerance to the inside of the profile can be indicated by placing a phantom line to the inside of the toleranced feature.

Unidirectional tolerances can also be applied to create the affect of a minus size tolerance. See Figure 11-10. The previously shown slot is repeated, but the unidirectional tolerance is indicated this time by drawing a phantom line to the inside of the slot.

Size and location dimensions on the shown slot are basic. A profile tolerance of .022″ is applied to the slot and referenced to three datums. The tolerance specification is modified by showing a phantom line on the inside of the slot.

With the phantom line located to the inside of the feature, the entire .022″ tolerance zone is to be located to the inside of the theoretical profile. A perfect profile sized and located by the basic dimensions forms one limit of the tolerance zone. The other limit is located .022″ to the inside of this one.

The effect on the shown slot is a plus .000″ and minus .044″ size tolerance. The location of the slot center relative to the referenced datums is similar to a ± .011″ tolerance.

Unequally Distributed Profile Tolerance

An unequally distributed profile tolerance is shown with a phantom line on each side of the controlled profile. See Figure 11-11. One phantom line is dimensioned relative to the theoretical profile of the part. The tolerance zone is understood to be the width specified in the feature control frame, with one boundary located by the dimension shown relative to the perfect profile.

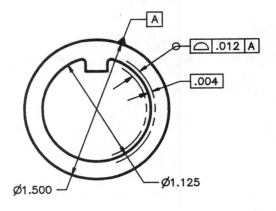

Figure 11-11. An unevenly distributed tolerance zone can be indicated by placing phantom lines on each side of the profile and dimensioning the offset to one of the phantom lines.

The given drawing shows a profile tolerance of .012″ applied to the inside surface of the part. Phantom lines are drawn on each side of the part profile. A dimension of .004″ is shown from the perfect profile to the outside phantom line. This requires that one tolerance zone boundary be located .004″ outside the perfect profile. The second boundary is .008″ inside the feature outline, to provide a total tolerance of .012″.

ACHIEVABLE LEVELS OF CONTROL

Profile tolerances can be used to establish a variety of controls on part features. When profile tolerances are applied to a single surface, then it is possible to control form, a combination of form and orientation, or a combination of form, orientation, and location. If the tolerance is applied all around a closed feature such as a slot, it is possible to achieve size control in addition to form, orientation, and location.

The feature control frame is one factor in determining the level of control established by a profile tolerance. See Figure 11-12. A profile tolerance without any datum reference in the feature control frame only controls form. If the feature control frame includes datum references, then more than form is controlled. The exact level of control must be determined from how the part is dimensioned.

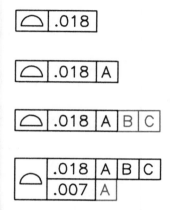

Figure 11-12. Various levels of control can be achieved through the composition of the profile tolerance specification.

Composite profile tolerances are specified using a two line feature control frame that is similar to the one used for composite position tolerances. Only one tolerance symbol is shown in the feature control frame. Composite profile tolerances were not illustrated in the 1982 and earlier dimensioning and tolerancing standards, but are included in the current standard. To avoid possible interpretation problems when working to the 1982 or earlier standards, a note should be placed on the drawing if composite profile tolerances are utilized. The note should indicate that datum references in the second line only provide an orientation control for the tolerance zone.

Options of line or surface profile tolerances combined with the available levels of tolerance specification provide a significant amount of flexibility in what can be achieved with profile tolerances. This flexibility can be used to express the required level of tolerance control to meet functional design requirements. Care must be taken not to overutilize the possible controls, because overspecification of tolerances can cause product costs to increase.

CONTROL OF SURFACE FEATURES

Proper application of profile tolerances is necessary to achieve the desired level of control. In addition, how feature dimensions are applied is important. The feature shape must always be defined by basic dimensions when using a profile tolerance, but location dimensions may or may not be basic. Whether or not location dimensions are basic depends on the level of control that is necessary.

Examples of how the various levels of control can be applied to surface features are provided in the following sections.

Form Only

Form of a feature can be controlled with a profile tolerance. Either profile of a line or profile of a surface may be used, depending on the design requirement. To control form only, the feature control frame for a profile tolerance must not include any datum reference. See Figure 11-13. Control of form alone can be used only when other means are employed to control location and orientation.

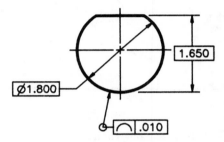

Figure 11-13. Control of form and size can be achieved with a profile tolerance that does not include any datum references.

The given figure shows a D-shaped hole. This type of hole is commonly used in control panels in which switches and electrical connectors are mounted. The purpose for the shape of the hole is to hold the inserted device so that it can't rotate. It is important that the shape and size of the hole be correctly made.

Basic dimensions define the diameter of the hole and the distance across the hole at the flat. Tolerances for basic dimensions must be defined in a feature control frame since title block tolerances do not apply to basic dimensions.

A line profile tolerance of .010″ is applied to the hole. It includes an all around symbol, and there are no datum references. There is no phantom line to indicate a unidirectional tolerance zone, therefore the tolerance zone is bilateral.

The tolerance zone is .010″ wide, and is centered on the perfect profile created by the basic dimensions. The location of the perfect profile is not defined by the feature control frame since there are no datum references.

Location dimensions for the hole must include coordinate tolerances since the profile tolerance does not include datum references. The coordinate tolerances provide the allowable location error for the hole. Making the location dimensions basic would be an error unless datum references are added to the profile tolerance specification.

Form And Orientation

Combining form and orientation control into one profile tolerance specification can be achieved by including one or more datum references in the feature control frame. See

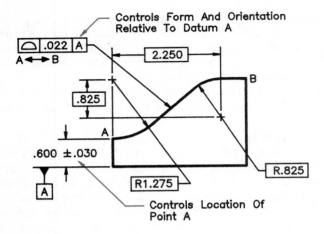

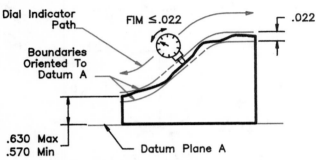

Figure 11-14. The addition of datum references to a profile tolerance establishes a requirement for orientation control.

Figure 11-14. The surface shape must be shown with basic dimensions.

The location of the surface must not be shown with basic dimensions if the tolerance is to control only form and orientation. Incorrectly using a basic dimension to show location would result in the profile tolerance also controlling location.

Basic dimensions are used to define the shape of the controlled surface in the given figure. There are two basic radii, and the locations of the center points relative to one another are also basic. A height dimension including a coordinate tolerance of ± .030″ is used to locate the height of the surface relative to the bottom of the part.

A surface profile tolerance specification of .022″ is attached to the surface. It includes a reference to datum A.

The tolerance zone established by the specification is .022″ wide, and it is oriented to the referenced datum. The location of the tolerance zone may float vertically, provided the orientation to datum A is maintained.

Location of the surface is controlled by the ± .030″ tolerance that locates one corner. The corner may be anywhere within the .060″ wide location tolerance band. Surface errors may go in any direction from the located corner, provided the .022″ profile tolerance zone can be moved into a position that contains all the surface errors. The profile tolerance zone can be moved to any position, provided the orientation to datum A is maintained.

Form, Orientation, and Location

Profile tolerance specifications that include datum references control form, orientation, and location if the surface

is located by basic dimensions. See Figure 11-15. This is the same figure as previously shown, except the .600″ location dimension for the corner has been made basic.

The tolerance zone on the given surface is .022″ wide. It is centered on the perfect profile created by the basic dimensions that locate and define the shape of the surface. The tolerance zone has a fixed location and orientation relative to the referenced datum.

The tolerance zone is centered on the perfect profile since no phantom lines are shown to indicate a unidirectional tolerance. Unidirectional tolerance could be indicated by placement of a phantom line on one side of the controlled profile.

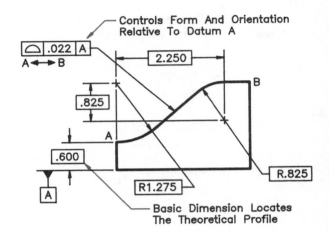

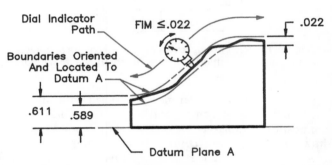

Figure 11-15. Location of the toleranced feature with basic dimensions results in the profile tolerance that controls location.

CONTROLS ON COPLANAR FEATURES

Large machined parts sometimes have raised areas, called *bosses*, machined away to create localized mounting pads. This prevents surface variation across large areas from causing distortion when the part is bolted to another part. The same technique is used on castings and forgings. Local bosses are built into the cast or forged parts and later machined to create coplanar surfaces that provide good mounting provisions.

Specification of the coplanarity requirements for multiple features is achieved through the specification of profile tolerances. See Figure 11-16. The level of control achieved can be varied by what is shown in the dimensions and tolerance specification. All the information shown in the given figure can be used to achieve the maximum level of control.

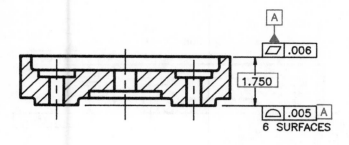

Figure 11-16. The level of control on coplanar surfaces is determined by the dimensions and tolerance specifications placed on the part.

Some or all of the shown tolerance specification and dimensions can be eliminated, depending on the desired control.

Flatness tolerances are not used to control coplanarity. Flatness is a control of individual features. Specification of a flatness tolerance on six bosses could result in six flat bosses, no two of which lie in a common plane.

Coplanarity Only

The amount of control placed on a feature or group of features should reflect the functional requirements of the part. In some designs, coplanarity of features is important to create a good mounting plane for a part. It is possible that the location and orientation of the mounting plane is not very important. For this type of situation, a profile tolerance can be applied to control only coplanarity.

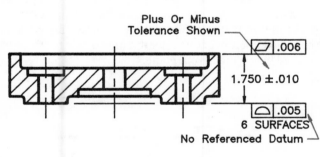

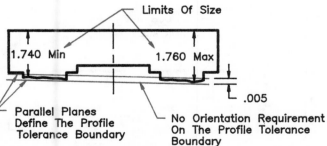

Figure 11-17. Coplanarity is required by the profile tolerance if no datum references are shown.

Figure 11-17 shows the section view from Figure 11-16. This figure shows a surface profile tolerance of .005″ applied to the six bosses on the part. A 1.750″ ± .010″ dimension is applied to locate the bosses. The profile tolerance specification of .005″ is on an extension line from the bosses.

The combination of the ± .010″ location dimension tolerance and the .005″ profile tolerance defines the coplanarity, location, and orientation requirements. Coplanarity is controlled by the .005″ profile tolerance. The profile tolerance zone may be in any location and orientation, but the surfaces of all six bosses must lie between two planes that are separated by .005″.

Location and orientation of the surfaces on the bosses are controlled by the ± .010″ tolerance on the location dimension. It is possible for the lowest point on one of the bosses to be at 1.740″, and the highest point on another boss to be at 1.760″. This is permitted so long as all surfaces are contained within two planes separated by .005″ as required by the profile tolerance.

Coplanarity And Orientation

Orientation of a part is sometimes as important as the required coplanarity for the mounting surfaces. The combined coplanarity and orientation requirement can be met by a profile tolerance that references a datum. See Figure 11-18.

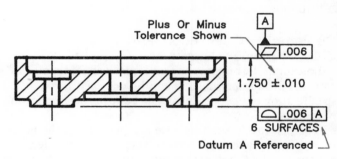

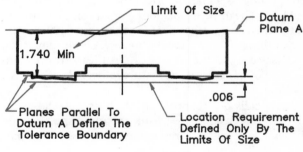

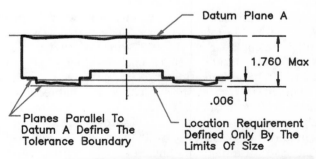

Figure 11-18. Coplanarity and orientation of the tolerance zone is required if a datum reference is shown.

Profile   229

The previously shown figure has been repeated with some changes to the dimensioning and tolerancing. A datum feature symbol has been added to one surface. The surface profile tolerance applied to the bosses is .006″ and is referenced to datum A. On the given part, there is no need to show a basic angle between the datum feature and the controlled surfaces since drawing convention permits parallelism to be assumed.

The profile tolerance zone must be properly oriented to the referenced datum at the angle shown on the drawing. In this case, it is parallel to datum A. The profile tolerance zone requires coplanarity within a .006″ zone, and also requires parallelism within the same zone.

Location of the bosses is allowed to vary within the ±.010″ tolerance on the 1.750″ dimension. The location tolerance is not reduced, but is refined by the allowable orientation error invoked by the profile tolerance. If a part is produced with the lowest point on one of the bosses at 1.740″, then the highest point on any other boss is controlled by the .006″ profile tolerance.

Another part can be produced with the highest point on one boss at 1.760″. If this occurs, the lowest point on any other boss is controlled by the .006″ profile tolerance.

The profile tolerance does, in effect, control size variation within a .006″ tolerance zone that is parallel to datum A, but the .006″ zone can fall anywhere within the 1.750″ ±.010″ tolerance zone.

Coplanarity, Orientation, and Location

Occasionally, the location of a group of features is as important as the orientation and coplanarity of them. Control of all three parameters can be achieved through the specification of one profile tolerance and proper application of dimensions.

The previously shown figure is repeated and the necessary changes have been made to result in the profile tolerance controlling coplanarity, orientation, and location. See Figure 11-19. The drawing includes one surface identified as datum feature A. A surface profile tolerance of .008″ referenced to datum A is applied to the bosses. A basic dimension of 1.750″ is used to locate the bosses. Profile tolerances control location of a feature if the feature is located with a basic dimension.

A theoretically perfect profile of the bosses is established at the 1.750″ dimension. The profile tolerance specification establishes a .008″ wide tolerance zone bounded by two parallel planes. This tolerance zone is centered on the 1.750″ location.

Surface variations on the bosses must lie between the two planes that form the tolerance zone. Surfaces that lie within the zone are coplanar within .008″, oriented to datum A within .008″, and located within a .008″ zone that is centered on the 1.750″ basic dimension.

The profile tolerance zone of .008″ permits bidirectional variation relative to the 1.750″ perfect profile location. Allowable variation is similar to a ±.004″ tolerance.

A unidirectional tolerance of .008″ could be established by placement of a phantom line on one side of the object profile. Direction of the allowable variation is determined by the side of the object on which the phantom line is drawn.

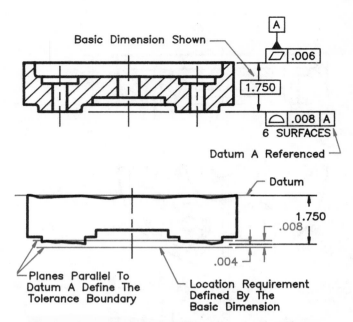

Figure 11-19. Coplanarity, orientation, and location are controlled by the profile tolerance if a datum reference is shown and the location dimension is basic.

Coplanarity Relative To A Datum Plane Established By The Controlled Features

Some controversy has existed over what the requirements are when a profile tolerance is applied to multiple features and referenced to a plane that is established by one or more of the controlled features. See Figure 11-20. The given figure shows a part with three lands on it. Two of the lands are identified as datum features A and B. A profile tolerance of .008″ is applied to all three surfaces, and referenced to datum A-B.

The interpretation that appears prevalent for the given tolerance specification is shown in the figure. Datum plane A-B is established from the surfaces of features A and B. The profile tolerance zone is centered on the datum plane. This creates a .008″ wide tolerance zone for all features other than the datum features. The datum features only get half the specified tolerance zone since it is impossible for them to be located below the datum plane.

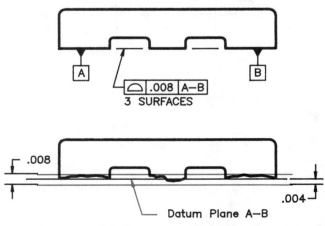

Figure 11-20. A profile tolerance can be referenced to a datum plane that is established by the controlled feature(s).

PROFILE APPLIED TO CONICAL SURFACES

Profile tolerances provide a means for controlling the form, orientation, and location of conical surfaces. No other tolerance permits control of these parameters for a conical feature. Circular runout can be used to control circular elements on a cone, but that does not control the overall form of the cone.

Control of cone form is specified with a surface profile tolerance that does not include any datum references. See Figure 11-21. In the given figure, all dimensions that define the cone are basic. A surface profile tolerance of .012″ is applied to the cone surface with a leader. Placement on an extension line from the surface is also permitted.

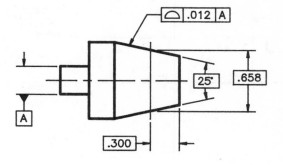

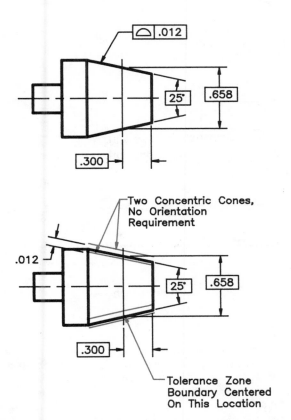

Figure 11-21. The form and size of a conical surface can be controlled through the use of a profile tolerance.

Surface variations on the cone must fall within two concentric cones that are separated by the specified .012″ tolerance. The two concentric cones control form and size of the cone since the cone dimensions are basic and the tolerance zone is centered on the given dimensions. Although the two cones forming the tolerance zone boundary must remain concentric, they can move together to any orientation or location. This is allowable since no datum references are shown in the tolerance specification.

Form and orientation of a conical surface can be controlled with a profile tolerance specification that includes a datum reference. See Figure 11-22. The previously shown figure is repeated, but a reference to datum A has been added to the profile tolerance specification.

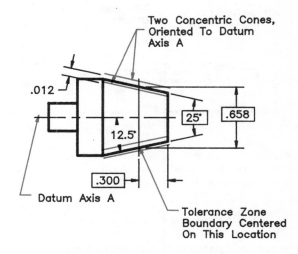

Figure 11-22. Orientation of a conical surface can be controlled by adding a datum reference to the profile tolerance specification.

Referencing datum A requires that the tolerance zone be oriented to that datum. The tolerance zone has perfect orientation relative to the datum axis on the given part. Surface variations may fall anywhere within the tolerance zone.

Location of a conical surface can also be controlled through a profile tolerance. It would be the same as Figure 11-22 except that a basic dimension would be used to locate the cone from a referenced datum.

COMPOSITE PROFILE TOLERANCES

Design applications often require tighter tolerances on the form of a feature than is needed for the orientation or location of the same feature. One example would be the mounting surfaces for a piece of optical equipment. The coplanarity of the surfaces is very important to prevent distortion of the optics when the parts are fastened in place. Orientation and location of the mounting surfaces may be less important because adjustment provisions can be included in the design of the optical equipment. The adjustments provide a means of compensation for orientation and location errors.

Composite profile tolerance permit specification of relatively large tolerances for location, orientation, and form,

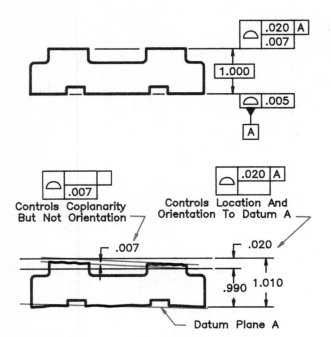

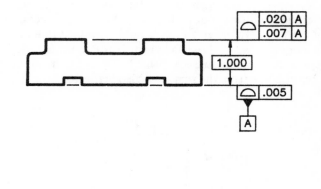

A composite profile tolerance can include a second line that controls coplanarity and orientation. See Figure 11-24. This requires datum references in the second line of the composite tolerance. Datum references in the second line only invoke an orientation requirement relative to the datums.

Figure 11-23. Composite profile tolerances permit large tolerances for less critical characteristics, and close control of the characteristics that need smaller tolerances.

with a second requirement that more closely controls form. See Figure 11-23. The given part has two lands that have a composite profile tolerance applied to them.

The first line of the composite tolerance creates a tolerance zone .020″ wide. It is bounded by two planes that are parallel to datum A. The tolerance zone is centered on the perfect profile that is located by the 1.000 basic dimension.

The second line of the composite profile tolerance establishes a coplanarity requirement that is bounded by two parallel planes separated .007″. Location and orientation of this coplanarity tolerance zone is not controlled since no datum references are shown. However, the surface variations must fall within the requirements of both tolerance zones.

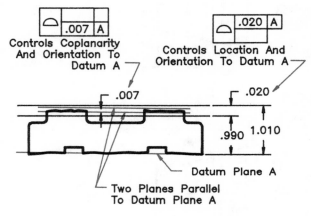

Figure 11-24. Datum references in the second line of a composite profile tolerance can be used to establish an orientation requirement.

CHAPTER SUMMARY

Profile of a line controls line elements that are parallel to the profile on which the tolerance is specified.

Profile of a surface simultaneously controls all points on a surface.

Limits of application for a profile tolerance must be specified, otherwise the tolerance controls the entire feature to which it is attached.

An all around symbol is required if a profile tolerance is to apply beyond the abrupt changes in direction that define the boundary of a feature.

Tolerance zones created by specified profile tolerances are bidirectional unless a phantom line is used to indicate otherwise.

A phantom line can be used to indicate that a profile tolerance applies in one direction relative to the profile defined by basic dimensions.

Levels of control that can be achieved include form, form and orientation, and form, orientation, and location. Form control also includes size control in some situations.

Coplanarity requirements can be specified with a profile tolerance.

Coplanarity, orientation, and location of multiple features can be achieved with profile tolerances. The level of control depends on how the part is dimensioned and whether or not datum references are used.

Profile is the only geometric tolerance other than straightness that can be used to specify form tolerances for a cone.

Composite profile specifications can be used to indicate a relatively large overall control and a smaller tolerance for more critical requirements such as coplanarity and orientation.

REVIEW QUESTIONS

Answer the following questions on a separate sheet of paper. Do not write in this book. Accurately complete any required sketches.

MULTIPLE CHOICE

1. Profile tolerance specifications may include _____ datum references.
   A. no
   B. one
   C. two or three
   D. All of the above.
2. Profile tolerance values are assumed to apply _____.
   A. RFS
   B. MMC
   C. LMC
   D. None of the above.
3. A profile tolerance with _____ datum references only controls form.
   A. no
   B. one
   C. two or three
   D. None of the above.
4. Errors measured on one surface element are independent of the errors on adjacent elements if a _____ tolerance is used.
   A. flatness
   B. surface profile
   C. line profile
   D. composite profile
5. When a profile tolerance is applied to a portion of a feature, _____ must be indicated if no abrupt changes in direction exist.
   A. datums
   B. surface profile tolerance
   C. line profile tolerance
   D. limits of application
6. Including a datum reference in a profile tolerance specification establishes, as a minimum, that the tolerance zone must be _____.
   A. oriented to the referenced datums
   B. located to the referenced datums
   C. Both A and B.
   D. Neither A nor B.
7. Coplanarity of flat surfaces should be achieved with a _____ tolerance.
   A. size
   B. surface profile
   C. line profile
   D. flatness
8. Profile tolerances can be applied to control the _____ of a conical surface.
   A. angle
   B. size
   C. location
   D. All the above.

TRUE/FALSE

9. References to datum features of size in a profile tolerance specification are assumed to apply RFS, but MMC modifiers are permitted. (A)True or (B)False?
10. In one respect, line profile is similar to a straightness tolerance applied to a surface, since the line profile tolerance controls individual surface elements. (A)True or (B)False?
11. Unless indicated otherwise, all profile tolerances are assumed to be totally distributed on one side of a theoretically perfect shape. (A)True or (B)False?
12. The combination of basic dimensions used to locate a feature and profile tolerances referenced to datums create a tolerance zone that is located and oriented relative to the datums. (A)True or (B)False?
13. Profile tolerances can be applied to only permit variation to one side–either inside or outside–of the basic feature outline. (A)True or (B)False?

FILL IN THE BLANK

14. Two types of profile tolerances are _____ and _____.
15. The shape of a curved feature must be defined by _____ dimensions if a profile tolerance is applied to the surface.
16. Surface profile establishes a tolerance zone across _____ of a surface.
17. A profile tolerance extends across a feature to _____ changes in surface direction.
18. A _____ line drawn on one side of a toleranced feature indicates that all the profile tolerance is in one direction from the basic shape.
19. The _____ line of a composite profile tolerance controls only orientation to any referenced datums.

SHORT ANSWER

20. Show the two profile tolerance symbols and label each symbol.
21. Explain how a profile tolerance can be specified to extend all the way around a feature.
22. When a profile tolerance references one or more datums, what difference does it make whether or not location dimensions for the feature are basic?

APPLICATION PROBLEMS

Each of the following problems require that a sketch be made of the figure. All sketches should be neat and accurate. Each problem description requires the addition of some dimensions for completion of the problem. Apply all required dimensions in compliance with dimensioning and tolerancing requirements. Show any required calculations.

23. Apply a bidirectional surface profile tolerance of .020″ relative to datum A primary and datum B secondary.

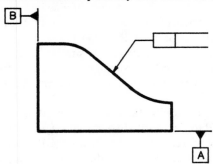

24. Apply a unidirectional surface profile tolerance of .015″ to the outside of the feature and related to datum A primary and datum B secondary.

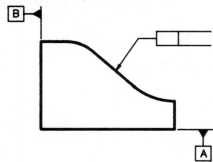

25. Apply a line profile tolerance of .018″ all around the cutout. Relate the tolerance to datums A primary, B secondary, and C tertiary. Datums B and C should be RFS.

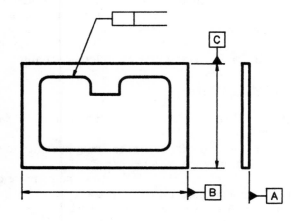

26. Apply a surface profile tolerance of .030″ relative to datum A primary and datum B secondary. Limit the application to the surface area between the indicated points.

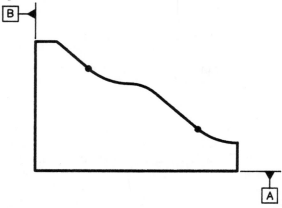

27. Control the three surfaces to be coplanar and oriented to datum A within .008″ and located to datum A within ±.010″.

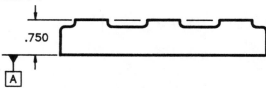

28. Control the three surfaces to be located, oriented, and coplanar within .012″ relative to datum A.

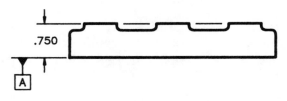

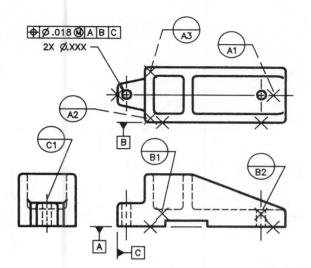

Chapter 12

Practical Applications and Calculation Methods

―――――――――――― CHAPTER TOPICS ――――――――――――

☐ Position tolerance effects on clearance holes in two and three stacked parts.

☐ Calculation of position tolerances for two clearance holes of different sizes.

☐ Distribution of the total allowable position tolerance for a floating fastener application.

☐ Calculation of position tolerances for two and three stacked parts when a fixed fastener condition exists.

☐ Proof of an advantage of projected tolerance zones for threaded holes.

☐ Accumulation of tolerances in an assembly and selection of datums to minimize tolerance accumulation.

☐ Increasing manufacturing freedom through the proper utilization of zero position tolerance at MMC.

☐ Paper gaging techniques for position tolerances.

INTRODUCTION

Floating fastener and fixed fastener conditions include two or more parts that are assembled together. Regardless of the number of parts in an assembly, tolerances can be calculated to ensure the parts fit together. This chapter shows how to calculate position tolerances for assemblies containing two or more parts stacked together. It also shows how to calculate tolerances when multiple parts are assembled in a manner that results in an accumulation of tolerances.

Application of the correct tolerance to properly specify the needed level of control requires that careful consideration be given to the design application. A logical process can be used to determine the proper type of tolerance, or groups of tolerances, to apply.

Application of tolerance zones that specify the required level of accuracy, but that permit the maximum manufacturing flexibility, require an understanding of how the tolerances affect the part. It also requires that calculations be completed on the basis of how the parts fit together. A proper understanding of tolerance types and concepts such as zero tolerancing at MMC will ensure that tolerances are maximized.

FLOATING FASTENER CONDITION

Floating fastener condition has been described to some extent in a previous chapter. A more in-depth look at the floating fastener condition makes it possible to see how calculations of tolerances can be completed for simple or complex designs that include floating fasteners.

A floating fastener condition exists anytime a bolt, pin, shaft, or other feature passes through clearance holes in an assembly. A floating fastener condition does not exist if any of the holes contain threads, a press fit, or any other feature that fixes the location of the fastener. The fastener must be free to float within all the holes.

TWO STACKED PARTS

The simplest assembly in which a floating fastener condition exists is one in which two parts are stacked together. See Figure 12-1. Each of the given parts is a simple rectangular plate with two clearance holes. A bolt passes through the clearance holes.

An assumed functional requirement for the assembled parts is to have two edges aligned. These edges are identified as datums, and dimensions for the holes originate from these datums. Position tolerances specified for the holes reference the datums. A single line feature control frame for position tolerances is used to ensure the edges of the parts will align.

Calculations are completed to ensure that bolts can be installed through the clearance holes. When clearance holes in two parts are located from common datums, the formula used to calculate the tolerance value is:

$$T = H - F$$

This formula results in a tolerance value that is applied to each of the parts. The letter T is the position tolerance value, H is the MMC of the hole, and F is the MMC of the fastener.

Figure 12-1 shows an enlarged section view of a bolt passing through two clearance holes. The datum features of the

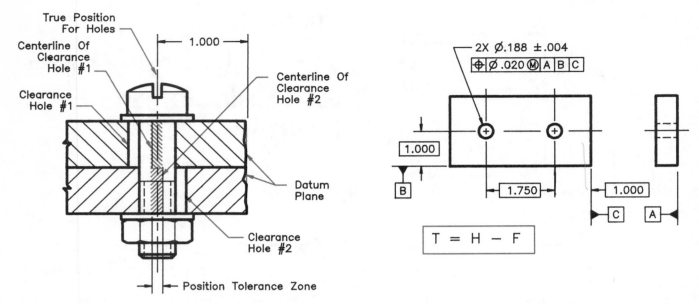

Figure 12-1. A floating fastener condition has a shaft passing through multiple parts that all contain clearance holes.

two parts are aligned with a common datum plane. A true position axis is located at the basic distance from the datum plane. A tolerance zone for each hole is centered on the true position axis. Each tolerance zone diameter is calculated using the $T = H - F$ formula.

The figure shows the axis of each hole moved to opposite extremes within the tolerance zone. This results in opposite sides of the hole making contact with the bolt. Even though the holes are in the worst possible location relative to one another, the bolt is still able to pass through the holes. Any other location of the holes would be better than the one shown, and would also permit the bolt to pass through.

The same size hole is shown in each part. The amount of clearance between each of the holes and the bolt is the same. This clearance is equal to the permitted position tolerance on each of the holes.

Position errors for holes are sometimes a combination of hole location and orientation variations. See Figure 12-2. Provided the axis of a hole falls anywhere within the tolerance zone permitted by properly completed calculations, the fastener will be able to pass through the clearance holes.

The given example shows two holes produced with orientation errors. Each hole axis falls within the permitted tolerance zone. The fastener does fit through the holes since the amount of clearance between the hole and fastener permits the axis to be at any location or orientation within the tolerance zone.

Alignment of the edges of parts in an assembly are not always required. See Figure 12-3. Any time an allowable mismatch between edges of an assembly exists, tolerances may be calculated for a composite position tolerance. A previous chapter explained that a composite position tolerance creates a relatively large pattern locating tolerance and a smaller feature relating tolerance.

The second line of a composite position tolerance specifies the feature relating tolerance. For a floating fastener condition, the tolerance value for a feature relating tol-

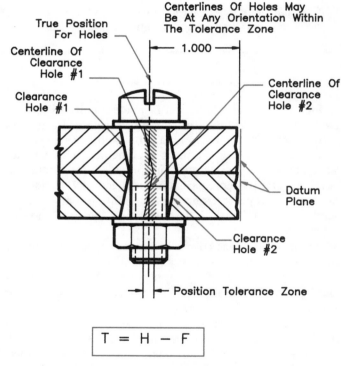

Figure 12-2. The axis of a hole may lie anywhere within the position tolerance specified for that hole.

erance (T_F) is usually calculated using the following formula:

$$T_F = H - F$$

This formula works when the datum references are to surfaces and the primary datum is repeated in the feature relating tolerance specification. Calculations completed in this manner ensure that hole locations within a pattern are adequately controlled to permit the fasteners to be installed.

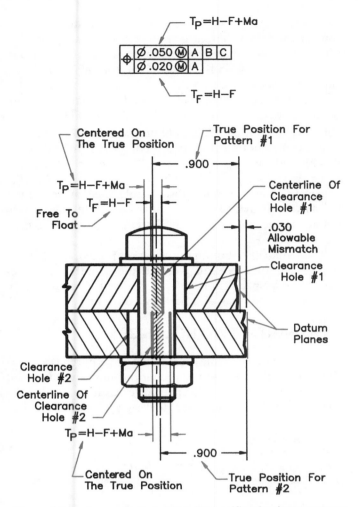

Figure 12-3. A composite tolerance specification has a pattern locating tolerance that permits relative movement of the assembled parts and a feature relating tolerance that ensures the fasteners can pass through the holes.

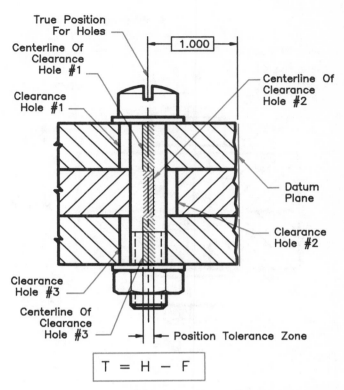

Figure 12-4. If the holes are all equal size, position tolerances for three stacked parts in a floating fastener application can be calculated in the same manner as for two stacked parts. Parallelism of surfaces is assumed to be exact.

This tolerance does not have any affect on the allowable mismatch at the edges of the assembled parts.

Calculation of the pattern locating value (T_P) of the composite position tolerance is completed using the following formula:

$T_P = H - F + M_a$

T_P = Tolerance, Pattern Locating

M_a = Mismatch, Allowable

Pattern locating tolerance values calculated this way permit the edges of assembled parts to be misaligned by a distance equal to the value M_a. Misalignment of any referenced datum features is permitted to be a maximum value equal to the value entered for the M_a value in the formula, provided those features are surfaces and referenced in the same order of precedence in the tolerance specification for both of the parts.

One factor that has not been considered in the preceding explanation is the effect of primary datum surface variations on the two parts. If surface variations are large, it is possible that high points on one surface could fit inside low points on the other part. This is unlikely, but if it happens, the primary planes on the two parts will not coincide.

Generally, the error between datum planes is so slight that it is not necessary to adjust specified position tolerances. If, however, the surface errors are large in relation to the surface size, errors need to be considered in the position tolerance calculations.

It is also possible to minimize the surface variations to ensure datum coincidence. A form tolerance can be applied to the datum features, or the form of the surface can be controlled by the size tolerance and Rule #1. The amount of form control expressed through form and size tolerances should not be overly restrictive.

THREE STACKED PARTS

Assemblies may have three parts stacked together. See Figure 12-4. When all three parts have clearance holes, a floating fastener condition exists.

The explanation of position tolerance calculations for three stacked parts assumes that parallelism of surfaces on the individual parts is perfect. As the surfaces depart from parallel, the resulting orientation of the holes requires an increase in hole diameter or a reduction in the allowable position tolerance.

Position tolerances for three stacked clearance holes are calculated in the same manner as for two holes. The same calculation methods work under the condition that the holes in all three parts have the same MMC size.

Figure 12-4 shows a position tolerance zone centered on

Practical Applications and Calculation Methods 237

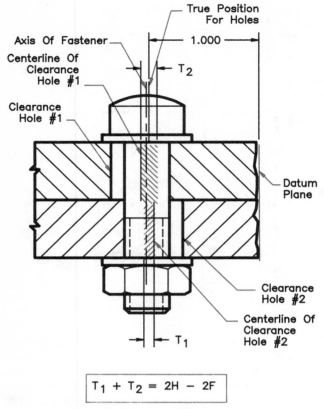

$$T_1 + T_2 = 2H - 2F$$

Figure 12-5. The total allowable position tolerance may be distributed unequally.

the true position for the three holes. The size of the tolerance zone is determined using the formula T = H - F. Each of the holes in the figure is offset to an extreme allowable position. Even with the holes in the shown positions, the fastener passes through the holes.

Provided the same edges on the three parts are identified as datum features and referenced in the same order of precedence, the edges of the parts can be aligned when the fastener is installed. Using different datum references can result in misalignment of the edges, however the hole patterns will align to permit installation of the fasteners.

TWO PARTS, UNEVENLY DISTRIBUTED TOLERANCES

Two parts containing holes of the same size may have the total amount of position tolerance unevenly distributed between the holes. See Figure 12-5. Two parts are shown stacked together in the given figure. The edges of the two parts are aligned. The aligned edges are used as datums to locate the holes.

Position tolerances for the two clearance holes are calculated using the following formula:

$$T_1 + T_2 = 2H - 2F$$

The resulting tolerances should be carefully applied on the drawing. It is advisable to apply the large tolerance as a projected tolerance zone. This is necessary because the position of the fastener is affected by the amount of tolerance in excess of the tolerance permitted by the hole

diameter. In effect, the hole acts as a fixed fastener condition for the tolerance that is in excess of the amount permitted by the hole diameter. The formula in Figure 12-5 can be proven to work on the basis of the formula used when calculating one tolerance value to apply on two holes. The formula T = H - F results in a value for T that is used on two holes. The total tolerance is therefore equal to T + T, which is the same as 2T. To obtain a value of 2T, the formula can be multiplied by two, giving the formula $2T = 2H - 2F$. Substituting $T_1 + T_2$ for 2T in the formula results in the previously given formula of $T_1 + T_2 = 2H - 2F$.

Any portion of the total position tolerance can be applied to one hole. Provided the remainder of the total tolerance is applied to the mating hole, the fastener will be able to pass through both holes.

If unequal distribution of tolerances is used, it should be done with consideration given to the manufacturing requirements. Specification of an extremely small tolerance on one of the two parts is generally a mistake.

TWO PARTS, TWO HOLE SIZES

Designs are sometimes completed using purchased standard (or catalog) parts in combination with newly designed parts. Since purchased parts may have existing holes and position tolerances, it is necessary to design parts with hole sizes

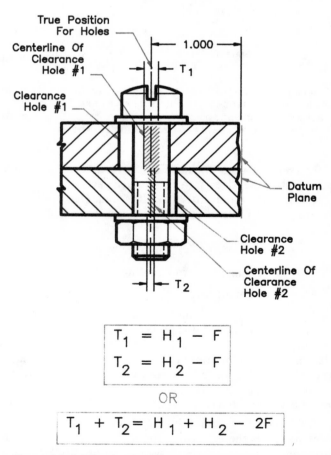

$$T_1 = H_1 - F$$
$$T_2 = H_2 - F$$

OR

$$T_1 + T_2 = H_1 + H_2 - 2F$$

Figure 12-6. The simple formulas shown may be used to calculate tolerances for individual parts provided the same methods are used for all parts in the assembly.

and tolerances that will assemble with the purchased parts.

Position tolerances can be calculated when clearance holes in a floating fastener condition are not equal in size. See Figure 12-6. The shown assembly has a large diameter hole in the top plate. The bottom plate has a smaller hole in it. Tolerance zones for the two holes are calculated on the basis of the hole sizes.

One method of calculating position tolerances permits the tolerance for each hole to be calculated separately. This method may only be used if the tolerances on both holes are calculated in the same manner. The two position tolerance values are calculated using the following formulas:

$$T_1 = H_1 - F$$

$$T_2 = H_2 - F$$

T_1 = Tolerance, Hole #1

H_1 = MMC, Hole #1

T_2 = Tolerance, Hole #2

H_2 = MMC, Hole #2

CAUTION: Do not use the above method unless tolerances on both parts are known to be calculated using this method.

A second method of calculation permits the total allowable tolerance, based on the two hole sizes, to be distributed in any desired manner. Distributing the tolerance also makes it possible to use a small tolerance zone on a large hole while using a larger tolerance zone on a small hole. The tolerance values may be calculated using the following formula:

$$T_1 + T_2 = H_1 + H_2 - 2F$$

This formula is similar to the one shown for distribution of tolerance on two equal size holes. The only difference is that $H_1 + H_2$ has been substituted for 2H.

When using this method, it is advisable to apply one of the position tolerances as a projected tolerance. The projected tolerance should be placed on the position tolerance that is larger than the hole size would normally permit.

FIXED FASTENER CONDITION

Fixed fastener condition has been described to some extent in a previous chapter. A more in-depth look at the fixed fastener condition makes it possible to see how calculations of tolerances can be completed for simple or complex designs that include fixed fasteners.

A fixed fastener condition exists anytime a bolt, pin, shaft, or other feature is fixed in position by a feature on a part. Examples of fixed fastener conditions include a threaded hole into which a fastener is threaded, a press fit for a shaft or dowel pin, and a tapered hole into which a tapered pin is inserted.

More than two parts can be stacked together in a fixed fastener condition. However, all the parts except one must have clearance holes if position tolerances are to be applied to establish interchangeable parts.

The following figures show fixed fastener conditions and calculation methods for determining position tolerances. The calculations shown are accurate. However, the tolerance values are calculated on the basis of using projected tolerance zones. As defined in an earlier chapter, a position tolerance applied to a threaded hole or press fit hole, should

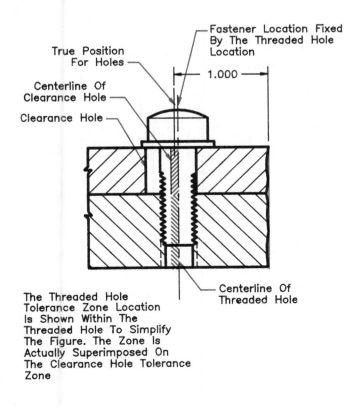

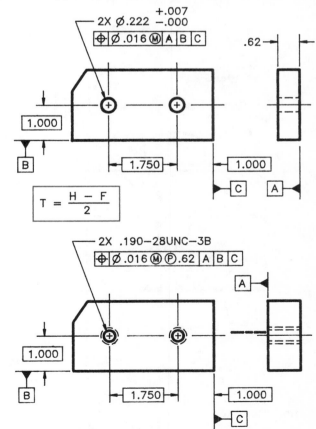

Figure 12-7. Fixed fastener calculations using the shown formula require the utilization of projected tolerance zones. The tolerance zone for the threaded hole is not shown projected to simplify the figure.

be applied as a projected tolerance zone. If projected tolerances are not used, calculations must be completed in a manner that compensates for the projection of the fixed location fastener. This results in a tolerance zone that is smaller than when projected tolerance zones are used. It is typically preferable to use projected tolerance zones to avoid reducing tolerance values.

Tolerances for Figures 12-7 through 12-10 are calculated as if though projected tolerance zones are to be specified. These figures are simplified and do not show projected tolerance zones. The tolerance zones for the threaded holes in these figures are shown within the threaded holes. This simplification is made possible by illustrating the hole position errors without any perpendicularity errors.

It is necessary to show the tolerance zones for both holes in order to provide an understandable explanation. Simplification of the figures is required to permit showing both tolerance zones. Without simplification, the tolerance zones would overlap, making the figure very difficult to understand. The true location of the projected tolerance zones lies outside the threaded hole, therefore they would be superimposed over the tolerance zone for the clearance hole. Showing this would make the figures difficult to understand.

TWO STACKED PARTS

A single line position tolerance specification results in tolerance zones that permit edges of the parts to be aligned. This is true if the same edges of the mating parts are identified as datum features and referenced in the same order of precedence.

The given figure shows a fixed fastener condition for two stacked parts. See Figure 12-7. A basic dimension is given to locate the true positions for the two holes. A tolerance zone for each hole is centered on the true position. Calculation of the diameter for the two tolerance zones is completed using the following formula:

$$T = (H - F)/2$$

Tolerance value T is applied to each of the holes. The previous formula is only used to determine the tolerance value if each part is to be assigned the same amount of tolerance. It may only be used if a projected tolerance zone is specified for the threaded hole.

Tolerance values for a particular clearance hole diameter used in a fixed fastener condition are only half the allowable tolerance value for the same size hole used in a floating fastener condition. This is true since the position error of the fixed fastener utilizes part of the clearance between the fastener and hole.

Figure 12-7 shows the threaded hole positioned at one extreme of the specified tolerance zone. The threaded hole location determines the fastener location. Movement of the

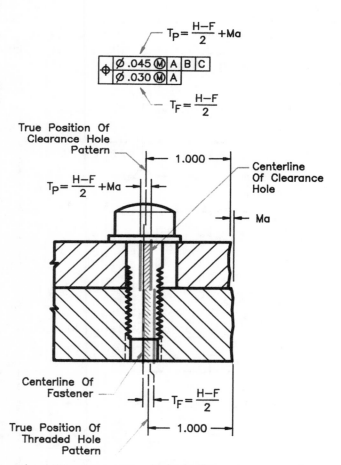

Figure 12-8. Composite tolerances may be used for fixed fastener conditions to allow relative part movement as was explained for floating fastener applications.

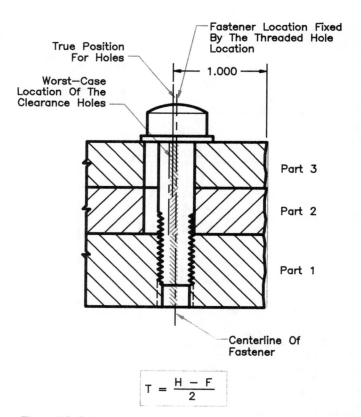

Figure 12-9. Position tolerances for multiple parts in a fixed fastener applications are calculated for pairs of parts, always using the fixed fastener feature as one of the parts in each pair.

fastener off the true position utilizes some of the clearance between the clearance hole and the fastener.

The given figure shows the clearance hole located at the opposite extreme of the tolerance zone relative to the fastener location. Assembly of the fastener through the clearance hole is possible if the fastener and clearance hole are displaced to opposite extremes of their allowable tolerance.

Figure 12-7 shows equal tolerance zones on both the threaded hole and the clearance hole. To simplify the illustration, the tolerance zone for the threaded hole is shown within the threaded part. (In reality, the tolerance zone should be projected for the threaded hole, and therefore would lie outside the threaded part.) Since the tolerance zone for the threaded hole is shown within the threaded part, a perpendicular orientation of the fastener is maintained in the figure to show the effect of the calculated tolerances.

Requirements for a design often permit the edges of parts to be misaligned to some extent. See Figure 12-8. Composite position tolerances can be used to increase tolerance zones when edge misalignment is permitted.

Feature relating tolerance (T_F) values are calculated using the same formula that is used to obtain edge alignment, assuming the datums are surfaces and the primary datum is repeated for the feature relating tolerance. Pattern

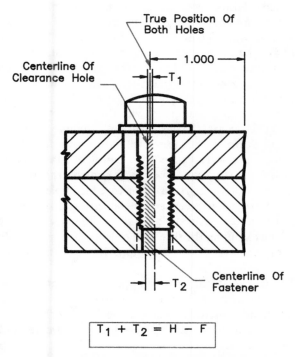

Figure 12-10. The total allowable position tolerance may be distributed between the features in a fixed fastener application.

locating tolerance (T_P) values are calculated using the following formula:

$$T_P = (H - F)/2 + Ma$$

Values calculated must be applied to the holes in each of the two parts.

THREE STACKED PARTS

Tolerances for a fixed fastener application with three parts stacked together can be calculated much in the same manner as for two stacked parts if the two clearance holes have the same MMC size. See Figure 12-9. The edges of the parts in the given figure are aligned. Calculating tolerances using the shown formula results in a tolerance value that must be specified in a single line feature control fame if alignment of the edges is to be maintained. The tolerance on the threaded hole must be specified to project a distance equal to the combined thickness of the parts containing the clearance holes.

The threaded hole in the shown figure is positioned at one extreme of the allowable tolerance zone. Both of the clearance holes are shown at the opposite extreme. This is a worst-case condition since any other position of either hole would result in clearance all around the fastener.

TWO PARTS, UNEVENLY DISTRIBUTED TOLERANCES

Holes in a fixed fastener condition may have the total allowable position tolerance unevenly distributed between the holes. See Figure 12-10. Two parts are shown stacked together in the given figure. The edges of the two parts are aligned. The aligned edges are used as datums to locate the holes.

Position tolerances for the two holes are calculated using the following formula:

$$T_1 + T_2 = H - F$$

This formula can be proven to work on the basis of the formula used when calculating a single tolerance value to apply on two holes. The formula $T = (H - F)/2$ results in a value for T that is used on both holes. Multiplying the formula by two results in:

$$2[T = (H - F)/2]$$

$$2T = H - F$$

Since 2T is equal to $T + T$, by substitution the formula can be rewritten as:

$$T_1 + T_2 = H - F$$

This formula permits any portion of the clearance between a hole and fastener to be applied to the fastener location and to the clearance hole location. Any portion of the total position tolerance can be applied to one hole, provided the remainder of the total tolerance is applied to the mating hole.

If unequal distribution of tolerances is used, it should be done with consideration given to the manufacturing requirements. Generally, threaded holes are permitted more of the total position tolerance than the clearance hole. One reason for this is that threaded holes include more manufacturing steps and are not as easy to locate as clearance holes. The other reason is that clearance holes normally have relatively large size tolerances, and the MMC modifier can be utilized to provide bonus tolerances for the location of the hole.

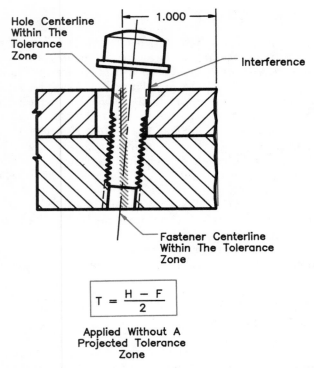

Applied Without A
Projected Tolerance
Zone

$$T = \frac{H - F}{2}$$

Figure 12-11. Failure to use a projected tolerance specification for a fixed fastener application can result in an interference condition.

Unevenly distributed tolerances can also be used for three stacked parts when a fixed fastener condition exists. Refer back to Figure 12-9. Calculations are completed for two parts at a time. Tolerances for Part 1 and Part 3 in the given figure can be calculated using the following formula:

$$T_1 + T_3 = H_3 - F$$

After tolerances for the first pair of parts (Parts 1 and 3) are calculated, a position tolerance for the remaining part (Part 2) must be calculated. The tolerance calculated for Part 1, in the previous step, must be utilized when calculating the tolerance for Part 2. This is necessary since Part 1 has the fixed fastener condition. The formula used is:

$$T_1 + T_2 = H_2 - F$$

Application of the position tolerance on Part 1 should show a requirement for a projected tolerance zone that extends a distance at least equal to the thickness of the two attached parts.

PROJECTED TOLERANCE ZONE

Position tolerances applied to features that result in a fixed fastener condition usually require a projected tolerance zone. Failure to specify a projected tolerance zone results in a tolerance zone that resides within the boundary of the feature to which the tolerance is applied. Failure to use a projected tolerance zone can result in an interference condition when the T = (H - F)/2 formula is used.

See Figure 12-11. The given figure shows the results of omitting the requirement for a projected tolerance zone. The position tolerance applied to the threaded hole and the mating clearance hole were calculated using the T = (H -

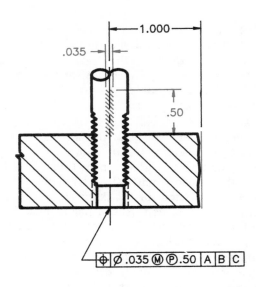

Figure 12-12. A projected tolerance specification controls the axis of the controlled feature only in the projected zone.

F)/2 formula, and applied without the requirement for a projected zone. The tolerance zone is therefore contained within the threaded hole.

The axis of a hole is permitted to lie anywhere within the specified tolerance zone. As shown in the figure, the axis of the hole may not be perpendicular. Since a bolt threaded into a hole will align on the axis of the threads in the hole, it will have the same orientation error that exists in the hole.

A clearance hole located at an extreme within the permitted tolerance zone is shown in the figure. When the hole is in an extreme location, it is possible for a fastener installed in the threaded hole to interfere with the edge of the clearance hole. If this occurs, the parts will not assemble.

Application of a projected tolerance zone on a threaded hole does, in effect, control the location of a fastener that is threaded into the hole. However, it controls the location of the fastener on the outside of the threaded hole. See Figure 12-12. A projected tolerance zone specification establishes a tolerance zone that resides outside the controlled feature. In the given example, the threaded hole includes a .035″ diameter tolerance zone that is to apply for a distance of .50″ outside the part.

The threaded hole must be located so that its axis lies somewhere within a tolerance zone that is projected outside the part and centered on the true position of the hole. The tolerance zone in the given figure is .035″ diameter and extends from the surface to a distance .50″ above the part. Controlling the location of the hole in this manner ensures that the fastener threaded into the hole has a known location outside the controlled part.

Tolerance specifications on fixed fastener features should include projected tolerance zone requirements. See Figure 12-13. Projected tolerance zone requirements result in superimposed position tolerance zones for the clearance hole and the fixed fastener feature. If the axis of the threaded hole falls within the projected tolerance zone, then the axis of the fastener will also fall within this zone. This ensures the fastener will pass through the clearance hole.

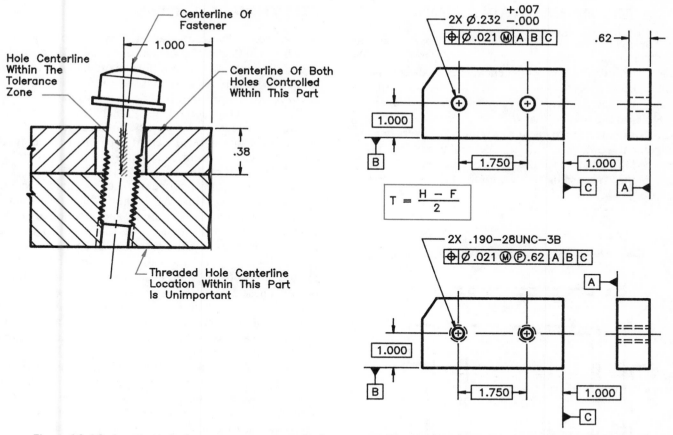

Figure 12-13. A projected tolerance zone ensures the fastener axis is controlled where it extends through the part containing the clearance hole.

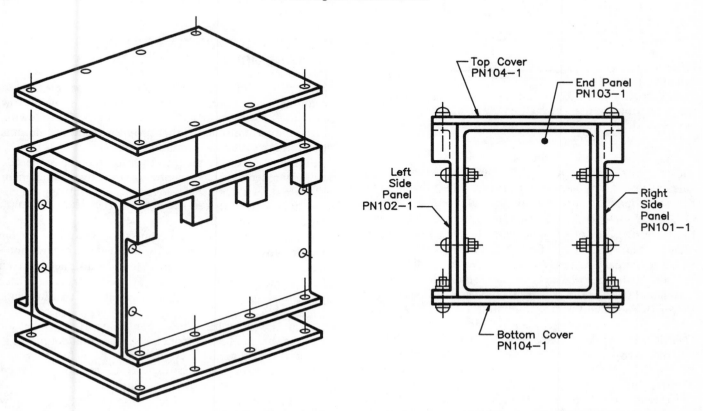

Figure 12-14. How tolerances in an assembly accumulate must be considered when calculating tolerances and applying them on the drawings.

MULTIPLE PARTS IN ASSEMBLY

A complete assembly is often made up of parts that stack together in more than one plane. See Figure 12-14. The shown assembly is a six-sided enclosure. Four side panels bolt together to form a complete set of side walls. A top and bottom cover are bolted to the side panels.

Holes in the covers must align with those in the side panels before bolts can be installed to hold the covers in place. Alignment is only possible if the accumulation of tolerances between the side walls and covers are controlled. The holes must be sized according to the amount of tolerance accumulation.

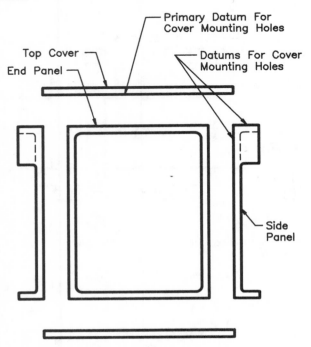

Figure 12-15. Datums are normally selected on the basis of how the parts in an assembly fit together.

Details of how to calculate and apply tolerances to the shown figure are described in this chapter. Completion of calculations for any assembly requires the assembly be thoroughly understood before starting calculations.

The shown assembly has two side panels with threaded holes for attachment of the covers. The left side panel is identified as part number PN102-1. The right side panel is identified as part number PN101-1. Two end panels identified as PN103-1 are placed between the two side panels.

Each of the side panels must include a position tolerance on the threaded cover attachment holes. The top and bottom covers must also include a position tolerance on the clearance holes. Tolerances on the side panel hole locations and on the cover clearance holes both affect assembly of the covers to the side walls. Also, there is a tolerance on the length of the end panels, and this tolerance affects the distance between side panels. The tolerance on the distance between side panels affects the relative distance between cover mounting holes on the assembled side panels.

SELECTION OF DATUMS

Total tolerance accumulation in an assembly such as the one shown in Figure 12-14 must be minimized to keep clearance hole sizes at a reasonable diameter and to permit assembly of the parts. One contributor to minimizing tolerance accumulation is the proper selection of datum features.

Datum selection should always include consideration of the functional features in the design. See Figure 12-15. The figure shows an exploded view of parts in an assembly. For an assembly such as this, identification of functional features is relatively simple.

The top cover only has one surface that comes in contact with the side walls. This surface should therefore be identified as a primary datum feature for specification of position tolerances applied to the holes on this part. Edges of the cover should be used as the secondary and tertiary datums.

The right side panel has one surface that is in contact with the top cover. It has another surface that contacts the end panel. These two surfaces should both be identified as datum features for locating the cover attachment holes. One of them will be selected as a primary datum, and the other will be used as a secondary datum. An end of the side panel will be used as a tertiary datum.

Datum selection for cover attachment hole locations on the end panels is not necessary since the end panels do not include cover attachment holes. It would be necessary to identify datums on the end panels for other purposes such as specification of orientation tolerances and side panel attachment hole position tolerances. The methods explained for cover attachment can be applied for control of side panel attachment.

CALCULATION OF TOLERANCES FOR AN ASSEMBLY

Selection of the proper datums makes it possible to keep accumulation of tolerances to a minimum. By showing how to calculate tolerances for the given assembly, the following paragraphs will also show that proper selection of datums does minimize tolerance accumulation.

Figure 12-16 shows the top cover located above the side walls subassembly. Centerlines for the cover attachment holes are shown. On the top cover, there is only one dimension that locates the distance between the clearance holes; that dimension is 3.876″.

The side walls subassembly shows the centerlines for the threaded cover attachment holes. These holes are nominally located 3.876″ apart, which is the same distance between the clearance holes in the cover. However, there are three dimensions that total the 3.876″ distance on the side walls subassembly.

Each side panel has a basic dimension of .438″ that locates the threaded holes from the surface that contacts the end panel. The dimension extends from the surface that contacts the end panel since locating surfaces are normally used as datum features when dimensioning parts.

A detail view of one side panel is shown in the figure to show the selected datum features on the side panels. Datum feature symbols are shown. These datums are referenced in the

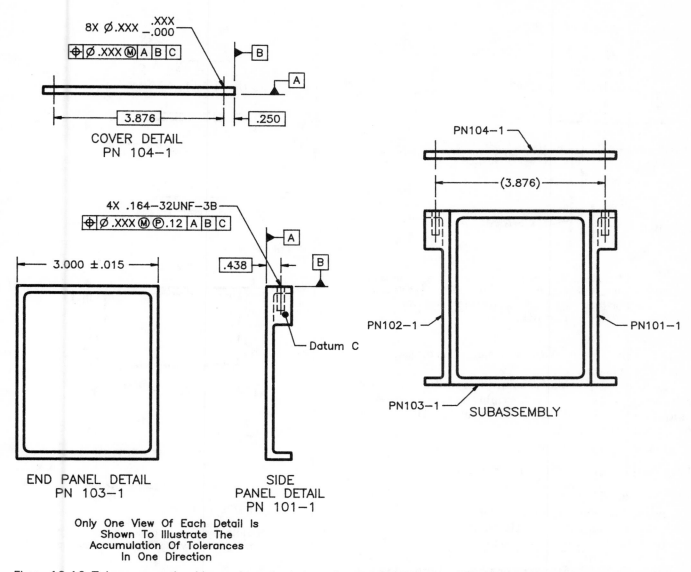

Figure 12-16. Tolerances on the side panels, end panels, and top cover affect the size clearance holes that must be produced in the cover.

position tolerance specifications for the attachment holes.

The figure shows that the end panel has a size dimension of 3.000″ ±.015″ applied to its length. Each of the side panels includes a .438″ hole location dimension. The total of the dimensions applied to the side panels and the end panel equal the 3.876″ distance between attachment holes on the cover.

An accumulation of tolerances does take place because of the dimensions applied to the three parts assembled together to establish the 3.876″ dimension. All tolerances must be carefully considered. There is a position tolerance on each of the side panels, and that tolerance permits location error relative to the side panel datums. There is also an end panel size tolerance that affects the distance between the side panels, which affects the distance between holes. To simplify this problem, the form and orientation tolerances on the end panel are initially ignored and therefore eliminated from the explanation of the tolerance accumulation. In fact, the end panels will initially be eliminated from the assembly to simplify the explanation for this assembly.

Position tolerance effects on the two side panels are considered first. See Figure 12-17. Elimination of the end panels from the assembly permits the datum features on the side panels to be brought together. For practical purposes, this creates a common datum reference frame for the two parts. The basic dimensions locating the holes on the two parts extend from a common datum reference frame. Since the position tolerance on the two groups of holes are, in effect, referenced to a common datum reference frame, the holes are treated as a single pattern. This means there is no accumulation of tolerances on the two side panels.

Selection of datum features to completely avoid a tolerance accumulation between two parts is not always possible, but it is always possible to minimize tolerance accumulation. Proper selection of datums and application of dimensions will reduce tolerance accumulation.

When the end panel is inserted between the two side panels, an accumulation of tolerance will take place because of the ±.015″ size tolerance on the end panel. See Figure 12-18. Calculation of position tolerances and clearance holes sizes

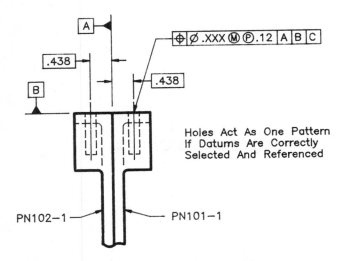

Figure 12-17. Selection of the appropriate datum features on the side panels will, for practical purposes, eliminate tolerance stackup (between the side panels) for the cover attachment holes.

for the assembly must take into consideration the size tolerance on the end panel.

The formulas previously defined for position tolerance calculation can be used for an assembly that has an accumulation of tolerances if the formula is expanded. The formula for a fixed fastener application is:

Total Tolerance = H - F

For a floating fastener condition the formula is:

Total Tolerance = 2H - 2F

To expand the formula, it is necessary to know how many tolerance values equal the total tolerance.

Figure 12-18 includes three tolerances that impact the assembly of the cover on the side walls. One tolerance is the position tolerance on the threaded holes. (Both side walls have the same position tolerance.) Another tolerance is the position tolerance on the clearance holes on the cover. The third tolerance is the size tolerance on the end panel. Since there are three contributors to the total tolerance and the application is a fixed fastener condition, the formula for this assembly is:

$$T_1 + T_2 + T_3 = H - F$$

Either the tolerances or the hole size may be assumed, leaving the other value to be calculated. Calculations for the given assembly are completed by assuming position tolerances to apply on the holes. Only the hole size must be determined. The given figure shows how the calculations are made.

Two of the tolerances applied in the calculation are the position tolerances of .012″ and .018″. These are values taken directly from the feature control frames. The third tolerance used in the calculation is the .015″ bilateral size tolerance. The value of .015″ is used since the size tolerance can only cause a worst-case difference in the nominal location of the holes of .015″. Although the ± .015″ tolerance is a total size tolerance of .030″, it can't cause a .030″ movement of the holes relative to nominal.

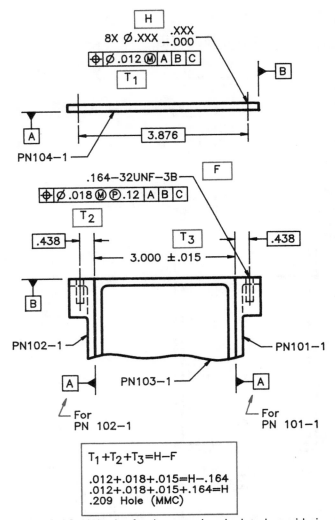

Figure 12-18. Hole size for the cover is calculated considering the tolerances on the cover, side panels, and end panels.

ZERO POSITION TOLERANCE AT MMC

Position tolerances when properly calculated and applied can provide increased tolerance zones and thereby improve the freedom in the manufacturing of parts. Previously shown position tolerance calculations have been based on the use of clearance holes that have an MMC larger than the fastener MMC. This is a common practice, and is far better than calculating plus or minus tolerances for hole locations. The advantages of position tolerances were explained in Chapters 8 and 9.

A further increase in manufacturing freedom can be obtained through the application of a concept known as *zero tolerancing at MMC*. Proper application of this concept does increase tolerances. Care must be taken not to misapply the concept or it will reduce the tolerances to zero.

The zero tolerance at MMC concept should only be applied when clearance conditions exist. It can only be used for tolerances that include the MMC modifier, and there must be a size tolerance on the located feature.

The position tolerances as already defined in previous segments of this text are significantly better than plus or

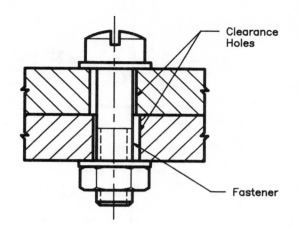

$\phi\ .282\ {}^{+.006}_{-.000}$

	PRODUCED HOLE DIA	ALLOWABLE POSITION TOLERANCE
MMC	.282	.032
	.283	.033
	.284	.034
	.285	.035
	.286	.036
	.287	.037
LMC	.288	.038

Figure 12-19. A functionally acceptable hole can be rejected if it is produced under the specified minimum size limit for the hole.

minus location tolerances. This is true because the properly calculated position tolerances provide more permissible variation. Understanding of the application of zero position tolerances at MMC can result in even greater permissible variation without any impact on the functional quality of the produced parts.

Using only the information defined to this point can result in tolerance specifications that require rejection or rework of some functionally acceptable parts. The parts would function as needed for the design application, but not meet drawing requirements. Rather than use tolerance specifications that may result in the rejection of parts that are functionally acceptable, it would be better to use tolerance specifications that accept all good parts. The following examples show a comparison of methods for applying position tolerances. The zero position tolerance at MMC method results in a specification that accepts all usable parts.

See Figure 12-19. A floating fastener application is shown. A position tolerance of .032″ diameter at MMC is calculated for the .282″ diameter MMC holes. A size tolerance of + .006″ is applied to the holes. The fastener diameter is .250″.

A table in the figure shows the acceptable position tolerance for the range of acceptable hole diameters. Any hole produced to the requirements of the table is acceptable.

Any hole produced outside the range shown in the table is not acceptable. As an example, a .280″ diameter hole produced at a position tolerance of .015″ diameter is not acceptable according to the hole specification. The hole does not meet the minimum size requirement of .282″. According to the drawing requirement, the part must be rejected. It is possible to increase the diameter of the hole to make it meet the size requirement, but that means an additional machining operation and another inspection cycle. These extra efforts cost time and money.

If the drawing requirements for the hole are ignored, it can be determined that the .280″ diameter hole with a position error of only .015″ diameter meets the functional requirements of the hole. In fact, a .250″ diameter fastener will fit through a .280″ diameter hole if it has a position error of as much as .030″. Why does the drawing specification require rejection of the part?

The zero tolerance at MMC concept permits specification of hole sizes and position tolerances that allow the full range of functionally acceptable parts to be produced. See Figure 12-20. This figure is the same as the previous one, except the hole specification and table have been revised. A .000″ diameter tolerance zone at MMC is specified, and the

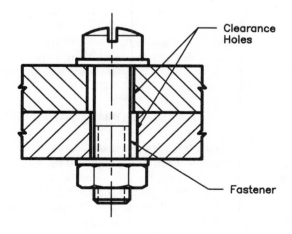

$\phi\ .250\ {}^{+.038}_{-.000}$

	PRODUCED HOLE DIA	ALLOWABLE POSITION TOLERANCE
MMC	.250	.000
	.251	.001
	.252	.002
	.282	.032
	.283	.033
	.287	.037
LMC	.288	.038

Figure 12-20. Properly applying zero position tolerances at MMC can increase manufacturing freedom and permit acceptance of all functionally acceptable holes.

hole has an MMC equal to the fastener diameter. The hole size has a tolerance of + .038″.

The hole size specification and zero tolerance at MMC permits the same maximum diameter hole (.288″) and the same position tolerance (.038″) on the maximum size hole as was permitted in the previous figure. In fact, the range of hole sizes and the associated position tolerance shown in the previous figure is permitted by the hole specification in this figure. Additionally, all hole diameters between .250 and .282″ diameter along with their associated position tolerance have been added by specifying the hole in the manner shown here.

A significant amount of manufacturing freedom is provided through the proper use of zero tolerancing at MMC. Any functionally acceptable hole can be accepted without rework of the parts. No functionally acceptable parts must be rejected since all functionally acceptable parts meet the hole specification.

There are a couple of precautions that should be taken when specifying zero position tolerances at MMC. The first is to make sure that all people working to the drawing understand the MMC concept. If the MMC concept is not understood, people may think the zero tolerance specification is an absolute value. Of course a zero tolerance is not possible to maintain, and may cause manufacturing to question the sanity of the designer that specifies zero tolerances.

Another precaution is not to specify small size tolerances when using the zero position tolerance at MMC. A zero tolerance at MMC is intended to increase manufacturing freedom by placing the position tolerance and size tolerance into one value. This lets manufacturing determine how to distribute the tolerance within their manufacturing methods.

POSITION TOLERANCE AT LMC

Distance between the edge of a part and a hole is sometimes a major consideration in the allowable location of a hole. Position tolerances can be specified including the Least Material Condition (LMC) modifier when the remaining material outside the hole is a primary concern. See Figure 12-21.

The LMC modifier indicates that the specified tolerance value applies when the controlled feature is at its least material condition. In the case of a hole, the least material condition is when the hole is at its maximum diameter. The specified tolerance increases as the controlled feature departs from its LMC size.

The given figure shows a hole located by a basic dimension of .500″ from the edge of the part. When the hole has a least material condition of .510″ diameter, the allowable position tolerance is .040″ diameter. A hole produced at a diameter of .510″ and at the maximum displacement allowed by the .040″ diameter position tolerance has an edge distance of .225″.

A hole produced at the minimum allowable diameter of .500″ is permitted a position tolerance of .050″ because of the LMC modifier. A hole produced at the .500″ diameter and at the maximum displacement allowed by the .050″ diameter position tolerance has an edge distance of .225″.

The minimum edge distance calculated for the least material condition is maintained when using the LMC modi-

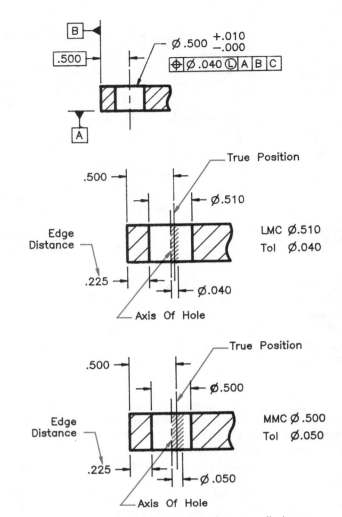

Figure 12-21. Edge distance can be controlled to one minimum value by using the LMC modifier on hole position tolerances.

fier. The two limits of size for the specified hole in the figure show that the edge distance is maintained.

Allowable position tolerance at the MMC size must be considered when sizing holes for installation of fasteners. If the LMC modifier is applied to a tolerance specification, the allowed tolerance at MMC must also be calculated. The allowable tolerance at MMC must be a value that will work with the size hole and fastener being used. The hole size and position tolerance value must be adjusted to result in the desired minimum edge distance and also permit fastener installation.

PAPER GAGING TECHNIQUES

Paper gaging is a process of recording inspection data on paper to determine if the measured part meets drawing specifications. This procedure can be used for verification of hole positions. It is not necessary to use this technique when using functional gages to verify feature locations or when using computer-driven measurement machines.

Verification of hole size and location are both required to determine if drawing requirements are met. See Figure 12-22. The hole size can initially be checked with go no-go

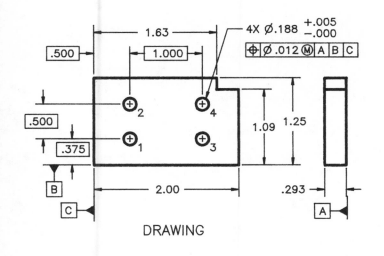

DRAWING

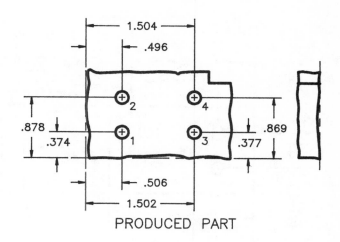

PRODUCED PART

MEASURED HOLE DATA

Hole #	1		2	
Diameter	.191		.192	
	X	Y	X	Y
Measured Location	.506	.374	.496	.878
Drawing Dimension	.500	.375	.500	.875
Error	.006	−.001	−.004	.003

Hole #	3		4	
Diameter	.191		.191	
	X	Y	X	Y
Measured Location	1.502	.377	1.504	.869
Drawing Dimension	1.500	.375	1.500	.875
Error	.002	.002	.004	−.006

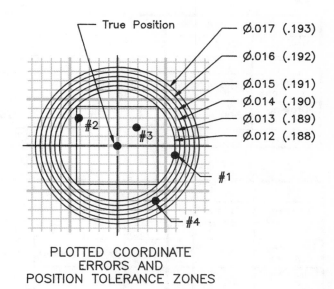

PLOTTED COORDINATE
ERRORS AND
POSITION TOLERANCE ZONES

Figure 12-22. Paper gaging of position tolerances is one method of verifying whether or not measured location errors are acceptable.

gages. These gages will determine if hole sizes are within the allowable size range, but they do not determine the actual hole size. If all holes are verified to be within the allowable size range, then hole locations are measured.

Hole locations must be measured relative to the datum reference frame. Failure to measure hole locations relative to the datum reference frame will result in inaccurate paper gaging conclusions. Location measurements for each hole is recorded in a table. See Figure 12-22. From these measured values, the coordinate errors can be calculated. Care must be taken to accurately determine whether the values are positive or negative.

Coordinate error values may be used to calculate the diameter position errors or they may be plotted on a grid to determine the diameter position errors. The given figure shows how the coordinate errors can be plotted to determine position errors.

Coordinate errors are plotted on a grid at a greatly enlarged scale. A minimum scale of 100:1 is recommended. A scale of 100:1 allows using a .100″ grid to represent .001″ of error.

Errors calculated in the table are plotted on the grid. A point is located for each hole and the associated hole identification number is placed by the point. After all points are plotted, a set of concentric circles is centered on the origin for the grid. A scale equal to that of the grid must be used to draw the circles. The smallest circle is drawn to represent the smallest permitted position tolerance. Larger circles are drawn at an increment to represent .001″.

Each of the concentric circles represent a tolerance zone diameter, and each tolerance zone diameter is associated with a hole size. Assuming the position tolerance is specified with an MMC modifier, the smallest diameter circle is associated with the MMC size of the hole. The next larger tolerance diameter circle is associated with a feature that has departed from the MMC value by .001″. In the given figure, the smallest circle represents a tolerance zone .012″ in diameter and is associated with a hole size of .188″ diameter. The

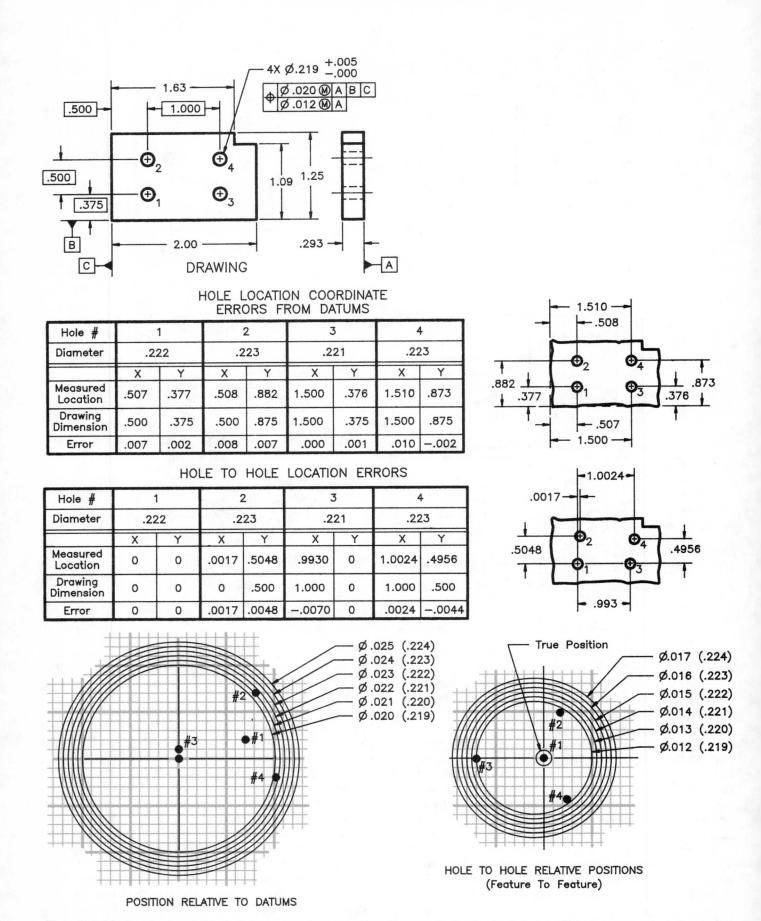

HOLE LOCATION COORDINATE ERRORS FROM DATUMS

Hole #	1		2		3		4	
Diameter	.222		.223		.221		.223	
	X	Y	X	Y	X	Y	X	Y
Measured Location	.507	.377	.508	.882	1.500	.376	1.510	.873
Drawing Dimension	.500	.375	.500	.875	1.500	.375	1.500	.875
Error	.007	.002	.008	.007	.000	.001	.010	−.002

HOLE TO HOLE LOCATION ERRORS

Hole #	1		2		3		4	
Diameter	.222		.223		.221		.223	
	X	Y	X	Y	X	Y	X	Y
Measured Location	0	0	.0017	.5048	.9930	0	1.0024	.4956
Drawing Dimension	0	0	0	.500	1.000	0	1.000	.500
Error	0	0	.0017	.0048	−.0070	0	.0024	−.0044

Ø.025 (.224)
Ø.024 (.223)
Ø.023 (.222)
Ø.022 (.221)
Ø.021 (.220)
Ø.020 (.219)

POSITION RELATIVE TO DATUMS

True Position

Ø.017 (.224)
Ø.016 (.223)
Ø.015 (.222)
Ø.014 (.221)
Ø.013 (.220)
Ø.012 (.219)

HOLE TO HOLE RELATIVE POSITIONS
(Feature To Feature)

Figure 12-23. Composite position tolerances can be paper gaged by separately plotting hole locations as measured relative to the true positions created by the pattern locating and the feature relating tolerances.

next larger circle represents a .013″ diameter tolerance zone. It is associated with a hole size of .189″ diameter.

Any hole produced at a diameter of .188″ diameter must have a position error within a .012″ diameter. Any hole produced at a .189″ diameter must have a position error within a .013″ diameter.

The given figure shows four points plotted on the grid in positions representing hole locations. Concentric circles located on the grid origin show the position error of each hole. Holes #2 and #3 are both within the smallest allowable position tolerance, therefore their actual hole size is unimportant, provided the hole diameters are somewhere within the allowable size tolerance.

Hole #1 is located outside the .012″ diameter circle, but inside the .013″ diameter circle. This means that this hole must have a diameter of at least .189″. The hole size must be checked if it is not already recorded in the inspection data table. In the figure, hole number one is shown to be .191″ diameter. It is therefore acceptable in the shown location.

Paper gaging techniques can also be applied to composite position tolerances. The pattern locating tolerances are verified in the same manner as described for the single line position tolerance feature control frame. The feature relating tolerance can be verified in a slightly different manner.

Figure 12-23 shows a paper gaging technique for composite position tolerances. The pattern locating tolerance for the given tolerance specification is verified in the manner described for the previous figure.

The feature relating tolerance shown in the second line of the composite tolerance specification is verified by measuring the relative locations of the holes. Measurements are made in a plane parallel to primary datum A, since datum A is referenced in the second line of the tolerance specification. Measurements between the holes are made with one of the holes acting as an origin. In the given figure, hole #1 is selected to serve as the origin.

Direction of measurements is established by selecting a second hole. Hole #3 in the given figure is selected for this purpose. After two holes are selected to locate the origin and orient the coordinate system axes, all hole locations must be measured from the coordinate system that is located by the two selected holes.

Measurements for the relative locations of the holes are recorded in a table and coordinate errors determined. Whether the coordinate errors are positive or negative is important and must be accurately recorded.

Coordinate errors are plotted on a grid. Care must be taken to plot the points in the proper locations. Positive and negative values must be properly plotted to obtain accurate results.

After all points are plotted, a set of concentric circles representing the allowable feature relating tolerances and the associated feature sizes is placed on the grid. The concentric circles may be placed in any position that encloses the plotted points. There is no requirement to center the circles on the origin of the grid. Forcing the circles to be centered on the grid would be incorrect since this would be the same as forcing location of the tolerance zones relative to the datum reference frame. There is no requirement for the feature relating tolerance zones to be located relative to the datums.

It is possible to make a quick acceptance check of the feature relating tolerance before measuring the feature-to-feature positions. This procedure may only be used to accept parts. Many good parts will pass this test. If a part passes this check method, it is not necessary to measure feature-to-feature positions. The time that would have been spent measuring feature-to-feature positions will be saved.

CAUTION: This quick check method must not be used to reject parts because some good parts will not pass this test.

One set of hole location measurements are taken, and the hole locations are plotted on a grid as illustrated in Figure 12-23. Two sets of concentric circles are then used on this one set of plotted points. The first set of circles is sized to the pattern locating tolerances. These circles must be centered on the origin of the grid. The second set of circles is sized to the feature relating tolerances. These circles may be located in any position encompassing all the plotted points. The feature relating tolerance circles will only encompass all the points if the entire hole pattern has shifted in one direction.

If both sets of circles enclose the points, then the hole locations are acceptable. In this case, the quick check has saved the time that would have otherwise been used to measure hole-to-hole locations.

If the pattern locating tolerance circles fail to enclose all the points, the part is not acceptable. Depending on the errors, it may be possible to rework the part.

If the feature relating tolerance circles fail to encircle the points, it does not mean that the part is bad. It only means that the feature-to-feature positions must be measured and plotted as illustrated in Figure 12-23.

SELECTION OF THE APPLICABLE TOLERANCE

Complex designs are usually a compilation of many relatively simple features. Application of tolerances can be made easier if the design is mentally broken down into the simple features and controls placed on them in the simplest possible means.

Once a feature is selected for application of a tolerance, it is necessary to determine which type or types of tolerances to apply. See Figure 12-24. Asking questions about the required level of control for a feature is a good method to use for determining what tolerances to apply. A flow diagram is shown that provides example questions that must be asked to determine what tolerance types to apply.

Typically, the first thing that must be determined is whether or not the feature needs to be controlled relative to other features. If no relationship to other features is required, then the tolerance type is going to be some type of form or profile tolerance. Additional questions will determine exactly what tolerance type is needed.

The given figure shows a flow diagram that permits selection of a tolerance control for many applications that require controls not referenced to datums. It shows the type of process that should be performed to select the proper tolerance type. The logical process shown can be expanded to include selection of tolerances for related features, the selection and referencing of datum features, and also the application of material condition modifiers.

The logic used to develop the shown flow diagram is explained in the following paragraphs.

Does the feature need to be controlled relative to any other feature? If the answer is no, then the tolerance must either be a form tolerance or a profile tolerance. Form toler-

ances do not reference other features (datums) and profile tolerances can be used without datum references to control the form of a feature. The NO path from this question must lead only to form and profile tolerances.

Is the required control applied to multiple coplanar features? If the answer is YES, all form tolerances are eliminated since form tolerances are applied to individual features. A NO answer leads to possible form or profile

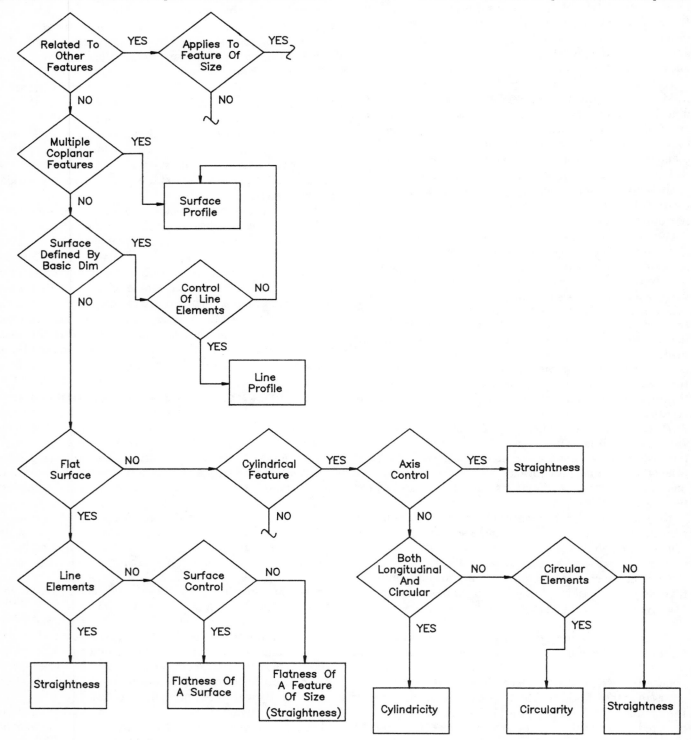

Figure 12-25. A complete flow diagram for determining tolerance application can be created using the format shown in this figure.

tolerances. Profile tolerances may be used on individual features or multiple coplanar features.

If the control is applied to an individual feature, a question is asked to determine if the feature profile is defined by basic dimensions. If the answer is YES, the required control is profile. If NO, the control is most likely a form tolerance.

Additional questions are asked to determine the type of form tolerance to apply to the feature.

MULTIPLE LINE TOLERANCE SPECIFICATIONS

Many tolerance types automatically include controls obtained by other tolerance types. As an example, perpendicularity applied to a flat surface also controls the flatness of the surface. If a perpendicularity tolerance of .015″ is applied to a surface, that surface must also be flat within .015″. It couldn't be perpendicular within .015″ if it had a flatness error greater than that value.

When applying a tolerance, all controls imposed by the specification should be considered. Sometimes, one specification can achieve all the needed control. Placement of unnecessary specifications on a drawing should be avoided. There is no need to apply a perpendicularity tolerance of .015″ on a surface and also apply a flatness tolerance specification of .015″. The flatness requirement is already included in the perpendicularity tolerance.

It is also important not to specify an overly restrictive, high-level control to avoid the specification of another tolerance. If a surface needs a flatness tolerance of .005″, a perpendicularity tolerance of .005″ should not be applied to obtain the needed flatness. It is better to specify only the needed level of perpendicularity control and specify a separate flatness requirement.

Orientation and Form

Multiple tolerance specifications may be applied to one feature. See Figure 12-25. A perpendicularity tolerance of .015″ and a flatness tolerance of .005″ are shown in the given figure. The two tolerances applied to one surface controls orientation to the .015″ value, and controls form within the .005″ value. The surface to which these tolerances are applied must meet both requirements.

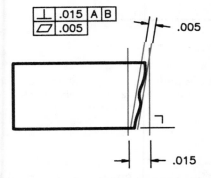

Figure 12-25. One tolerance specification can be used to refine the requirements of a higher level tolerance specification.

A similar specification may be applied to a hole. See Figure 12-26. A perpendicularity tolerance of .017″ is applied to a hole. The hole also has an axis straightness tolerance of .007″. The axis may be out of perpendicularity within a 0.17″ diameter zone, but it must be straight within a .007″ diameter zone.

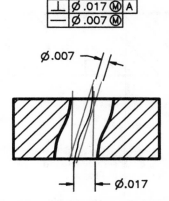

Figure 12-26. Tolerances applied to features of size may also include multiple levels of control.

Position, Orientation, and Form

Application of a position tolerance should only require the level of control needed for position. If further refinement of feature orientation or form is needed, then that control should be achieved with the correct tolerance specification, not be overly restrictive position tolerances. See Figure 12-27. The given figure shows an allowable position tolerance of .042″. An orientation tolerance of .012″ is applied to the same feature to prevent the orientation error from being the full .042″ permitted by the position tolerance. A straightness tolerance of .006″ is also applied to the feature to further control the form of the feature.

Multiple levels of control should be used when increased producibility of the parts is achieved by them. However, multiple levels of control should be applied with caution. The amount of effort necessary to verify two or three tolerance requirements must be balanced against the effort required to meet one tolerance specification that achieves all the needed controls within one specified tolerance value.

Position And Profile

Composite profile tolerances provide a means to specify a location requirement that is relatively large, with a smaller tolerance used to refine the size and form requirement. It is also possible to use profile to control size and form in conjunction with a position tolerance to control location. See Figure 12-28. When this is done, a note of BOUNDARY is placed beneath the position tolerance. The profile tolerance is interpreted exactly the same as if no position tolerance were shown. A maximum material condition boundary for the feature is created at one limit of the profile tolerance. The position tolerance creates a boundary that constrains the location of the MMC boundary.

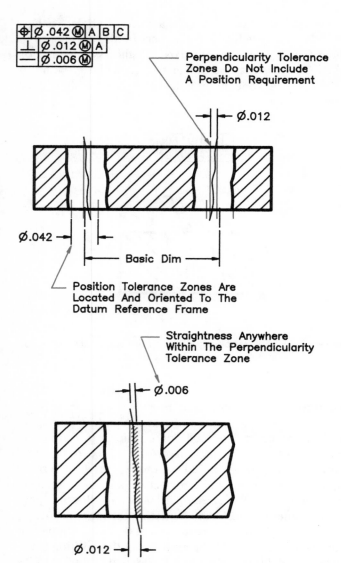

Figure 12-27. Only the level of control required to make the part functionally acceptable should be applied to any feature.

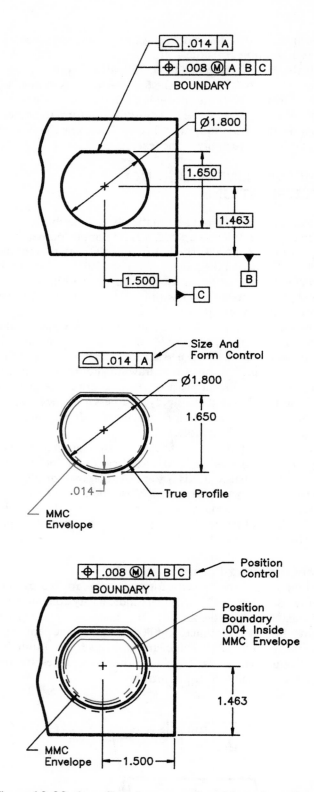

Figure 12-28. A profile tolerance and position tolerance may be used to control the size, form, and location of irregular features.

The Figure 12-28 shows a D-shaped hole on which a surface profile of .014″ has been applied. A position tolerance of .008″ has also been applied to the hole. Basic dimensions define the hole diameter as 1.800″ and the distance across the flat as 1.650″. Basic dimensions are given to define the hole's location.

Size and form of the hole are controlled by a .014″ wide profile tolerance zone that is centered on the true profile defined by the 1.800″ and 1.650″ dimensions. Location of the hole is controlled by a tolerance zone boundary that is .008″ smaller than the MMC of the hole. The location tolerance zone is located at the basic locations defined by the drawing.

The actual surface of the hole may move to any location that does not violate the location tolerance boundary. The surface must also fall within the profile tolerance boundary, which is free to move as long as its orientation to datum A is maintained.

Profile and position tolerances may also be applied to external features. When applied to an external feature, the position tolerance boundary is outside the MMC boundary created by the profile tolerance.

The total allowable tolerance for both the floating fastener and fixed fastener conditions may be distributed when hole sizes are equal or when they are unequal.

Tolerances for a floating fastener condition with three parts containing equal-size holes stacked together may be calculated in the same manner as when two parts are stacked together.

Tolerances for a fixed fastener condition with three stacked parts can be calculated considering the parts two at a time provided the fixed fastener feature be included in the calculations for each pair of parts.

Tolerance accumulation can be minimized through proper selection of datum features.

Zero position tolerances applied at MMC can increase manufacturing freedom and reduce rejection of functionally good parts.

Position tolerances including the LMC modifier are sometimes used to control edge distances.

Paper gaging techniques permit manually obtained inspection data to be evaluated for compliance with position tolerances.

REVIEW QUESTIONS

Answer the following questions on a separate sheet of paper. Do not write in this book. Accurately complete any required sketches.

MULTIPLE CHOICE

1. A _____ tolerance zone is typically used when applying a position tolerance to a threaded hole.
    A. small
    B. large
    C. profile
    D. projected
2. _____ datum features are identified and referenced in position tolerances when alignment of edges in an assembly is required.
    A. Random
    B. Similar
    C. Unrelated
    D. Opposite
3. Selection of the proper _____ can reduce tolerance accumulation.
    A. datums
    B. hole sizes
    C. fastener sizes
    D. None of the above.
4. The position tolerances applied to the holes in a floating fastener application _____ applied to each of the parts.
    A. may be one value
    B. may be two individual tolerances
    C. must be one value
    D. Both A and B.
5. Stacking three parts that all contain clearance holes of one size requires completion of calculations _____.
    A. using the same formula as is used for two parts
    B. using the same formula as for fixed fastener applications
    C. considering only two parts at a time
    D. None of the above.
6. The _____ datum reference in a position tolerance specification on holes is typically the datum feature that makes contact with the attached part.

    A. primary
    B. secondary
    C. tertiary
    D. Either A or C.
7. If edges of parts in an assembly must align, position tolerances are specified using a _____ position tolerance.
    A. composite
    B. single line
    C. Either A or B.
    D. Neither A nor B.
8. The _____ tolerance in a composite position tolerance controls relative location of holes in a group and ensures the fasteners can pass through the holes.
    A. pattern locating
    B. feature relating
    C. group holding
    D. MMC-specified
9. The _____ tolerance in a composite tolerance affects the amount of misalignment between features in an assembly when the parts are bolted together.
    A. pattern locating
    B. feature relating
    C. group holding
    D. MMC-specified
10. Control of edge distance sometimes warrants the use of the _____ on position tolerances.
    A. small values
    B. RFS modifier
    C. MMC modifier
    D. LMC modifier

TRUE/FALSE

11. The formula T = H - F may only be used to complete calculations when a maximum of two parts are stacked together. (A)True or (B)False?
12. If a .010″ diameter position tolerance is used on two stacked parts, the clearance hole diameter for a fixed fastener condition must be larger than the clearance hole diameter for a floating fastener condition. (A)True or (B)False?
13. It would be extremely restrictive, if not impossible to meet, a tolerance of .000″ RFS. (A)True or (B)False?

FILL IN THE BLANK

14. A _____ position tolerance at MMC in combination with a large size tolerance can increase manufacturing freedom when properly utilized.

15. A projected tolerance zone is normally specified to prevent _____ between the fastener and the clearance hole in a fixed fastener application.

16. _____ _____ are used to represent position tolerance zones when paper gaging.

17. Paper gaging should be completed at a scale of at least _____ : 1 .

SHORT ANSWER

18. List the two general fastener conditions for which position tolerances must be calculated.

19. Describe a floating fastener condition.

20. Show the formula for a fixed fastener condition in which two parts are stacked together and a projected tolerance zone is to be used on the fixed feature.

21. What is the formula for calculating position tolerances for a floating fastener condition in which there are two hole sizes and the tolerances are to be distributed unevenly?

APPLICATION PROBLEMS

Each of the following problems require that calculations be completed. Show any required calculations.

22-25. Use the data from the given table to calculate a position tolerance for each problem. Assume that two parts are stacked together and that the tolerance is for a floating fastener condition.

Problem Number	Fastener MMC	Hole Diameter At MMC	
		Part #1	Part #2
22	∅.164	∅.193	∅.193
23	∅.190	∅.218	∅.218
24	∅.190	∅.226	∅.226
25	∅.250	∅.282	∅.282

26-29. Use the data from the given table to calculate a position tolerance for each problem. Assume that two parts are stacked together and that the tolerance is for a fixed fastener condition. Assume that a projected tolerance zone is used.

Problem Number	Fastener MMC	Hole MMC
26	∅.164	∅.188
27	∅.190	∅.212
28	∅.250	∅.282
29	∅.250	∅.312

30-32. Use the data from the given table to calculate a clearance hole diameter for each problem. Assume that two parts are stacked together and that a fixed fastener condition exists. Assume that a projected tolerance zone is used.

Problem Number	Fastener MMC	Position Tolerance
30	∅.138	∅.018
31	∅.190	∅.024
32	∅.250	∅.029

Appendix A1
COORDINATE-TO-DIAMETER CONVERSION TABLE

Any measured pair of coordinates less than or equal to .020″ relative to true position may be converted to an inscribing position tolerance zone diameter using this table. The delta X error can be located along the top of the table, and the delta Y error can be located along the left column. Extending a line vertically from the delta X value and another line horizontally from the delta Y value results in an intersection of the lines on one number value in the table. The value at the intersection of the lines is the diameter of the circle that inscribes the X and Y values.

Example:

The measured location error for one hole is:

X = .005″

Y = .007″

The required diameter position tolerance zone to permit this measured location is .01720″.

CONVERSION TABLE

Y \ X	.001	.002	.003	.004	.005	.006	.007	.008	.009	.010	.011	.012	.013	.014	.015	.016	.017	.018	.019	.020
.001	.00283	.00447	.00632	.00825	.01020	.01217	.01414	.01612	.01811	.02010	.02209	.02408	.02608	.02807	.03007	.03206	.03406	.03606	.03805	.04005
.002	.00447	.00566	.00721	.00894	.01077	.01265	.01456	.01649	.01844	.02040	.02236	.02433	.02631	.02828	.03027	.03225	.03423	.03622	.03821	.04020
.003	.00632	.00721	.00849	.01000	.01166	.01342	.01523	.01709	.01897	.02088	.02280	.02474	.02668	.02864	.03059	.03256	.03453	.03650	.03847	.04045
.004	.00825	.00894	.01000	.01131	.01281	.01442	.01612	.01789	.01970	.02154	.02341	.02530	.02720	.02912	.03105	.03298	.03493	.03688	.03883	.04079
.005	.01020	.01077	.01166	.01281	.01414	.01562	.01720	.01887	.02059	.02236	.02417	.02600	.02786	.02973	.03162	.03353	.03544	.03736	.03929	.04123
.006	.01217	.01265	.01342	.01442	.01562	.01697	.01844	.02000	.02163	.02332	.02506	.02683	.02864	.03046	.03231	.03418	.03606	.03795	.03985	.04176
.007	.01414	.01456	.01523	.01612	.01720	.01844	.01980	.02126	.02280	.02441	.02608	.02778	.02953	.03130	.03311	.03493	.03677	.03863	.04050	.04238
.008	.01612	.01649	.01709	.01789	.01887	.02000	.02126	.02263	.02408	.02561	.02720	.02884	.03053	.03225	.03400	.03578	.03758	.03940	.04123	.04308
.009	.01811	.01844	.01897	.01970	.02059	.02163	.02280	.02408	.02546	.02691	.02843	.03000	.03162	.03329	.03499	.03672	.03847	.04025	.04205	.04386
.010	.02010	.02040	.02088	.02154	.02236	.02332	.02441	.02561	.02691	.02828	.02973	.03124	.03280	.03441	.03606	.03774	.03945	.04118	.04294	.04472
.011	.02209	.02236	.02280	.02341	.02417	.02506	.02608	.02720	.02843	.02973	.03111	.03256	.03406	.03561	.03720	.03883	.04050	.04219	.04391	.04565
.012	.02408	.02433	.02474	.02530	.02600	.02683	.02778	.02884	.03000	.03124	.03256	.03394	.03538	.03688	.03842	.04000	.04162	.04327	.04494	.04665
.013	.02608	.02631	.02668	.02720	.02786	.02864	.02953	.03053	.03162	.03280	.03406	.03538	.03677	.03821	.03970	.04123	.04280	.04441	.04604	.04771
.014	.02807	.02828	.02864	.02912	.02973	.03046	.03130	.03225	.03329	.03441	.03561	.03688	.03821	.03960	.04104	.04252	.04405	.04561	.04720	.04883
.015	.03007	.03027	.03059	.03105	.03162	.03231	.03311	.03400	.03499	.03606	.03720	.03842	.03970	.04104	.04243	.04386	.04534	.04686	.04841	.05000
.016	.03206	.03225	.03256	.03298	.03353	.03418	.03493	.03578	.03672	.03774	.03883	.04000	.04123	.04252	.04386	.04525	.04669	.04817	.04968	.05122
.017	.03406	.03423	.03453	.03493	.03544	.03606	.03677	.03758	.03847	.03945	.04050	.04162	.04280	.04405	.04534	.04669	.04808	.04952	.05099	.05250
.018	.03606	.03622	.03650	.03688	.03736	.03795	.03863	.03940	.04025	.04118	.04219	.04327	.04441	.04561	.04686	.04817	.04952	.05091	.05235	.05381
.019	.03805	.03821	.03847	.03883	.03929	.03985	.04050	.04123	.04205	.04294	.04391	.04494	.04604	.04720	.04841	.04968	.05099	.05235	.05374	.05517
.020	.04005	.04020	.04045	.04079	.04123	.04176	.04238	.04308	.04386	.04472	.04565	.04665	.04771	.04883	.05000	.05122	.05250	.05381	.05517	.05657

X COORDINATE (across top) · Y COORDINATE (down left side)

Appendix A2

COORDINATE-TO-DIAMETER CONVERSION CHART

This figure can be used to approximate the required diameter circle to inscribe a coordinate value. Grid lines are drawn on a .001″ increment, and circles are drawn on a .002″ diameter increment. Coordinate values in the X and Y axes are equal. Coordinate values are only labelled along the X axis. Diameters are labelled along the Y axis.

The chart is used by plotting a coordinate value on the grid. A circle passing through the plotted point, or the next larger circle, is traced counterclockwise to the Y axis to determine the diameter.

Any coordinate value can be plotted on this chart. If coordinate values extend beyond the chart boundaries, the scale of the chart can be increased. Simply multiply all values on the chart by the same number to increase its scale.

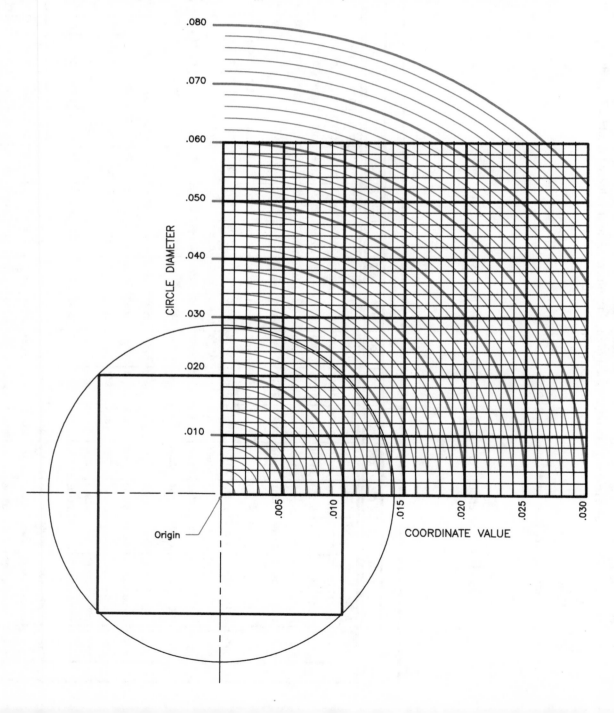

Appendix B1

CURRENT SYMBOLS

The majority of the current dimensioning and tolerancing symbols are the same as those used in the previous standards. A few new symbols have been added to more closely match international practices and to reduce the number of words placed on drawings.

DIMENSIONING SYMBOLS		
CURRENT PRACTICE	ABBREVIATION IN NOTES	PARAMETER
Ø	DIA	Diameter
SØ	SPHER DIA	Spherical Diameter
R	R	Radius
CR	CR	Controlled Radius
SR	SR	Spherical Radius
⌴	CBORE	Counterbore
⌴	SF or SFACE	Spotface
⌵	CSK	Countersink
↧	DP	Deep
○	—	Dimension Origin
□	SQ	Square
( )	REF	Reference
X	PL	Places, Times
⌒	—	Arc Length
⊿	—	Slope
⌲	—	Conical Taper
[2.38]	—	Basic Dimension
⟨ST⟩	—	Statistical
↔	—	Between
▲	—	Datum Feature Triangle

DIMENSIONING SYMBOLS		
CURRENT PRACTICE	ABBREVIATION IN NOTES	PARAMETER
□	—	Datum Feature Symbol
⊖	—	Datum Target Symbol
Ⓢ	RFS	Regardless Of Feature Size
Ⓜ	MMC	Maximum Material Condition
Ⓛ	LMC	Least Material Condition
Ⓟ	—	Projected Tolerance Zone
—	—	Straightness
▱	—	Flatness
○	—	Circularity
⌭	—	Cylindricity
⊥	—	Perpendicularity
//	—	Parallelism
∠	—	Angularity
⌖	—	Position
⌯	—	Symmetry
◎	—	Concentricity
↗	—	Circular Runout
⌰	—	Total Runout
⌒	—	Line Profile
⌓	—	Surface Profile

Appendix B2

Standard symbols have changed slightly as improvements to the dimensioning and tolerancing standard have been made. Symbols in the 1982 and 1973 standards are shown. Optional symbols contained in the 1982 standard are highlighted.

PAST DIMENSIONING SYMBOLS		
1982	1973	PARAMETER
⌀	DIA	Diameter
S⌀	SPHER DIA	Spherical Diameter
R	R	Radius
SR	SR	Spherical Radius
⊔	CBORE	Counterbore
⊔	SF or SFACE	Spotface
⌄	CSK	Countersink
↧	DP	Deep
○	—	Dimension Origin
□	SQ	Square
( )	REF	Reference
X	PL	Places, Times
⌒	—	Arc Length
◁	—	SLOPE
◁	—	CONICAL TAPER
2.38	2.38	Basic Dimension

PAST DIMENSIONING SYMBOLS		
1982	1973	PARAMETER
⊡	⊡	Datum Feature Symbol
⊖	⊕	Datum Target Symbol
Ⓢ	Ⓢ	Regardless Of Feature Size
Ⓜ	Ⓜ	Maximum Material Condition
Ⓛ	Ⓛ	Least Material Condition
Ⓟ	Ⓟ	Projected Tolerance Zone
—	—	Straightness
▱	▱	Flatness
○	○	Circularity
⌀/	⌀/	Cylindricity
⊥	⊥	Perpendicularity
//	//	Parallelism
∠	∠	Angularity
⌖	⌖	Position
⌖	=	Symmetry
◎	◎	Concentricity
↗	↗	Circular Runout
↗↗	↗ TOTAL	Total Runout
⌒	⌒	Line Profile
⌓	⌓	Surface Profile

GLOSSARY

A

aligned dimensions: A system of dimensioning in which dimension values are aligned with the dimension lines. Values are oriented to read from the bottom or right side of the page. This system is seldom used and is no longer illustrated in the standard.

all around symbol: A circle drawn at the corner in a leader that extends from a feature control frame. It indicates that the tolerance applies all the way around the profile of the feature to which the leader is attached.

allowance: The difference between the maximum shaft size limit and the minimum hole size limit.

Anglo-Saxon foot: A distance standard previously used in Great Britain. It was based on the 11.65″ long Roman foot.

angularity tolerance: The amount of permitted error on the angular relationship between features. It is specified on angles other than 90° or parallel.

ANSI: The abbreviation for American National Standards Institute.

apex offset, cone: The distance from the apex of a cone to a line extending perpendicular to and from the center of the cone base.

arc: Any segment of a circle other than a full circle.

arrowhead: A symbol placed at each end of the dimension lines and on the end of a leader to indicate the point of application.

ASME: The abbreviation for the American Society of Mechanical Engineers.

arc length: The circumferential distance measured along an arc.

B

base diameter, cone: The distance across the bottom of a cone.

baseline dimensioning: A system in which all dimensions in one direction originate from the same feature.

basic dimension: A theoretically exact value applied to the size or location of a feature. Title block tolerances do not apply. Tolerance must be applied by feature control frames or other means.

basic hole system: A system for calculating the limits of size for a mating hole and shaft. The size limits for the hole are calculated first, and the size limits for the shaft are based on the hole dimensions.

basic shaft system: A system for calculating the limits of size for a mating hole and shaft. The size limits for the shaft are calculated first, and the size limits for the hole are based on the shaft dimensions.

bidirectional profile: A profile tolerance that extends equally in each direction relative to the perfect profile of the feature.

blind hole: A hole that does not go all the way through the material in which it exists.

bonus tolerance: Any increase in allowable tolerance that is permitted by the departure of a feature from the material condition specified in the feature control frame.

boss: A local protrusion on a part that is produced to provide a small area to machine. This prevents the need to machine an entire surface when a small locally machined area is adequate.

British Imperial Yard: The permanent standard established in Great Britain in 1855.

C

CAD: The abbreviation for Computer-Aided Design.

caliper: A tool used to make measurements.

chain dimensioning: A system in which dimensions go point-to-point to locate each consecutive feature from the previous one.

chamfer: A small inclined surface cut on the edge of a part. It may be required for the purpose of removing sharp edges, and is commonly used to make assembly of parts easier.

centerdrill: A tool used to produce a combined drilled hole and countersink. The center-drilled hole produced with this tool is used to mount the workpiece on a machine center.

circular runout: Surface variation measured relative to an axis of rotation at a circular cross section.

circularity tolerance: A form tolerance used to control the circular cross section of features such as cylinders and cones.

clearance fit: A condition in which an internal part is smaller than the external part it fits into.

clearance hole: A hole that has an MMC larger than the MMC of the fastener that goes into the hole.

clocking: Control of the rotational position of one or more features relative to another feature on a part or in an assembly.

CMM: The abbreviation for Coordinate Measuring Machine. See *coordinate measuring machine.*

coaxiality: The condition of two or more feature being centered on the same centerline.

coaxial hole pattern: Two or more holes that lie on the same axis. An example is the set of coaxial holes for a hinge pin.

composite position tolerance: A position tolerance specification that provides a pattern locating tolerance and a feature relating tolerance.

compound datum features: Two datum features that act together to establish a single datum plane or axis.

compound datum reference: Two datum letters separated by a dash and placed in one cell of a feature control frame or shown in a note. Datum reference B-C indicates that features B and C are to be used to establish one datum.

concentricity tolerance: An allowable variation of the location of an axis for one feature relative to the axis of another feature. This is a very difficult tolerance to verify. Runout and position tolerances are normally preferable.

cone angle: The angular measurement across cone elements in a right circular cone with the apex acting as the vertex of the angle.

cone height: The distance from the base to the apex of a cone.

coordinate measuring machine: A machine that electronically records coordinates for features on a part. It may also calculate feature variations from nominal and report any existing discrepancies.

coplanarity: Two or more feature that lie in the same plane.

corner radius: Machine cutters typically include a small radius at the end of the cutter. This produces a corner radius at locations such as the bottom of a counterbore. Corner radii are typically desired to reduce stress concentrations.

counterbore: A stepped increase in the diameter of a hole. Counterbores are located at the ends of holes.

counterdrilled hole: A stepped hole made of two diameters located on the same centerline. The transition between the two diameters is conical and is formed by the drill angle on the large diameter drill.

counterdrill depth: The distance from the entrance of the hole to the bottom of the large diameter.

countersink: An angular increase in the size of a hole. It has the shape of a right circular cone and is coaxial with the hole.

countersink angle: The cone angle of the countersink. Countersink angles for screw heads must match the screw head angle.

cubit: An ancient distance standard, equivalent in length to the Egyptian Pharaoh's forearm.

cylinder: A geometric shape formed by revolving a line around an axis that is parallel to the line. It has a diameter and a length.

cylindricity tolerance: The allowable amount of variation from the shape of a perfect cylinder. The tolerance boundary is formed by two perfect cylinders.

D

datum: Theoretical point, line, or plane established from datum features on a part.

datum axis: An axis established by a datum feature such as a hole or shaft.

datum feature: Physical feature on a part that is used to establish the location of a theoretically perfect datum.

datum feature of size: A physical feature that has size, such as a hole, that is identified with a datum feature symbol.

datum feature symbol: A rectangle drawn around a letter with a dash on each side of the letter. The symbol is used to identify datum features on a drawing.

datum reference: Placement of a datum letter in a feature control frame or in a note that specifies requirements relative to datums.

datum reference frame: Three planes that establish a coordinate system. It is created by datum references in a feature control frame or note. One part may have multiple datum reference frames.

datum simulation: The use of a tool to contact a datum feature and establish the location of the part relative to the datum.

datum simulator: A tool used to contact a datum feature. The tool probably contains some error and therefore does not establish a perfect datum, but adequately simulates the datum for practical purposes.

datum targets: Points, lines, or areas on a feature that are identified to establish datums.

datum target area: An area on a surface that is contacted to establish a datum.

datum target line: A line along a surface that is contacted to establish a datum.

datum target point: A point on a surface that is contacted to establish a datum.

datum target symbol: A circle containing a letter and number to identify datum targets and associate them with a specific datum.

diameter symbol: A circle with a diagonal line through it. It is placed in front of numbers to indicate the value is a diameter.

decimal places: The number of digits to the right of the decimal point.

departure from MMC: The amount of size variation from the maximum material condition of the feature.

dial indicator: A device used to measure variations relative to a desired condition. It does not measure size.

diameter: The distance across a circle measured on a line that passes through the center.

digit: A distance standard used over 500 years ago. It is based on the distance across a finger.

dimension line: A line used to indicate the direction of the dimension. It is drawn parallel to the dimensioned feature.

dogleg: An offset in an extension line that can be used to provide clearance between the extension line and other lines or symbols.

double dimensioning: A condition in which the size, location, or tolerance on a feature can be determined in more than one way.

drill depth: The distance from the penetrated surface to the bottom of the full diameter.

E

extension line: A line used to extend dimensioned features to the dimension lines. Extension lines make it possible to define part size without placing dimensions on the view.

equalizing datum targets: Targets located such that when they are contacted the forces on the part result in a centering effect.

extrusion: A shape formed by pushing or pulling material through a die.

F

feature control frame: A tolerance specification in a format that shows the tolerance type, tolerance value, any applicable material condition modifiers, and any applicable datum references.

feature relating (locating) tolerance: The tolerance value in the second line of a composite position tolerance. It controls locations between features in a pattern.

feature relating tolerance zone framework: The relative true positions of the features in a pattern.

feature of size: Generally, a feature that affects the weight of a part if its dimension is changed. As an example, changing a hole diameter affects how much material is in the part, thereby affecting the weight of the part.

fixed fastener condition: An assembly condition in which two or more parts are stacked together with a fastener passed through them. One of the parts has a hole that in some manner fixes the location of the fastener. The hole may be threaded with the fastener threaded into it, or the hole may be sized for a press fit.

flatness tolerance: A tolerance control that defines the allowable amount of variation relative to a perfect plane.

floating fastener condition: An assembly condition in which two or more parts are stacked together with a fastener passing through them. All of the parts have clearance holes in them.

force fit: A combination of sizes on a shaft and hole that result in an interference fit. The shaft is larger than the hole.

foreshortened radius: A radius seen in a view where the line of sight is not perpendicular to the plane in which the arc is located.

form: The condition of a single feature such as a straight rod, flat surface, round hole, or cylinder.

form tolerance: The amount of allowable variation of a single feature from the theoretically perfect form of that feature.

free state variation: The condition of a part that permits its dimensions to change after it is removed from the manufacturing equipment.

full indicator movement (FIM): The difference between the minimum and maximum reading obtained with a dial indicator. Full indicator movement is also referred to as total indicator reading (TIR).

functional gage: A tool that can be used to check dimension limits by placing it on or in the workpiece. An example is a plate with four pins that checks the locations of four holes.

G

general note: Information that applies to an entire drawing is placed in a general note. General notes are precise in specifying requirements.

I

implied datum: The practice of specifying only the tolerance type and tolerance value, leaving the datum selection to the judgment of the manufacturer and inspector. This practice is now prohibited by the standard.

inch: A standard unit of measurement approximately 25.4 millimeters long. In 1324 AD, it was a distance equal to the length of three barley corns placed end-to-end.

inclined surface: A surface at any angle other than parallel or perpendicular to another surface.

interference fit: A condition in which an internal part is larger than the external part into which it is assembled.

irregular curve: For purposes of this text, an irregular curve is any curve that doesn't have a constant radius.

K

keyseat: A rectangular slot cut in a shaft or hub for the purpose of installing a key to lock rotational relationships between parts.

L

lay: The direction of surface irregularities caused by machining processes.

leader line: A line used to connect specific information to a point on a drawing.

least material condition (LMC): The size of a feature when the least material exists in the part. Maximum hole diameter and minimum shaft diameter are examples of the least material condition.

limit dimension: A dimension that shows the minimum and maximum allowable values for a feature size or location.

line fit: A calculated set of limits for two parts that result in a zero allowance.

line profile: Contour of line elements in one direction on a surface.

LMC: The abbreviation for Least Material Condition. See *least material condition.*

location tolerance: The allowable variation on where a feature is supposed to be in relationship to other features.

location clearance fit: A condition in which there is clearance between mating parts. The amount of clearance is based on the desired accuracy of location. This type of fit is not used for rotating parts.

location dimension: A value that indicates where one feature belongs relative to another.

location interference fit: A condition in which the shaft is larger than the hole. The amount of interference is only meant to provide a known relative position for the parts. It is not meant to hold the two parts together.

location transition fit: A condition that may result in either a slight interference or clearance. It is used to provide a good relative location between the two parts.

M

machine relief: A feature produced in a part to provide machine cutter clearance. Also called a machining relief.

mate drill: An operation in which two parts are placed together and both parts drilled in one operation.

mated assemblies: Parts mate drilled or mate machined may not be interchangeable with parts from other assemblies. If they are not interchangeable, then they are mated assemblies.

material condition modifier: A symbol used in tolerance specifications to indicate at what material condition the specified tolerance value is applicable.

maximum material condition (MMC): The size of a feature when the maximum material exists in the part. Minimum hole diameter and maximum shaft diameter are examples of the maximum material condition.

meter, standard: A standard established in 1798 based on one ten-millionth of the distance from the north pole to the equator when measured on a line passing through Paris. This standard has been replaced by the U.S. prototype meter #27.

micrometer: A tool used for making measurements.

MIL-STD-8: A military standard issued in the 1940s to specify dimensioning and tolerancing methods. The current standard evolved from this one.

minor radius, countersink: The shortest distance between the axis of a hole and the intersection of the countersink with an irregular curved surface.

MMC: The abbreviation for Maximum Material Condition. See *maximum material condition*.

multiview drawing: An orthographic drawing that shows the required views to describe a part.

N

normal: At an angle generally thought of as perpendicular. In the case of an ellipse, a line extending from a point on the ellipse is normal to the ellipse if it is perpendicular to a tangent line that passes through the point.

note flag: A symbol and number placed on a drawing to indicate that a note in the notes list is applicable. Note flags are often attached to specific features with a leader.

O

Olympic foot: The official name for the 16 digit foot established by the Greeks. Its length is 12.16″.

order of precedence: The order in which datum references are made to indicate which datum is primary, secondary, and tertiary.

ordinate dimensioning: A system of dimensioning in which the dimension values are shown at the ends of the extension lines. No dimension lines are shown.

orientation: The angular relationship of one feature to other features.

orientation tolerance: An allowable variation expressed in a perpendicularity, parallelism, or angularity tolerance.

origin of measurement: The point, line, or plane from which measurements are made.

P

palm: The distance across the palm of a hand. It was used as a distance standard more than 5000 years ago.

paper gaging: A method of analyzing collected inspection data to see if the part meets tolerance requirements.

parallelepiped: A six-sided prism where all surface intersections form 90° angles.

parallelism tolerance: The allowable orientation error of a feature relative to a parallel feature. It does not control location.

pattern locating tolerance: The tolerance value in the first line of a composite position tolerance. It controls locations and orientation of the features relative to the referenced datums.

pattern locating tolerance zone framework: The true positions of the controlled features relative to the referenced datums.

perpendicularity tolerance: The amount of acceptable variation on a 90° angle between a feature and a referenced datum.

pitch diameter: The diameter of the pitch cylinder created by threads. It is approximately midway between the root and crest of the threads.

plus and minus tolerance: Values that show the allowable variation above and below the nominal dimension value. It is possible for one of the tolerance values to be zero.

polar coordinate: A point defined by a combination of an angle dimension and a distance from an origin point.

position tolerance: A value that defines the acceptable amount of variation in the location of a feature relative to the true position specified for the feature.

primary datum: The first datum reference in a feature control frame. If the primary datum is a surface, at least three points on this surface must be contacted to establish the primary plane.

profile tolerance: A tolerance on a feature that can specify any needed control ranging from a singular control of form to a simultaneous control of form, orientation, and location.

profile of a line: A control of line elements in only one direction on a surface.

profile of a surface: A simultaneous control of all points over the full area of a surface

projected tolerance zone: A tolerance zone that extends to the outside of the controlled feature.

R

radial hole pattern: A pattern of holes located on radial lines.

radial distance: A distance measured along a line extending from the center of a feature such as a cylinder.

radial line: A line extending from the center of a round feature such as a cylinder.

radius symbol: The letter "R" placed in front of the radius dimension value.

reference dimension: A dimension that provides information about location or size but which does not indicate a requirement. A dimension value is enclosed in parenthesis to indicate that it is reference.

reference value: A number shown on the drawing for information purposes. It does not establish a firm requirement for the acceptance or rejection of parts.

right square pyramid: A pyramid with its apex located on a line that is perpendicular to and centered on the base.

roughness: The small peaks and valleys on a surface that are caused by manufacturing processes.

round off: Creating a value with a predetermined number of decimal places from a value that has a greater number of decimal places.

round off error: The error in answers obtained when calculating with rounded off numbers.

Royal Cubit: A permanent distance standard made of black granite and based on the cubit length (20.67″). It was used to calibrate measuring sticks.

Rule #1: Any dimensioned feature of size must have perfect form when the feature is produced at its maximum material condition.

Rule #2: Applicable material condition modifiers must always be shown on position tolerance specifications.

Rule #3: Material condition modifiers are assumed to be regardless of feature size on all tolerances except position.

running and sliding clearance fit: A condition in which the shaft is always smaller than the hole and the parts are free to move.

runout: Variation of a surface relative to an axis of rotation.

S

secondary datum: The second datum reference in a feature control frame. It must be established after the primary datum.

scale: The proportion of the drawing relative to the actual size of the part.

sharp diameter, flathead screw: The diameter created by extending the screw head cone all the way to the top of the head.

simultaneous datum features: See compound datum features.

simultaneous datum references: See compound datum references.

sine bar: A device that has a well-controlled distance between mounting feet, and is used to set up angles. The length is fixed at a distance that permits stacking blocks under one foot to produce a height equal to a multiple of the sine of the angle to be established.

single limit dimension: A dimension that only shows one acceptable limit. The other limit must be controlled by the geometry of the part.

sixteen digit foot: A distance standard established by the Greek's (12.16″). It is approximately two-thirds the length of the Royal Cubit.

size dimension: A value that defines how large a feature is.

slope: A ratio of the rise over the run of an inclined surface.

span: A distance standard used approximately 5000 years ago. It was based on the distance from the end of the thumb to the end of the small finger when the hand is spread.

specific note: Information attached to one or more specific features on a drawing.

spherical radius: The distance from the surface of a spherical feature to its center point.

spotface: A stepped increase in the diameter of a hole. It is similar to a counterbore but generally is not used to recess a screw head or other part. It is generally used to provide a good bearing surface for a screw head or to establish a controlled material thickness.

square symbol: A small square placed in front of a dimension value to indicate that the dimension applies across both sets of flats on a square part.

station point: Point located by tabulated coordinates.

stepped datum: A datum established by two features that lie in different planes.

stock shape: A geometric cross section of stock material that may commonly be purchased. Examples are round, square, rectangular, and hexagonal bar stock.

straightness tolerance: The acceptable amount of variation relative to a straight line.

surface feature: Any flat or curved feature that is not defined by a size dimension.

surface profile: The simultaneous control of all points across an entire surface contour.

symbol size: Symbol sizes are proportional to the size of dimensioning characters used on a drawing.

symmetrical features: A group of features that are equally located and shaped relative to a center plane or axis.

T

tabulated dimensions: Coordinate and/or size values contained in a table and referenced to the drawing by symbols or letters placed at the locations of the dimensioned feature.

tangent plane: A plane that contacts the high points on a surface.

taper: A ratio of the change in diameter per unit of length.

target area: A portion of a surface that is identified as a target. It is to be contacted with a tool that is sized to the dimensions of the target area.

target line: A line on a surface that is to act as a target. The side of a dowel pin can be positioned to contact the line and establish the datum.

target point: A single point on a surface that is to act as a target. A spherical tool located to contact the point can be used to establish the datum.

tertiary datum: The third datum reference in a feature control frame or note. It is established after the primary and secondary datums are already established.

times/places: The number of occurrences of a feature. It is indicated by a number followed by an "X".

title block tolerance: Allowable dimensional variations associated with untoleranced dimensions on the drawing. The tolerance values may be shown in the title block or notes list.

tolerance: The acceptable dimensional variation that is permitted on a part.

tolerance accumulation: The addition of tolerances associated with features on a part. Tolerance accumulation is also referred to as tolerance buildup and tolerance stackup.

tolerance buildup: The addition of tolerances associated with features on a part. Tolerance buildup is also referred to as tolerance accumulation and tolerance stackup.

tolerance value, feature control frame: The number following the tolerance symbol. It indicates the permitted amount of variation.

total runout: The amount of variation measured across an entire surface when the part is rotated on an axis of rotation.

transition fit: Limits of size applied to two parts that can result in either a clearance or interference fit.

true position: A theoretically exact location for a feature. A specified tolerance defines the allowable variation relative to the true position.

true radius: The actual distance between an arc center and the arc. This is different than the apparent radius seen when viewing an arc at any angle other than perpendicular to the plane of the arc.

U

unidirectional dimensions: A system of dimensioning in which all numbers are written horizontally and read from the bottom of the page.

unidirectional profile: A profile tolerance that extends in one direction relative to the perfect profile of the feature.

U.S. prototype meter #27: The standard against which all U.S. units are established. Its length is established relative to a specific wavelength of light.

V

virtual condition: The combined effect of the maximum material condition and any applicable tolerance.

W

waviness: Large surface variations on which roughness variations may be superimposed.

workpiece: The part on which work is being performed.

Y

yard: A distance standard originally established in the 12th century. It was equivalent to the distance from King Henry's nose to his thumb.

Z

zero position tolerance at MMC: A position tolerance that shows no tolerance in the feature control frame, but instead places all the tolerance on the size of the located feature. This permits maximum flexibility in manufacturing if properly done.

INDEX